Biostatistics with

Apply Python for biostatistics with hands-on biomedical and biotechnology projects

Darko Medin

Biostatistics with Python

The author acknowledges the use of cutting-edge AI, such as ChatGPT, with the sole aim of enhancing the language and clarity within the book, thereby ensuring a smooth reading experience for readers. It's important to note that the content itself has been crafted by the author and edited by a professional publishing team.

Associate Group Product Manager: Niranjan Naikwadi
Publishing Product Managers: Sanjana Gupta and Yasir Khan
Book Project Manager: Shambhavi Mishra
Senior Editor: Tiksha Lad
Technical Editor: Rahul Limbachiya
Copy Editor: Safis Editing
Proofreader: Tiksha Lad
Indexer: Hemangini Bari
Production Designer: Prashant Ghare
Senior DevRel Marketing Executive: Vinishka Kalra

First published: November 2024

Production reference: 1221124

Published by Packt Publishing Ltd.

Grosvenor House

11 St Paul's Square

Birmingham

B3 1RB, UK.

ISBN 978-1-83763-096-7

`www.packtpub.com`

Contributors

About the author

Darko Medin is a researcher and a biostatistician who graduated from the Faculty of Mathematics and Natural Sciences, Experimental Biology and Biotechnology, University of Montenegro. Darko is an expert biostatistician, especially in the fields of research and development in the biotech and pharma industries. He is a Python-based data scientist with more than 10 years of experience in the areas of clinical biostatistics and biomedical research. As a biologist and data scientist, he has worked with many research companies and academic institutions around the world and is an experienced machine learning and AI developer.

About the reviewers

Meghal Gandhi is currently a software engineer and machine learning researcher at Charles R. Drew University of Medicine and Science based in Los Angeles. He holds a master's in computer science from California State University, Fullerton. While working in AI in healthcare, he built machine learning and deep learning models to predict the risk of getting diseases based on medical records. His research work has been published in prestigious medical journals and conferences. Prior to this, he worked as a software engineer on telecommunication and performance engineering projects at AT&T.

Russell Reeve, PhD, is VP and global head of biostatistics at Syneos Health and was formerly VP and global head of biostatistics at IQVIA. He provides advice on clinical trial design and analysis, is responsible for delivery, and leads machine learning initiatives. At IQVIA, Dr. Reeve led the development of three successful SaaS products, including a machine learning solution for subgroup identification and synthetic patient generation. Dr. Reeve has provided trial design advice for hundreds of trials, including for adaptive and platform trials, and has over 30 peer-reviewed papers. He has presented over 50 conference presentations on his research. Dr. Reeve received his doctorate in statistics from Virginia Polytechnic Institute and State University.

Table of Contents

2

3

4

Part 2: Introduction to Python for Biostatistics – Methodology and Examples

5

Introduction to Python for Biostatistics 91

6

Biostatistical Inference Using Hypothesis Tests and Effect Sizes 107

7

Predictive Biostatistics Using Python 133

8

Part 2 Exercise – T-Test, ANOVA, and Linear and Logistic Regression 151

9

Biostatistical Inference and Predictive Analytics Using Cardiovascular Study Data 171

Part 3: Clinical Study Design, Analysis, and Synthesizing Evidence

10

Clinical Study Design 193

14

Part 3 Exemplar Project – Meta-Analysis of Survival Data in Clinical Research 269

Part 4: Biological and Statistical Variables and Frameworks, and a Final Practical Project from the Field of Biology

15

Understanding Biological Variables 285

16

Data Analysis Frameworks and Performance for Life Sciences Research 301

Preface

Unlike other books aimed at biostatisticians, this book is written for all those who wish to use the power of Python for biostatistics in their respective fields. Python has become the most used programming language, adopted in almost all segments of the biotech industry, the medical and pharma research industries, and academia, due to its ability to be integrated with artificial intelligence and machine learning. For this reason, implementing biostatistics with Python is one of the most important fields of research today. This book capitalizes on Python's strengths and scalability to augment and improve the researcher's toolbox, helping anyone in the life sciences and biostatistics fields. This book is one of the rarest resources for this book's audience.

This book provides a comprehensive guide to combining Python programming with biostatistics for applications in life sciences, biotech, and AI-driven fields. It offers real-world projects and examples from oncology, cardiology, biology, and biotech, making learning practical and relevant. The book integrates the biological, data science, and statistical domains with coding, catering to both novices and experienced programmers. Python's scalability and efficiency make it an invaluable tool for biotech, clinical, life sciences, and bioinformatics professionals, enabling the automation of data processing and analysis tasks while significantly reducing time and effort.

You will gain the modern programming skills necessary to perform complex statistical analyses and connect your Python expertise to cutting-edge fields such as artificial intelligence, machine learning, and digital product creation. The book equips you with domain-specific biostatistics knowledge tailored to life sciences and biotech, eliminating the need to learn additional programming languages. It also empowers AI developers, software engineers, and digital product creators to evaluate AI models, test results, and deliver insightful analyses within the Python ecosystem.

By bridging Python programming and biostatistics, *Biostatistics with Python* offers a well-structured approach to mastering essential statistical concepts, unlocking powerful applications across numerous scientific and technological domains.

Who this book is for

This book is designed for everyone in the fields life sciences, biodata science, biotech, and Python programming fields. Here are the main audience groups that may be interested in this book:

- **Biologists with an interest in using Python capabilities**: Biology researchers who require a robust statistical programming language and are looking to integrate biology, data science, and statistics to analyze experimental data and Python's capabilities

- **Python programmers entering life sciences**: Software developers, engineers, data scientists, and analysts who want to use Python for biostatistics, as well as academics and researchers in computational fields
- **Python-based data analysts interested in biostatistics**: Analysts using Python who wish to specialize in biostatistics and life sciences
- **Doctors and medical researchers**: Medical professionals involved in clinical research, cardiology, and oncology who need to perform complex analyses, study disease patterns, and evaluate treatment efficacy in Python
- **Data scientists in biotech**: Individuals engaged in drug target discovery and drug development who utilize statistical methods to design clinical trials, analyze pharma data, and optimize biostatistics that could be integrated with machine learning and AI in the future
- **AI and machine learning specialists in life sciences**: Professionals from the AI and machine learning sectors in life sciences research who use biostatistical approaches to evaluate the effectiveness of AI/machine learning products in Python
- **Bioinformaticians with an interest in biostatistics**: Experts handling bioinformatics data who need biostatistical methods to interpret complex datasets and derive meaningful biological insights in Python
- **Computational biologists with an interest in biostatistics**: Computational biologists who require Python proficiency in biostatistics to deal with complex datasets and use efficient, scalable, and reproducible methods for data analysis in Python
- **Hobbyists and enthusiasts**: Anyone with a passion for Python programming and biology who is seeking to expand their knowledge and apply Python to biostatistical concepts and projects

What this book covers

Chapter 1, Introduction to Biostatistics, introduces the field of biostatistics and its use cases. You will learn why biostatistics is important for biomedicine, clinical trials, biology, biotechnology, and life sciences. You will also understand why it's important to use computational programming languages such as Python to process biological and biomedical data and try to answer the research questions we may have.

This chapter lays the theoretical foundation needed to proceed with the use of biostatistics in life sciences fields and understand how Python can be used for biostatistical analysis within the biotech and life sciences research fields. You will need this understanding to proceed with the hands-on projects in the following chapters.

Chapter 2, Getting Started with Python for Biostatistics, is about facilitating Python installation and getting started with Python such as Spyder IDE and Jupyter Notebook. You will learn how to install Python and its IDEs using the open source Anaconda distribution. Finally, you will learn how to navigate the interfaces of IDEs.

Chapter 3, Exercise 1 – Cleaning and Describing Data Using Python, will help you learn more about the basics of data science, including data types, how to load data in Python, and more. Practical exercises for loading the famous Iris dataset, cleaning data, and describing data are among the topics in this chapter. The chapter prepares you for the next exercise on diabetes data. It introduces the concept of **Exploratory Data Analysis** (**EDA**), which will be used in many other chapters of the book.

Chapter 4, Part 1 Exemplar Project – Load, Clean, and Describe Diabetes Data in Python, is where you will apply what you learned in *Chapter 3*. The dataset is the Pima Indians Diabetes dataset. EDA, cleaning, and data visualizations as output are the practical goals of this chapter. The chapter also covers theoretical aspects associated with the dataset, such as the theoretical foundation of diabetes mellitus biomarkers.

Chapter 5, Introduction to Python for Biostatistics, covers the libraries used for specific biostatistical methods and how those methods work. You will learn about the libraries for hypothesis tests, effect size analysis, predictive analysis, and more. Toward the end of the chapter, you will learn how to select specific hypothesis tests and biostatistical implementations for different research questions. The main goal of the chapter is to introduce you to the Python framework for biostatistical analysis.

Chapter 6, Biostatistical Inference Using Hypothesis Tests and Effect Sizes, is on biostatistical inference. How to apply hypothesis tests such as Student's t-test, the Wilcoxon test, and the Chi-square test is covered. Another topic covered is finding the associations between variables using correlation analysis. ANOVA and Kruskal–Wallis tests are explored, as is how to analyze multiple groups using Python.

Chapter 7, Predictive Biostatistics Using Python, looks at predictive biostatistics and its uses in different areas of biology, biomedicine, and other life sciences fields.

You will learn about different types of variables in relation to predictive analysis, such as dependent variables, independent variables, and latent variables. You will learn how to implement linear regression and logistic regression in Python. Finally, you will learn how to create and interpret multivariable regression models.

Chapter 8, Part 2 Exercise – T-Test, ANOVA, and Linear and Logistic Regression, is mostly about practical exercises in Python hypothesis testing and predictive analysis. You will learn how to implement Student's t-test for comparing two groups and **Analysis of Variance** (**ANOVA**) to compare multiple groups in biological data. In this chapter, you will also learn how to practically define, create, and implement linear and logistic regression models using Python. At the end of each analysis, you will also learn how to create a publication-ready and intuitive data visualization using Python's data visualization libraries.

Chapter 9, Biostatistical Inference and Predictive Analytics Using Cardiovascular Study Data, contains an exemplar project based on the Cleveland Heart Disease dataset. The main focus of this chapter is the practical implementation of biostatistical inference and predictive analytics in cardiology. The chapter includes both biological and statistical aspects of a practical cardiology project in the field of biostatistics. Hypothesis tests and linear and logistic regression are this time applied to a cardiovascular dataset, including cardiovascular disease modeling.

Chapter 10, Clinical Study Design, looks at how one of the most important aspects of any biostatistics project is the study design. In this chapter, the main topic is understanding clinical studies from the design perspective. You will understand the principles for observational studies, including cohort and case control studies, but also different designs of clinical trials. Furthermore, you will learn how to add sample size calculation for study planning and design. Finally, you will learn how to define protocol documentation for clinical studies.

Chapter 11, Survival Analysis in Biomedical Research, will see you start by loading and understanding an oncology dataset (Veterans Oncology dataset) using scikit-learn. Then, you will how to use survival analysis and **Kaplan-Meier** (**KM**) curves to visualize and analyze survival in different groups of oncology patients. You will learn how to implement Cox Proportional Hazards regression models to perform survival analysis inference and identify the appropriate oncology survival models in the data.

Chapter 12, Meta-Analysis – Synthesizing Evidence from Multiple Studies, shows you how to synthesize evidence from multiple studies or analyses. This chapter lays the theoretical foundation to help you understand how to use meta-analysis to synthesize evidence from multiple studies and create overall estimates of treatment effects in biostatistics. You will learn the differences between random and fixed-effects meta-analysis models and when to use them. You will learn how to reason about and interpret forest and funnel plots, which are often the main focus of meta-analysis interpretation and data visualization.

Chapter 13, Survival Predictive Analysis and Meta-Analysis Practice, is about the practical implementation of meta-analysis code in Python. You will be using the `PythonMeta` package and the DerSimonian & Laird inverse variance method. You will learn about **Overall Survival** (**OS**), **Progression-Free Survival** (**PFS**), **Disease-Free Survival** (**DFS**), and **Recurrence-Free Survival** (**RFS**) metrics in oncology meta-analysis. Finally, the main outcome of the chapter is being able to practically implement meta-analysis and visualize and interpret results using Python.

Chapter 14, Part 3 Exemplar Project – Meta-Analysis of Survival Data in Clinical Research, starts with **Non-Small Cell Lung Cancer** (**NSCLC**) dataset and the treatment used to target a specific molecule associated with this cancer called **Tyrosine Kinase Inhibitors** (**TKI**). The project involves performing a real-world meta-analysis with data from real studies. This exemplar project is a simulation of a real-world oncology meta-analysis, all done using the powerful Python programming language.

Chapter 15, Understanding Biological Variables, looks at simplifying the complexity of biological systems by focusing on the observation and analysis of key variables. It explores latent variables and provides detailed guidance on selecting significant variables for biodata analysis. You will learn how to connect biological questions with observable variables, ensuring the meaningful interpretation of data. The chapter concludes with techniques for validating the biological relevance of data, reinforcing the connection between theory and practical application.

Chapter 16, Data Analysis Frameworks and Performance for Life Sciences Research, focuses on learning to differentiate different statistical data analysis frameworks. We discuss the frequentist and Bayesian frameworks, their differences, when to use them, and how to apply them to different research problems.

You will also learn how to choose the correct statistical framework for your analysis. Finally, you will learn how to connect an experiment design with the statistical aspects of the analysis and perform in-depth interpretation of the results based on the statistical framework you choose.

Chapter 17, Part 4 Exercise – Performing Statistics for Biology Studies in Python, contains a state-of-the-art biology research exemplar project for you to use Python programming and advanced statistical approaches. You start with the mice proteomics dataset, in which we explore the biological aspects of neuroscience and proteins associated with different conditions. The approaches included are **Principal Component Analysis** (**PCA**), **Random Forest** (**RF**) for feature selection, and **Structural Equation Modeling** (**SEM**). By the end of the chapter, you will know how to create protein association SEM models and how to relate biological domain knowledge with latent variables to perform protein SEM analysis and statistically test biological pathways understood using the theoretical domains of molecular biology.

To get the most out of this book

To get the most out of this book, you don't need any previous knowledge about biostatistics or Python. What's most important is to install Python, Spyder IDE, and Jupyter Notebook using the Anaconda navigator. Other ways of installation may also work but will probably be more complex.

Software/hardware covered in the book	Operating system requirements
Python 3.8 or above	Windows, macOS, or Linux
Spyder IDE	Windows, macOS, or Linux
Anaconda Distribution (contains all of the above)	Windows, macOS, or Linux

If you are using the digital version of this book, we advise you to type the code yourself or access the code from the book's GitHub repository (a link is available in the next section). Doing so will help you avoid any potential errors related to the copying and pasting of code.

Remember, anyone can learn the topics in this book, as long as an interest in biotech, biomedical research, other life sciences areas, and the use of Python for biostatistics is present.

Download the example code files

You can download the example code files for this book from GitHub at `https://github.com/PacktPublishing/Biostatistics-with-Python`. If there's an update to the code, it will be updated in the GitHub repository.

We also have other code bundles from our rich catalog of books and videos available at `https://github.com/PacktPublishing/`. Check them out!

Conventions used

There are a number of text conventions used throughout this book.

`Code in text`: Indicates code words in text, database table names, folder names, filenames, file extensions, pathnames, dummy URLs, user input, and Twitter handles. Here is an example: "We will be looking at `CLASS=='Y'` versus `CLASS=='N'`".

A block of code is set as follows:

```
#Load Iris dataset as a csv file
data=pd.read_csv(r'C:\Users\MEDIN\Desktop\Iris.csv')
```

The # sign is used to comment code; such lines of code will not be run by Python. The paths in the book may be replaced by the paths on your computer. In this example, `C:\Users\MEDIN\Desktop` may be replaced by `\Users\path\Desktop` or another path where files of interest for the book may be located.

> **Tips or important notes**
> Appear like this.

Get in touch

Feedback from our readers is always welcome.

General feedback: If you have questions about any aspect of this book, email us at `customercare@packtpub.com` and mention the book title in the subject of your message.

Errata: Although we have taken every care to ensure the accuracy of our content, mistakes do happen. If you have found a mistake in this book, we would be grateful if you would report this to us. Please visit `www.packtpub.com/support/errata` and fill in the form.

Piracy: If you come across any illegal copies of our works in any form on the internet, we would be grateful if you would provide us with the location address or website name. Please contact us at `copyright@packtpub.com` with a link to the material.

If you are interested in becoming an author: If there is a topic that you have expertise in and you are interested in either writing or contributing to a book, please visit `authors.packtpub.com`.

Share Your Thoughts

Once you've read *Biostatistics with Python*, we'd love to hear your thoughts! Scan the QR code below to go straight to the Amazon review page for this book and share your feedback.

`https://packt.link/r/1837630968`

Your review is important to us and the tech community and will help us make sure we're delivering excellent quality content.

Download a free PDF copy of this book

Thanks for purchasing this book!

Do you like to read on the go but are unable to carry your print books everywhere?

Is your eBook purchase not compatible with the device of your choice?

Don't worry, now with every Packt book you get a DRM-free PDF version of that book at no cost.

Read anywhere, any place, on any device. Search, copy, and paste code from your favorite technical books directly into your application.

The perks don't stop there, you can get exclusive access to discounts, newsletters, and great free content in your inbox daily

Follow these simple steps to get the benefits:

1. Scan the QR code or visit the link below

`https://packt.link/free-ebook/978-1-83763-096-7`

2. Submit your proof of purchase
3. That's it! We'll send your free PDF and other benefits to your email directly

Part 1: Introduction to Biostatistics and Getting Started with Python

In *Part 1* of the book, you will be introduced to biostatistics and Python. This part is on learning about biostatistics use cases in biomedical, biotech, and pharma research. Furthermore, you will learn how to install Python using Anaconda and use it along with Spyder IDE and Jupyter Notebook. Finally, you will start using Python with a hands-on exemplar mini-project using diabetes data.

This part has the following chapters:

- *Chapter 1, Introduction to Biostatistics*
- *Chapter 2, Getting Started with Python for Biostatistics*
- *Chapter 3, Exercise 1 – Cleaning and Describing Data Using Python*
- *Chapter 4, Part 1 Exemplar Project – Load, Clean, and Describe Diabetes Data in Python*

1
Introduction to Biostatistics

Welcome to the world of biostatistics. This book will guide you through the principles and practical examples of biostatistics and you will go through a portfolio of exemplar projects with real-world data and learn how to use one of the most advanced programming languages today: Python.

Biostatistics is one of the most important science disciplines today; it enables research, is the foundation of most life sciences, and is growing as a key factor in many industries today, from pharmaceuticals to medicine, biology, and many other life sciences. This chapter explains why biostatistics is important for different areas of biomedicine, clinical trials, biology, and life science areas.

In this chapter, we're going to cover the following main topics:

- Understanding the need for biostatistics in life sciences
- Formulating the scientific questions in life sciences and research
- How statistics and computation can help answer different questions in life sciences

At the end of this chapter, you will have a better understanding of the principles that make biostatistics the foundation of life science and what the advantages of using Python exemplar projects for biostatistics are.

Why do we need biostatistics in life sciences?

Life sciences are some of the most important fields of science today. Throughout the disciplines of biology, biomedicine, and pharmaceutical sciences in pharmaceutical and biotech companies, biostatistics plays a key role. We use it to analyze the data from experiments, improve study designs, interpret the results of studies, and make decisions within all these areas of life science. Biostatistics is applicable in all of these areas, and more, because it allows us to understand the underlying processes that you may be investigating.

While biostatisticians are essential in many areas, from biology and medicine to public health, understanding biostatistics is critical for other professionals in these areas, too.

If you are performing an experiment, conducting a study, or are interested in life science analytics, you will need to analyze the data to make conclusions or get insights from it. Biology and biomedical professionals will encounter biostatistics in most areas of their careers.

When reading almost any life science research publication, you will need to understand how to read biostatistics to understand the results. This is essential for both biologists and biomedical professionals who want to stay current with the latest research statistics for the pharmaceutical industry to discover biomarkers or therapies for patients.

Biostatistics enables us to understand and analyze the data or results we get from experiments, research, or observations. This is one of the reasons why the biostatistical field is important not only for biostatisticians, but also for doctors, biologists, epidemiologists, public health decision-makers, bioinformaticians, health data scientists, and other professionals from most life science branches.

The next subsection will help you understand the specific areas of life science where biostatistics is used. This is very important as every life science area is different and requires a different approach to resolve the research problems

Biostatistics in human life sciences

Biostatistics is essential in many human life sciences. Epidemiologists heavily rely on different types of data to infer their insights. Understanding statistical concepts is essential to understanding population-level biological events and helps both doctors and public health professionals in their work. One such example is the past SARS-CoV-2 pandemic. You must have heard about concepts such as reproductive number (R) or SARS-CoV-2 cumulative incidence of infections, mortality, lethality, excess deaths, and other similar terms. All these concepts are derived using biostatistical concepts and formulas.

Epidemiology is predominantly used in biomedical science areas by public health professionals to make decisions for disease response and to keep the population as safe as possible.

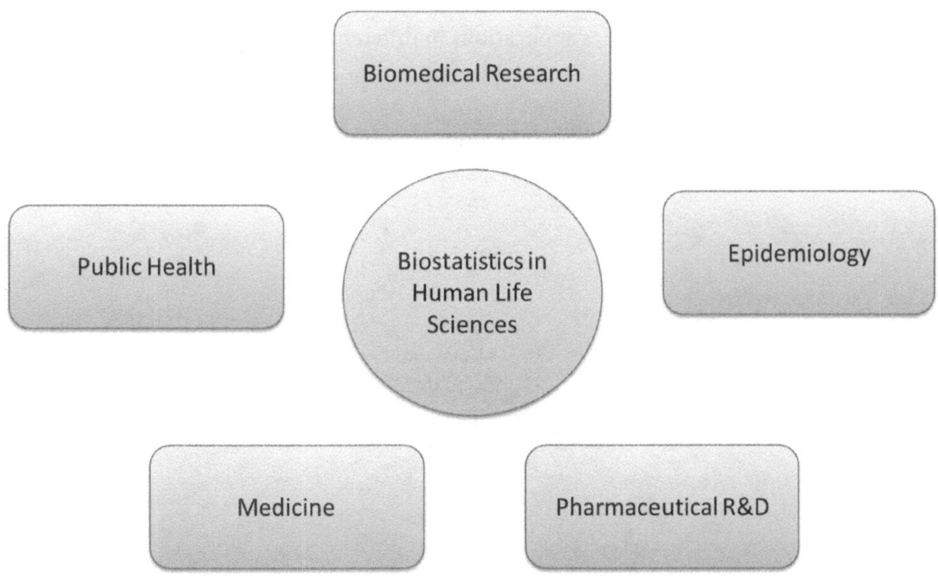

Figure 1.1 – Areas of human life science where biostatistics is used

Medical doctors need biostatistics, not only for their everyday work but also for publishing their academic work, which generally utilizes statistics to summarize and analyze the data of studies and to understand novel discoveries in their profession by interpreting study results from novel publications.

Biomedical research is one of the areas which is heavily reliant on biostatistics and knowing biostatistics is important, not only for statisticians but also for biomedical researchers. Even with access to expert biostatisticians, it is helpful to understand biostatistical thinking and analysis methodology to help with discussions on study design, analysis, and interpretation.

Pharmaceuticals, research, and the development of medications are among the largest industries today. The majority of advanced research in these areas is vital for the biomedical industry.

Biostatistics for biology

While statistics itself is used in many different areas of science, its application in biology has evolved in a specific way due to the nature of different biological domains. Statistics cannot be effectively applied without knowing the basic principles of these biological disciplines.

The following figure shows biostatistics applications in different areas of biology:

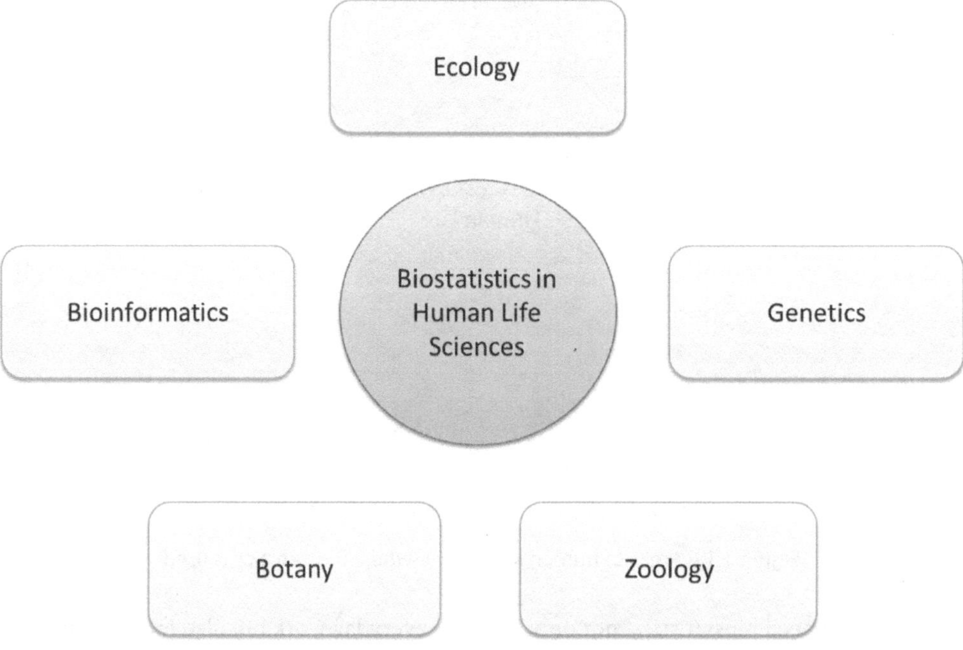

Figure 1.2 – Areas of biology where biostatistics is used

Bioinformatics relies on different statistical methods and algorithms combined with computational tools to process and analyze large amounts of biological data, such as RNA (ribonucleic acid) sequencing data (transcriptomics), DNA (deoxyribonucleic acid) data (genomics), and many other data types. Bioinformatics is specifically focused on genetics and molecular biology but implements methods such as biostatistics and machine learning.

Ecological studies are one of the examples where biostatistics is one of the main biological research drivers. Analyzing plant and animal populations, trends, dynamics, and relations between organisms and their environments would not be possible without biostatistics. Next, we will discuss biostatistics applications in different fields in more detail.

Biostatistics in epidemiology and public health

Epidemiologists and public health professionals answer some of the most important public health questions but also make decisions in different communities. They investigate diseases and events in smaller groups of people, cities, and countries, or even worldwide phenomena, such as pandemics. All this would not be possible without the use of biostatistics facilitating the process of analyzing the biomedical and population data.

Epidemiologists often create different statistical models to try to relate infectious outbreaks to causes and then prevent future infections and isolate the infection source. One such example is studying types of food ingested by infected individuals and identifying a potential bacterial or viral food source, or a location, such as a hotel or restaurant, as a source. Biostatistical models are often used in identifying the sources of infectious agents, which will be discussed in more detail in later chapters.

A few biostatistical concepts used in epidemiology and public health are as follows:

- Prevalence
- Cumulative incidence
- Identifying causes for infectious outbreaks
- Characteristics of microorganisms causing outbreaks in a population
- Epidemiological monitoring of populations
- Decision-making based on biostatistics

Biostatistics in medicine and biomedical research

Medicine and biomedical research are very active sciences today, as they directly or indirectly impact almost everyone's life today. These two disciplines rely heavily on the use of biostatistics. It is of the essence not only for medical doctors but also for biomedical researchers.

Medical doctors' understanding of the probability of different diseases or outcomes is highly dependent on understanding the statistical concepts and how these apply to groups of patients.

Here are some of the most important concepts used in biomedical research:

- Understanding of incidence of safety for treatments
- Making conclusions about the symptom and disease relations
- Creating biomedical studies
- Analyzing biomedical data
- Interpreting novel research
- Becoming specialized in biomedical data analysis

Biostatics in zoology and botany

A significant portion of the research in biological disciplines, such as zoology and botany, depends on quantifying different aspects of their behavior, life cycles, relations with their environmental factors, and many other aspects.

Some examples of areas in zoology and botany that apply biostatistical methods are as follows:

- Animal behaviors
- Plant growth
- Relations between animals and their environment
- Relations between plants and their environment
- Biochemical composition of different tissues in animals
- Biochemical composition of different tissues in plants
- Identifying feeding patterns in animals

Biostatistics in ecology

Ecology is one of the life science disciplines significantly based on biostatistics. Understanding the population's diversity and the relationships between organisms, as well as the relationships between organisms and their environments, is facilitated using different biostatistical methods.

Some important areas of the use of biostatistics in ecology are as follows:

- Relationships between animals and their environment
- Relationships between plants and their environment
- Studying biochemical and molecular aspects in zoology and botany
- Studying relations between humans, ecology, and environmental protection

Biostatistics in pharmaceutical research and design

The pharmaceutical industry is one of the main drivers of research and innovation today. Biostatistical analyses enable pharmaceutical companies to design, conduct, and make decisions based on different analyses and insights. In fact, almost any high-quality research project in the pharmaceutical industry consults biostatisticians to make sure that the design is statistically sound and that it can answer the research questions to drive forward the development of assets and to conform with regulatory requirements. Biostatistics is also the key to analyzing adverse events from the data collected during a study, which is essential for any pharmaceutical product. All medications are required to have a list of adverse effects and this is something that can be seen in everyday life. Biostatistical calculation of incidence rates is one of the ways to assess those adverse effects.

Biostatistics is used to assess the efficacy of different therapies and, as such, is a key element in selecting the candidate drugs for diseases such as diabetes or cancer, which are then further evaluated in clinical trials using different biostatistical methods.

Calculating required sample sizes for pharmaceutical studies is a common task of biostatisticians within the pharmaceutical industry, but this is also intertwined with trial design and endpoint selection.

Here is a summary of the uses of biostatistics in pharmaceutical R&D:

- Creating R&D studies
- Evaluating drug safety
- Selecting drug candidates through biostatistical screening
- Designing clinical trials
- Evaluating results
- Research publications
- Meta-analyses of therapy effects
- Regulatory submission

Biostatistics in bioinformatics and genetics

Molecular biology is one of the biological branches that is very specific in terms of using statistical analyses. From structural biology to analyzing gene expression, biostatistics plays one of the most important roles in bioinformatics. Statistical bases form many genetics areas, such as inheritance genetics and population genetics. Here are some of the areas of bioinformatics and genetics where biostatistics plays a pivotal role:

- Differential gene expression
- Structural biology
- Mutation biology
- DNA analytics
- Mendelian inheritance
- Mendelian randomization studies
- Population genetics

Formulating the scientific questions in life sciences and research

To be able to perform statistical analyses in life science and research, you will first need to learn how to address scientific questions in these areas. Scientific questions are a way to define what it is that we are trying to understand or what goal to achieve. In this chapter, you will learn by example how to formulate scientific questions related to various fields related to biostatistics, such as biomedical research, before any relevant statistical analysis is made. One of the first questions to answer is, "What is the goal of a statistical analysis?" This goal is closely related to different life science aspects, therapies, biological processes, or genetic characteristics, and in this section, those will be covered in more detail.

Once scientific questions are made, they are then used to formulate different scientific hypotheses. The main characteristic of any hypothesis is that it can be tested and there is an alternative (opposite) hypothesis to the main one. So, the baseline scenario assumption can be that there is no statistically significant result, and we can test the alternate scenario: that there is a significant result against the baseline or null scenario. We can call the null hypothesis *H0* and the alternate hypothesis *Ha*.

How to formulate scientific questions related to diabetes

The effect of different lifestyles on the outcomes of type 2 diabetes mellitus has been debated for decades.

Let's pose a couple of scientific questions about diabetes. We will use the letter *Q* for scientific questions:

- Q1. Is body weight related to type 2 diabetes mellitus?
- Q2. Are there other risk factors for type 2 diabetes mellitus among those studies?
- Q3. Which of the lifestyle factors is the most important risk factor in type 2 diabetes mellitus?

Now, let's formulate these questions even better. We will mark formulations using the letter *F*:

- **F1. Null hypothesis (H0)**: Body weight is not related to type 2 diabetes mellitus.

  Alternate hypothesis (Ha): Body weight is related to type 2 diabetes mellitus.

- **F2. Null hypothesis (H0)**: There are no other risk factors for type 2 diabetes mellitus among those studied.

  Alternate hypothesis (Ha): There are other risk factors for type 2 diabetes mellitus among those studied.

- F3. This question will not have a null hypothesis as it is already assumed there are risk factors in the questions. So, the goal of answering this question is to compare the risk factors and identify the most important one. This would be an observational scientific question.

So, why do we usually formulate the null hypothesis as a negation of what's being tested? Well, we want to know the following: Can I show evidence that contradicts that baseline negative assumption? If I can, then I can reject the null hypothesis. If there isn't enough evidence to negate the null hypothesis, I can say that I cannot reject the null hypothesis (avoid the mistake of saying that no evidence is evidence of a null hypothesis).

How to formulate scientific questions related to cardiovascular disease

Is ST (the last wave on the electrocardiogram of the heartbeat) elevation closely related to heart disease? With this, we move to the following questions:

- Q4. Do cigarettes increase the risk of cardiovascular diseases?
- Q5. Is an ECG closely related to cardiovascular disease?
- Q6. Are there any other risk factors for cardiovascular disease among the studied parameters?

  Let us make a more structured formulation as follows:
- **F4. Null hypothesis (H0)**: Cigarettes do not increase the risk of cardiovascular diseases.
- **Alternate hypothesis (Ha)**: Cigarettes increase the risk of cardiovascular diseases.
- **F5. Null hypothesis (H0)**: ECG is not closely related to cardiovascular disease.
- **Alternate hypothesis (Ha)**: ECG is not closely related to cardiovascular disease.
- **F6. Practice yourself!**

How to formulate scientific questions in biology

Here are a few examples for formulating questions in biology:

- Q7. Learn to explore which genes are highly suppressed in lung cancer.
- Q8. How similar are the genomes of mice and humans?
- Q9. What are the differences in plants and minerals collected from localities A and B (Ca, Mg, K)?
- Q10. Does water temperature affect plankton?

Practice formulating these questions as hypotheses or concrete study questions!

You may find the answers at the end of *Chapter 1*.

How computation can help answer different questions in life sciences

It is generally believed that biostatistics is mostly about numbers and graphs. The reality is quite different. Biostatistics is also about understanding life science problems and finding ways to resolve those using statistical methods. There are six main problem-solving skills in biostatistics:

- Helping life science professionals resolve research problems in these domains through the use of data
- Helping life science professionals interpret the results of their research
- Making sure the published research is both statistically and biologically valid
- Helping R&D professionals make decisions in the projects
- Revealing objective truths about different phenomena through the use of data
- Explaining the abstract features of mathematics and biology in an intuitive and easy-to-understand way

One of the most important impacts of biostatistics is transitioning from statistical knowledge to actual problem solutions in life sciences. This will be discussed in more detail in the rest of this chapter.

Biostatistics is needed to derive insights from life science experiments and convert measurements and observations to life science solutions.

Professionals in life science and biostatisticians, working together, design different types of experiments, measurements, and observations. All these can be written or stored as data. Data is a source of information from those experiments, measurements, and observations.

Data can originate from observations, too. One example of observation is the diagnosis by a dermatologist or the identification of species by biologists.

Biostatisticians are there to help make sure this data is valid and make it meaningful. Further, data should be organized and structured, often presented in the form of tables to be prepared for further analysis and interpretation.

To make the data useful, we must understand all the details about the data and how these are related to domains where biostatistics is applied. One of the most important aspects of biostatistics is the context around the data. This context can significantly affect the results and is one of the reasons why biostatisticians are more specialized in life science domains than general statisticians.

One of the main goals of biostatistics is to take all available inputs in the form of data and process them in such a way as to produce meaningful insights, answers, and conclusions and provide information to make decisions in life science.

Here is the biostatistics workflow:

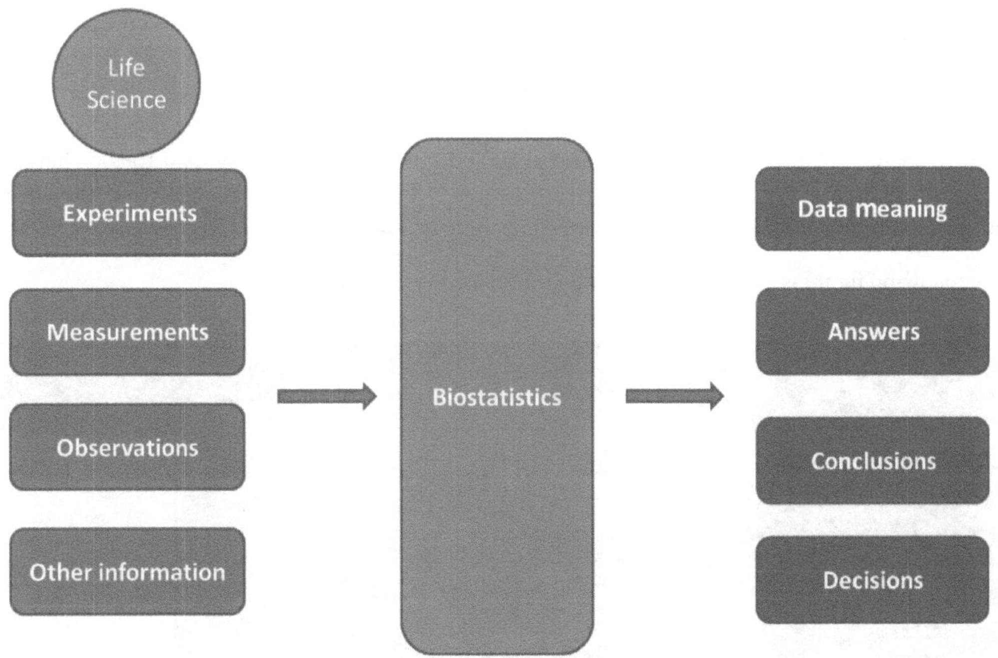

Figure 1.3 – Biostatistics workflow

There are two main types of data: numerical (for example, the measurement of the hemoglobin level in blood in which we are using numerical values such as grams per liter or g/L) and categorical, such as a doctor's diagnoses of their patients in a form; "Yes" for a positive diagnosis or "No" for a negative diagnosis. These types of data can be further divided into subcategories, which will be discussed in detail in the next chapters.

Understanding data sources is essential for biostatistics. Biostatistics is focused on statistical models but also on domain knowledge and, as such, has evolved as a separate branch of both statistics and life sciences.

This book will provide many different examples that will show you how to use biostatistics specifically for different domains, such as diabetes research, cardiology, and biostatistical studies. Further, in this chapter, we will discuss how the Python programming language can facilitate the implementation of biostatistical methods.

Biostatistics and Python

Most biostatistical analyses today are implemented in some form of software or a programming language. I chose Python as a programming language for this book for several reasons. Python is one of the most advanced languages for data science and biostatistics. As programmers today are moving toward using Python, keep in mind that it is one of the most wanted skills in most areas that have to do with analytics. Libraries such as Biopython and SciPy are among the more than 100,000 libraries that make Python so versatile, meaning that almost any biostatistical analysis can be performed using this programming language. It is open source, meaning it is transparent and free for anyone to use.

The following figure is an example of using Python for biostatistics:

Figure 1.4 – Biostatistics and Python

Its integration with advanced machine learning and bioinformatics algorithms gives a biostatistician a whole new spectrum of approaches and provides the most advanced frameworks for using biostatistical algorithms at this time.

Finally, the most important part – learning Python through a portfolio of practical projects provides you, as a reader, with two important qualities: being able to use one of the most wanted programming languages out there can be beneficial for your career, and having a portfolio of more than 10 practical projects using biostatistics and Python provides significant resources for your portfolio as someone who plans to use or advance your career by using biostatistics.

Answers for Chapter 1

(A stands for answer)

- **A6. Null hypothesis (H0)**: There are no risk factors for cardiovascular disease among the studied parameters.

  Alternate hypothesis (Ha): Risk factors are present among the studied parameters.

- A7. This question would have no concrete hypothesis. Instead, the overall goal of the study is to identify the genes that are highly expressed in lung cancer tissues..

- A8. We can re-formulate this question into three potential options based on different levels of similarity:

  - The mouse-human genome similarity is low (0-50%).
  - The mouse-human genome similarity is medium (50-90%).
  - The mouse-human genome similarity is high (>90%).

- A9. To answer this research question, we can formulate it as follows:

  Are Ca, Mg, and K concentrations higher in locality A compared to locality B?

- **A10. Null hypothesis (H0)**: Water temperature does not affect plankton.

  Alternate hypothesis (Ha): Water temperature affects plankton organisms.

Keep practicing yourself!

Summary

In this chapter, you learned about the needs and uses of biostatistics in life sciences, how to formulate the research questions, and how the Python programming language can help you with that. You also learned about the specific applications of biostatistics in biology, clinical research, and the pharmaceutical industry.

In the next chapter, you will learn in detail how to install and get started using the Python programming language.

2
Getting Started with Python for Biostatistics

Welcome to this second chapter! Here, you'll learn how to install and get started using Python. The main goal of this chapter is for you to learn how to install and get started using Python for biostatistics. To facilitate the use of the Python programming language, you'll learn how to use Jupyter Notebook and Spyder's **integrated development environment** (**IDE**), which are used to make Python programming easier.

In this chapter, we're going to cover the following main topics:

- Installing and navigating Python software on your platform
- Launching the Spyder IDE and using its interfaces
- Installing and using packages in Python
- Installing Python software on your platform

The Python programming language is open source and available for anyone to use. It can be used for academic, research, or even commercial purposes. Additionally, many types of IDEs facilitate and make it easier for us to learn Python programming. Some of the most popular IDEs for Python are the Spyder IDE and Jupyter Notebook, both of which will be used in this book. The easiest way to install Python along with its IDEs is by installing Anaconda Navigator, a data science platform.

Here are the steps for installing Anaconda Navigator:

1. Go to `www.anaconda.com`, Anaconda Navigator's official website.

2. Under the **Products** tab, go to **Anaconda Hub**, then **Distribution**, as shown in the following screenshot:

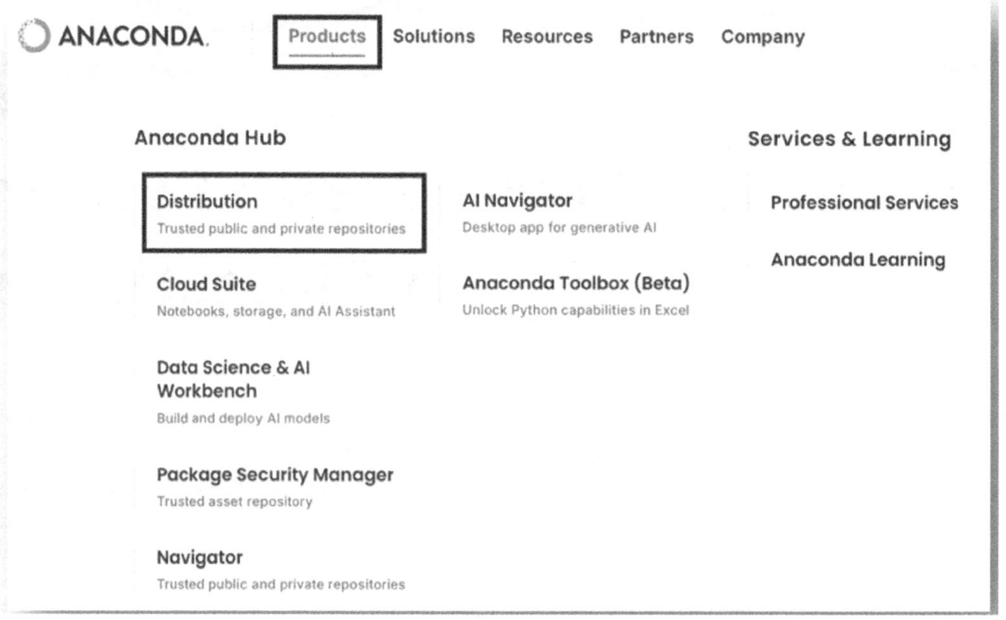

Figure 2.1 – Anaconda Hub – Distribution

3. Then, follow the necessary steps, such as providing an email and choosing your operating system. There are three options for this: Windows, MacOS, and Linux. Upon choosing **Distribution**, you'll be redirected to the download page for Anaconda Navigator.

You can use the **Download** button if your operating system is suitable. If not, choose the option that matches the operating system on your computer, as shown here:

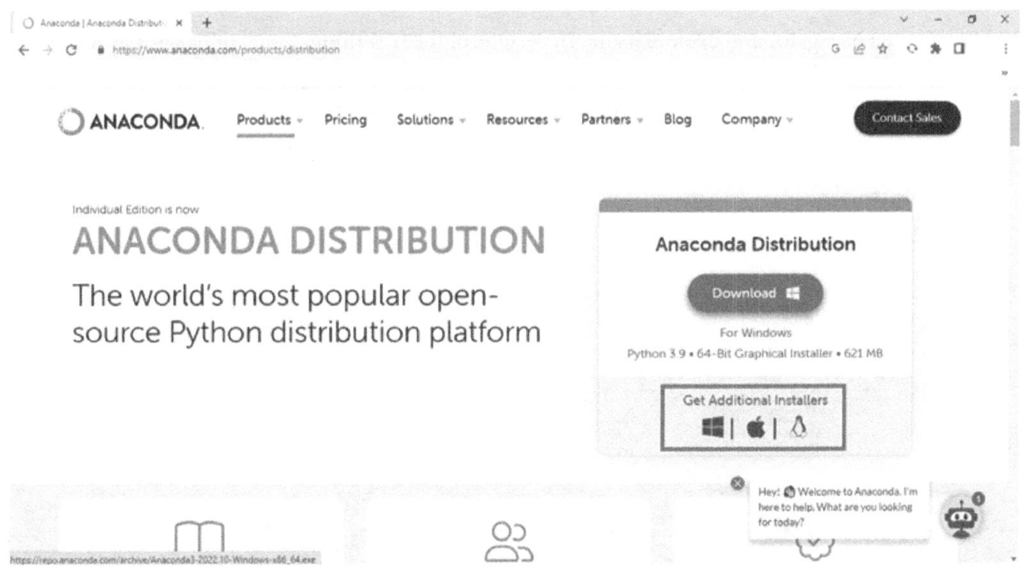

Figure 2.2 – Downloading the Anaconda Navigator

At this point, we can install it. Here are the steps:

1. Double-click on the downloaded file.
2. Click **Next** until you're redirected to the **Advanced options** window.
3. Select **Register Anaconda as my default Python** and click **Install**.

Make sure that Anaconda is installed in your system. For this book, we'll be using Windows.

Once Anaconda Navigator has been installed, it's installed automatically alongside Python. This means that Python will be part of the Anaconda framework and doesn't have to be installed separately. You may open it either from the start menu of your operating system or using the desktop shortcut.

Open Anaconda Navigator and explore its interface. One of the first things you'll notice is that Anaconda Navigator has many different tabs and boxes so that you can use different software. Jupyter Notebook, Visual Studio, PyCharm, RStudio, and Spyder IDE are among the available options, as shown here:

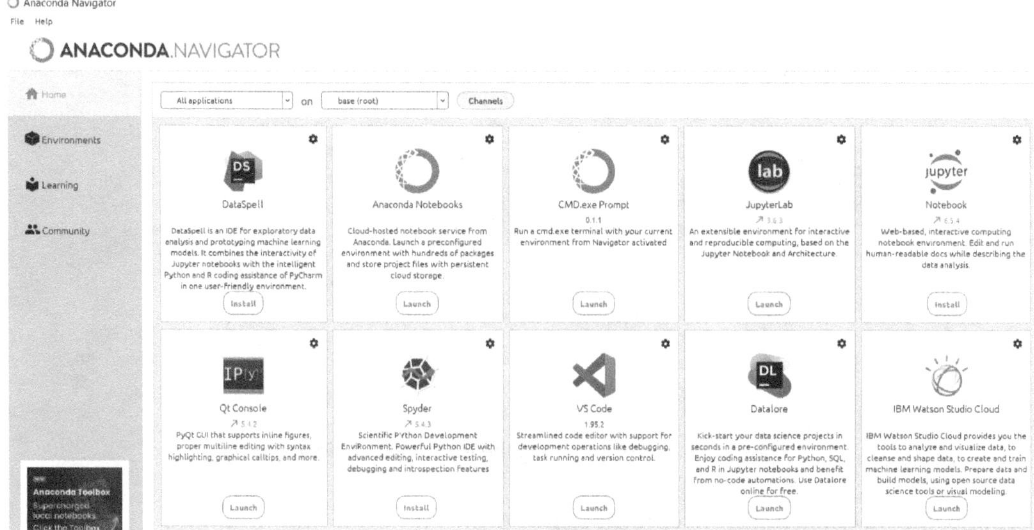

Figure 2.3 – Exploring Anaconda Navigator's interfaces

For this book, we'll be using Jupyter Notebook and Spyder as they are one of the best combinations for biostatistics. To proceed, we'll need to install them, something that Anaconda Navigator makes very easy. Any of the required software can be installed by simply clicking the **Install** button located below the icons. In newer versions of Anaconda, the installations for Spyder and most other software are automatic and already installed. If this isn't the case for you, use the **Install** button, as explained previously:

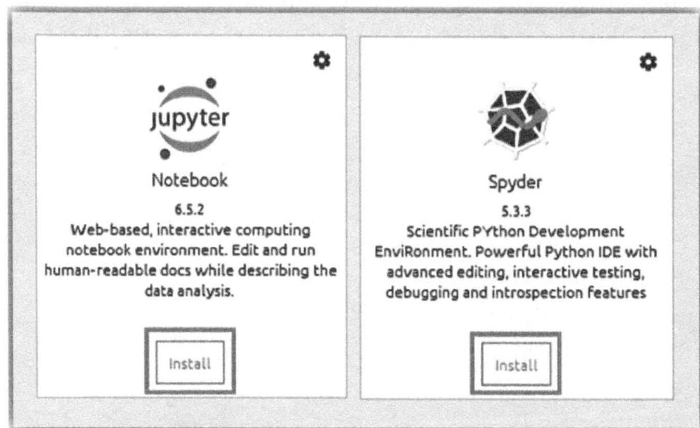

Figure 2.4 – Installing Jupyter Notebook and the Spyder IDE

Once you've initialized the installation process for Jupyter Notebook, repeat this process for the Spyder IDE. Anaconda Navigator will install the software, a process that may take a few minutes. I recommend that you restart your computer after the installation process. After restarting, you can launch and use both Jupyter Notebook and Spyder.

In this section, you learned how to download Anaconda Navigator, choose your operating system, and follow the installation steps. In the next section, you'll learn how to open and explore the different functionalities and interfaces of Anaconda Navigator, such as Jupyter Lab and the Spyder IDE.

Launching Jupyter Notebook and navigating its interfaces

First, we'll learn how to launch Jupyter Notebook. To do so, simply navigate to the Anaconda home panel and find the **Launch** button. Once you've installed Jupyter Notebook, instead of the **Install** button, a **Launch** button will appear, as shown here:

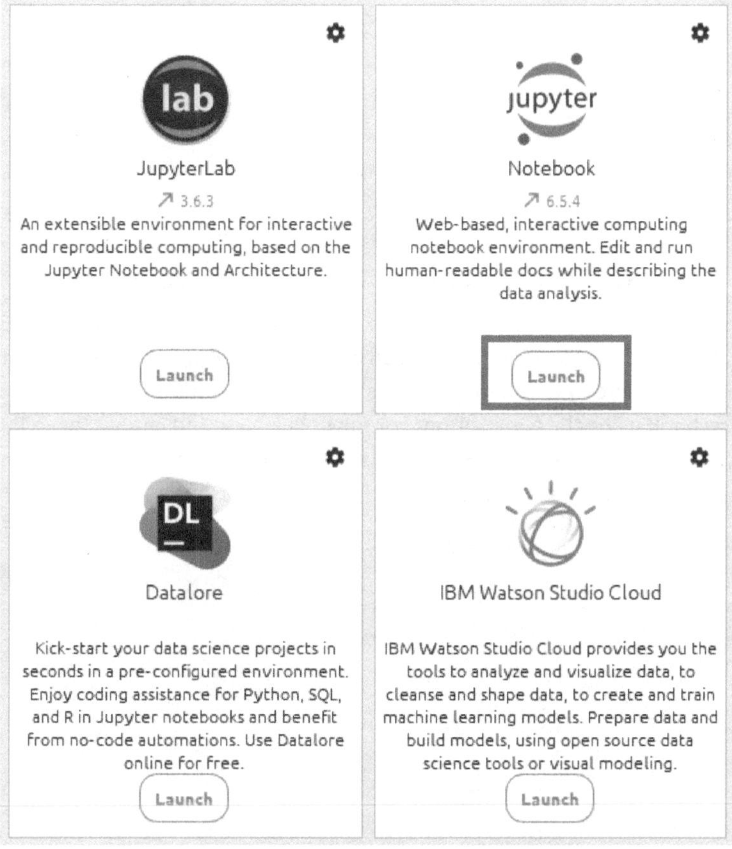

Figure 2.5 – Navigating Anaconda's windows and panels

Click on the **Launch** button and wait a couple of seconds. Upon launching Jupyter Notebook, it will automatically open in one of your internet browsers.

To avoid problems with launching Jupyter Notebook, make sure you have an internet browser installed on your computer. Jupyter Notebook will run on most standard internet browsers, including Google Chrome, Microsoft Edge, and Mozilla Firefox.

Here's the first interface you'll encounter once you launch Jupyter Notebook:

Figure 2.6 – Jupyter Notebook's management interface

As you can see, the Jupyter logo, various tabs, such as **Files**, **Running**, and **Clusters**, and buttons such as **Upload** and **New** are present. You may use this interface to load files using the **Upload** button or create a new Notebook using the **New** button. For now, click **New** and choose the Python 3 option (in some versions, this may be **New** | **Notebook**). This will create your new Python 3 Jupyter Notebook, after which it will be launched, as shown in the following screenshot:

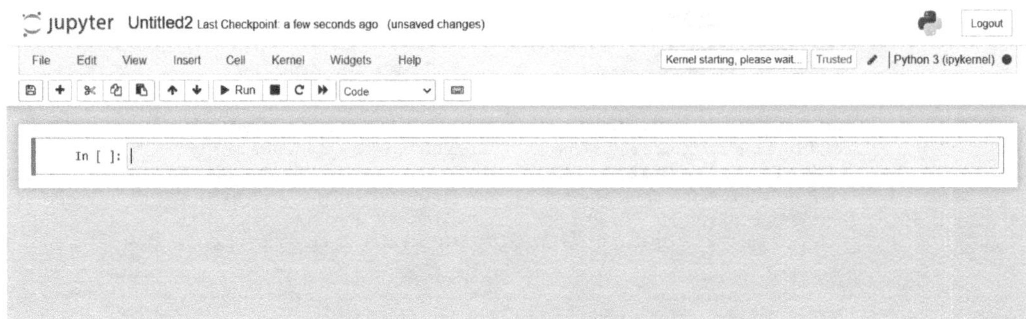

Figure 2.7– Jupyter Notebook has been launched

Using Jupyter Notebook to write Python code and a brief introduction to programming

Python programming, like most other forms of programming, is based on writing code. This code is a set of instructions for Python to perform. One of the most basic instructions is a statement to print certain text.

As an example, we're sending an instruction to Python to print the sentence, "*My first Python code.*" How can we send this instruction to Python?

Type this into your first empty Jupyter Notebook cell and click **Run**:

```
print("My first Python code")
```

Here's the output you'll see:

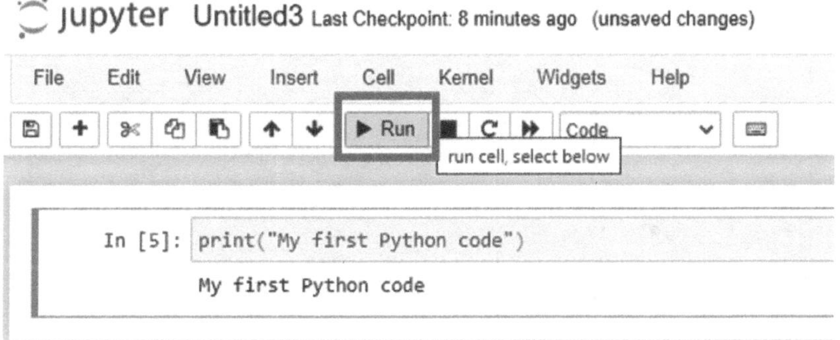

Figure 2.8 – Output after trying out some code in Python

As you can see, Python printed out `My first Python code`, as instructed.

Notice how text is defined between quotation signs and the `print` statement served as a function to print the text we wanted. The `print` statement is followed by brackets and the text to print out is written between those brackets. Quotation signs tell Python that we want the text to be printed. If we don't specify text between quotation signs, Python will print an error. Here's an example of such an error:

```
print(My first Python code)
  File "<ipython-input-4-776f56cf3b0f>", line 1
    print(My first Python code)
             ^
SyntaxError: invalid syntax
```

This syntax error occurs when the code isn't written well and because Python now assumes that our text is code, instead of being text. This is the reason why we must define the text we wish to print out using quotation marks. Single quotation marks, `'`, and double quotation marks, `"`, can both be used to define text in Python programming.

Once the first cell of code has been executed, Jupyter Notebook will automatically add a new cell so that we can write more code. You can also add as many cells as you want by using the plus sign (+). Now, you can use the new cell to perform a basic mathematical operation by typing `8+9`. The output will simply be their sum – that is, 17 – as shown here:

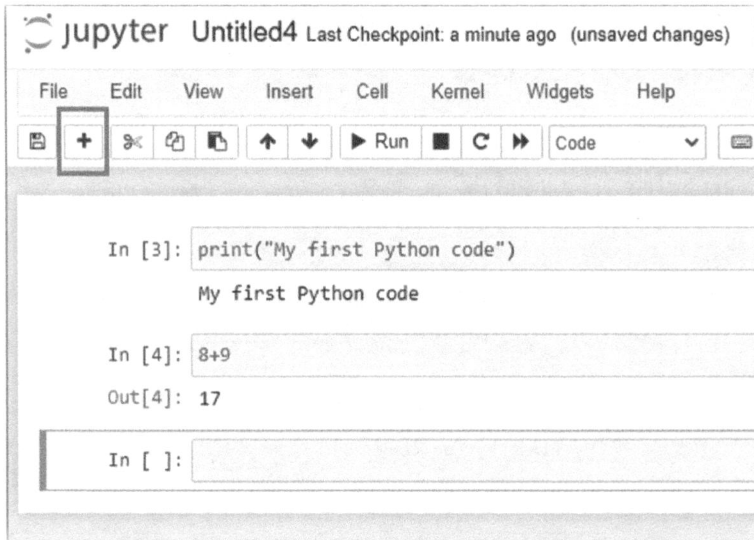

Figure 2.9 – Adding new cells and a new line of code to perform a basic mathematical operation

As you can see, no quotation marks were needed for numbers. Python can automatically recognize numbers and functions (such as print statements or print functions), while text needs to be specifically quoted to be recognized as text and differentiated from numbers and the rest of the code.

Now, let's try writing basic mathematical instructions (code) for Python to perform. We'll write the code to add 8 and 9 together.

Here's the code so far:

```
print("My first Python code")
My first Python code

8+9
17
```

Congratulations! You've successfully run your first lines of code and learned how to differentiate text, functions, and numbers in code.

Once Python code has been written, it's always good to document and explain what the code means. This can be done by adding comments to your code. But how can you add comments without affecting the code and producing errors? This can be done using the # sign:

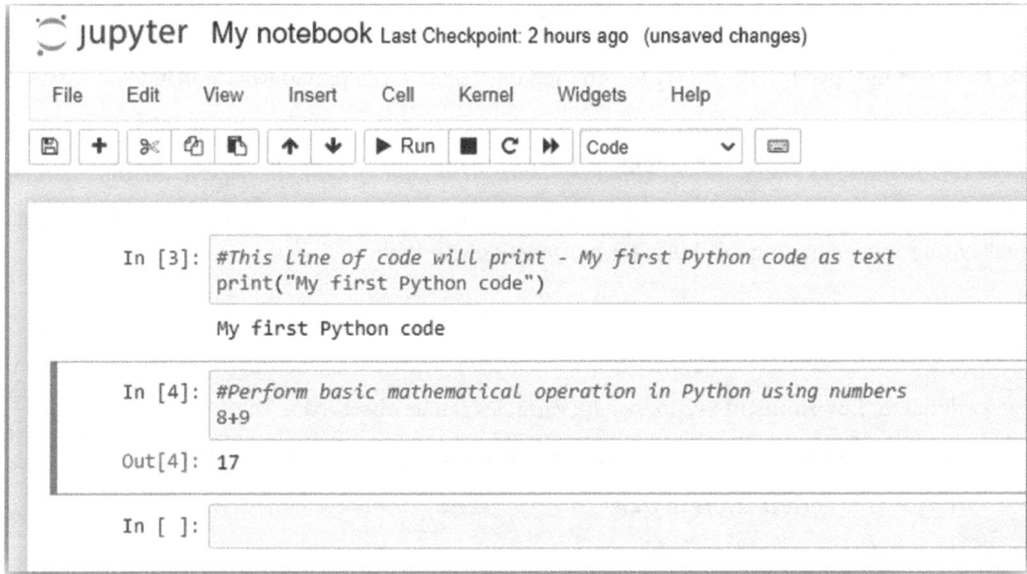

Figure 2.10 – Adding comments to code using the hash sign

Adding the hashtag will make the comments unreadable to Python. This means they won't affect the execution of the code but will remain as comments for us to explain and document our code. Every good biostatistician documents their code in detail, so keep this in mind while you're learning biostatistics with Python.

So far, we've uncovered the following coding principles:

- The `print()` function can be used to print textual statements
- Statements to be printed are written between brackets
- Quotation signs are used to define text and differentiate it from numbers and functions such as `print()`
- Hashtag (#) signs are used to differentiate comments from code
- Documenting code using hashtags is recommended

In the next section, we'll learn how to launch the Spyder IDE and use its interfaces. Here, we'll learn how to navigate the different panels of the interface and explore their functionalities.

Launching the Spyder IDE and using its interfaces

Using Anaconda Navigator is one of the easiest ways to launch Python applications. In this case, the Spyder IDE can be launched using Anaconda Navigator's panels.

So, let's open the Spyder IDE and try to run the same code we ran previously using it.

If you scroll through the Anaconda Navigator panes, you'll find the Spyder IDE icon. Note that Spyder stands for **Scientific Python Development Environment**. This is where its name is derived.

The Spyder IDE makes working with Python code much easier. Here are some of the features that make using Spyder an optimal choice for biostatistics with Python:

- It's easier to write code
- It provides detailed settings and preferences tabs
- It has an **Environment** section where variables can be observed
- It provides point-and-click options to facilitate loading and examining data
- It's easier to correct errors in code
- It provides an intuitive interface

These will be discussed further in the remaining part of this chapter.

Next, we'll open the Spyder IDE.

There are three main interfaces in Spyder:

- A section where code can be written as a script
- A section where the environment and help can be viewed
- The Python console

These are depicted in the following figure:

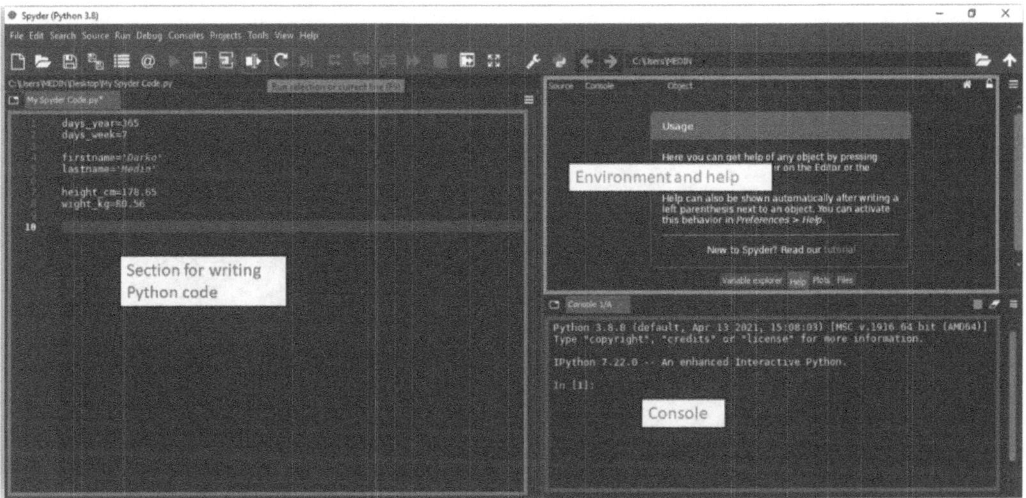

Figure 2.11 – The Spyder IDE

In the preceding screenshot, on the left-hand side, you'll find the section for writing Python code. This section isn't organized into cells, as was the case with Jupyter Notebook), but rather as a whole, so you may type code wherever you want in this section. The code is organized as lines of code, so you may see that each line is enumerated from 1 to 10. This is the section where the majority of coding operations will be performed. Here, you can write code, edit it, and correct any errors. Once the code is ready, you can run it by pressing *F9* or the **Run** selection button. But before we start running some code, let's look at the other sections.

In the top-right part of the interface, you'll find the **Environment** section.

There are a few buttons in this section:

- You can use the **Variable Explorer** button to view your variables and Python objects with a single click of a button. This is what makes Spyder one of the best IDEs for Python programming and facilitates its use for areas such as biostatistics.
- At the top right of the interface, you'll find the **Help** button. This button provides explanations about different packages and functions in Python. We'll discuss this in more detail in the next few chapters.
- Next to the **Help** button, we have the **Plots** button. We often use this to view plots that are outputted using Python programming.
- Finally, we have the **Files** button, which can be used to view, access, and load different files to the Spyder environment.

Now that you're familiar with Spyder's basic interfaces, let's try to run some code.

Selecting and running code in Spyder

The button highlighted on the right will run the selected code. *F9* is the default shortcut in Spyder for running such lines of code:

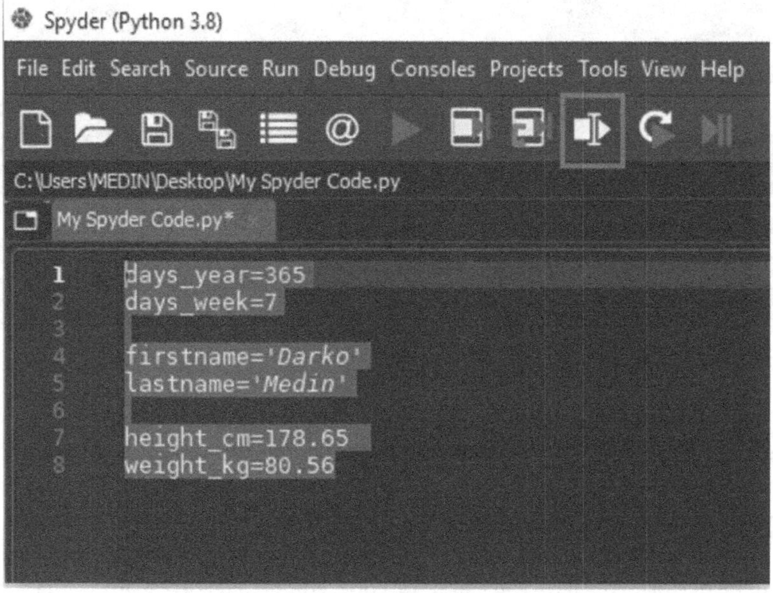

Figure 2.12 – Running code in Spyder

You may select as much code as needed and then run it by clicking the **Run** button or pressing the *F9* key on your keyboard. Now, write some code in the coding section to store some numbers and strings in Python objects. How can you do that? Here's the code:

```
days_year=365
days_week=7

firstname='Darko'
lastname='Medin'

height_cm=178.65
weight_kg=80.56
```

Here, we used the = sign to add data such as numbers and strings to a Python object. This way, we can store different values and save them in a Python environment, something we can check using the **Variable Explorer** button.

To check that our code runs well, we must check the console output. Navigate to the lower right, where **Console** is located, and check the output:

```
In [1]: days_year=365
   ...: days_week=7
   ...:
   ...: firstname='Darko'
   ...: lastname='Medin'
   ...:
   ...: height_cm=178.65
   ...: weight_kg=80.56

In [2]:
```

The output produced no errors, which means our code is OK and that the values that were defined were stored in objects.

Notice how no spaces are present in the names of Python objects. If the names contain more than one word, then either using the underscore (_) sign or merging two words is recommended. This is because having a space or other special characters in the names of Python objects would produce errors.

Here's an example of what would happen if any spaces were present in the names of objects:

```
days year=365
  File "C:\Users\MEDIN\AppData\Local\Temp\ipykernel_25844\4014293106.
py", line 1
    days year=365
         ^
SyntaxError: invalid syntax
```

The output will contain a syntax error. This error is caused by having a special character such as a space in the object's name.

In addition to special characters, syntax errors can occur when a number is the first character in the object's name (note that numbers can be in other parts of the variable's name). Here's the error that would be produced if we were to use a number:

```
1stname='Darko'
  File "C:\Users\MEDIN\AppData\Local\Temp\ipykernel_25844\2929313869.
py", line 1
    1stname='Darko'
     ^
SyntaxError: invalid syntax
```

Now, switch your view to the **Environment** section and click the **Variable Explorer** button. You'll see all the Python objects you created:

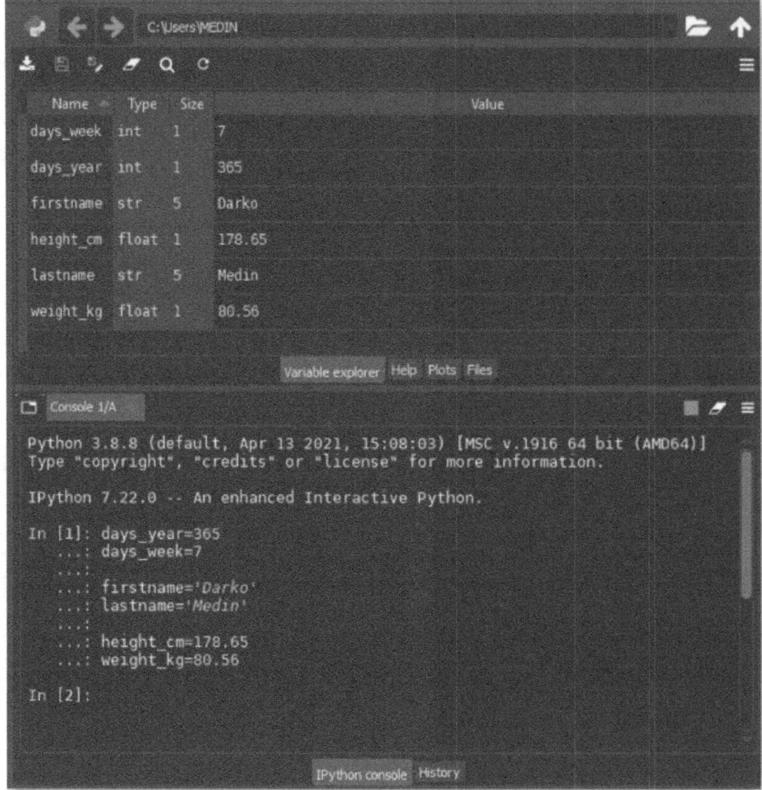

Figure 2.13 – Viewing all the Python objects that have been created

Notice how variables have certain tabs, such as **Name**, **Type**, and **Size**. We defined the names previously, so it makes sense that those are the names that are outputted. Now, using these names of objects, we can access the values and strings stored in them.

There are three basic data types in Python.

- **String**: Composed of sequences of letters and marked as text using quotation marks. Examples include **Darko**, **Medin**, **Car**, and **This is a text variable**.
- **Integer**: A discrete number with no decimal places. Examples include 365 (days in a year), 7(days in a week), and 100 (subjects in a study).
- **Float**: A continuous number that often contains decimal places. Examples include 178.65 (cm), 80.56 (kg), and 36.5 (Celsius degrees).

Note that more data types are available, such as Booleans, lists, tuples, ranges, dictionaries, and many more, all of which will be discussed later in this book.

The Python version we've installed is 3.8, though it may be higher. If any dependency errors occur, you can always downgrade the version to any version by using the following code:

```
conda create -n py38 python=3.8
```

This will create a specific environment (in this case, `py38`). It is advised that you use the later versions of Python for certain functionalities, such as 3.11 or above. However, for this book, 3.8 works optimally.

Installing packages in Python

One of the ways you can install packages is by using Anaconda's command prompt. Simply open the Anaconda Prompt from the start menu and type the following command:

```
pip install pandas
```

Here's the output:

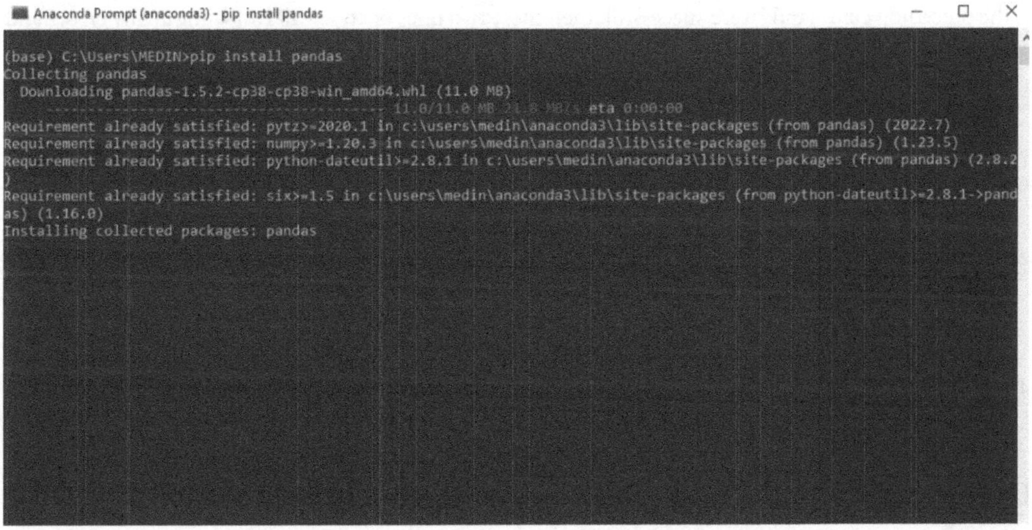

Anaconda Prompt (anaconda3) - pip install pandas

```
(base) C:\Users\MEDIN>pip install pandas
Collecting pandas
  Downloading pandas-1.5.2-cp38-cp38-win_amd64.whl (11.0 MB)
     ---------------------------------------- 11.0/11.0 MB eta 0:00:00
Requirement already satisfied: pytz>=2020.1 in c:\users\medin\anaconda3\lib\site-packages (from pandas) (2022.7)
Requirement already satisfied: numpy>=1.20.3 in c:\users\medin\anaconda3\lib\site-packages (from pandas) (1.23.5)
Requirement already satisfied: python-dateutil>=2.8.1 in c:\users\medin\anaconda3\lib\site-packages (from pandas) (2.8.2
)
Requirement already satisfied: six>=1.5 in c:\users\medin\anaconda3\lib\site-packages (from python-dateutil>=2.8.1->pand
as) (1.16.0)
Installing collected packages: pandas
```

Figure 2.14 – Installing packages in Anaconda

You can repeat the same process for the `numpy` package. To do so, run the following code in your command prompt:

```
pip install numpy
```

Here's the output:

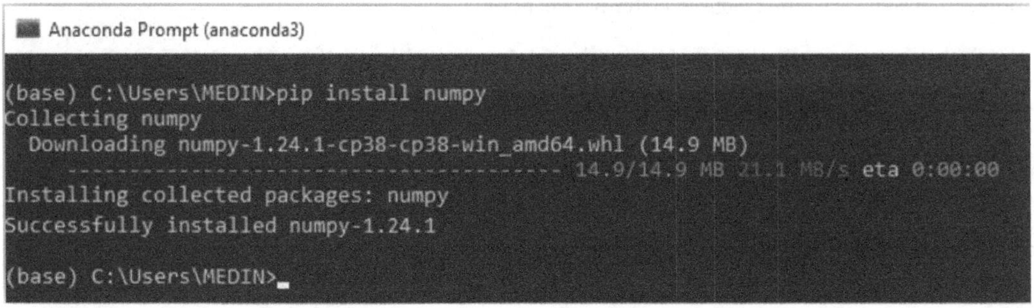

```
Anaconda Prompt (anaconda3)

(base) C:\Users\MEDIN>pip install numpy
Collecting numpy
  Downloading numpy-1.24.1-cp38-cp38-win_amd64.whl (14.9 MB)
     ---------------------------------------- 14.9/14.9 MB 21.1 MB/s eta 0:00:00
Installing collected packages: numpy
Successfully installed numpy-1.24.1

(base) C:\Users\MEDIN>_
```

Figure 2.15 – Installing the numpy package

Now, try running the following lines of code:

```
import pandas as pd
import numpy as np
```

If the preceding commands were successful, then these two lines of code will run without any errors and you can now use the `pandas` and `numpy` packages. Note that you may run into the following error:

```
import pandas as pd
Traceback (most recent call last):

  File "C:\Users\MEDIN\AppData\Local\Temp\ipykernel_13244\2886829073.
py", line 1, in <module>
    import pandas as pd

ModuleNotFoundError: No module named 'pandas'
```

This means the `pandas` package wasn't installed. In this case, make sure you install it by running `the pip install`, as described in this section.

Loading data in Python – how to load the Iris dataset

The next step is to learn how to load data in Python. One of the simplest ways to load data is to use the data already present in Python packages and load it directly. We can use the `sklearn` package to load the dataset using the `datasets.load` function. This function can be used to load the Iris dataset (Ronal Fisher, 1936) directly into the Python environment using the `sklearn` package.

The simplest way to load the Iris dataset is by running the following code:

```
#load the libraries needed to load Iris dataset
import pandas as pd
from sklearn import datasets

#Load Iris dataset using the sklearn datasets directly
iris = datasets.load_iris()
```

The preceding code will load the `iris` dataset from the `sklearn` or `scikit-learn` package directly. Notice how the `load_iris()` function was used to load the dataset and `iris = datasets.load_iris()` was used to assign the Iris dataset to the `iris` object.

Now, to explore the loaded data, use the **Variable Explorer** button in the **Environment** section:

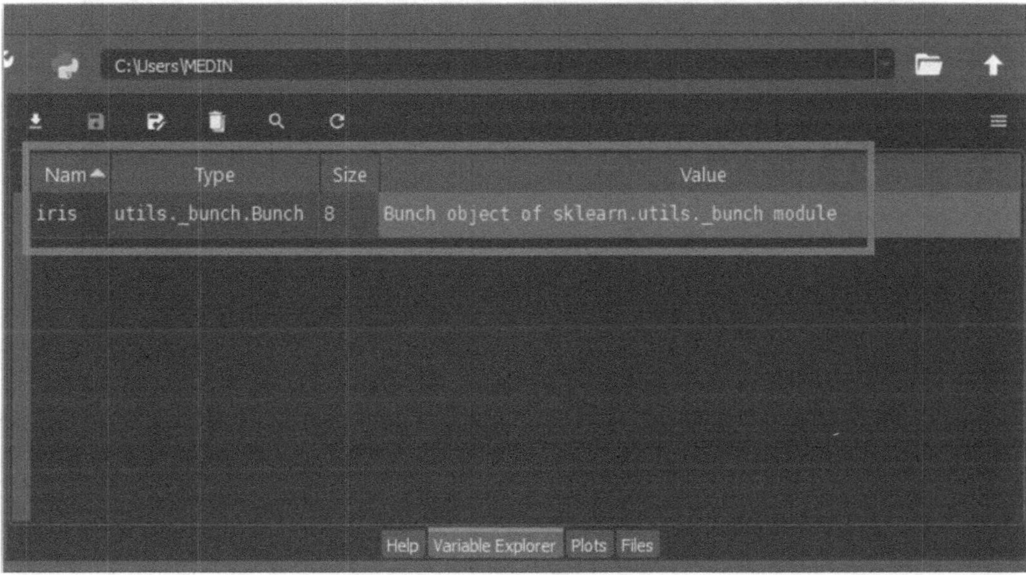

Figure 2.16 – Exploring objects using Variable Explorer

As you can see, the `iris` object is created. Note that it's a bunch object. A bunch object can store data but also different attributes about the data, such as variable names and explanations about the data. Just double-click anywhere on the Bunch object to start exploring the Iris dataset.

Exploring the Iris dataset

Click on the `iris` object shown in *Figure 2.15*. You'll see the structure of the Iris data, which is in the form of a dictionary. In Python, a dictionary is a collection of key-value pairs. In this case, each key on the left is associated with the data part. In this example, we're interested in the `data`, `feature_names`, and `target` keys. These keys can be used to access the values from the data (flower measurements), the names of the variables, and species categories:

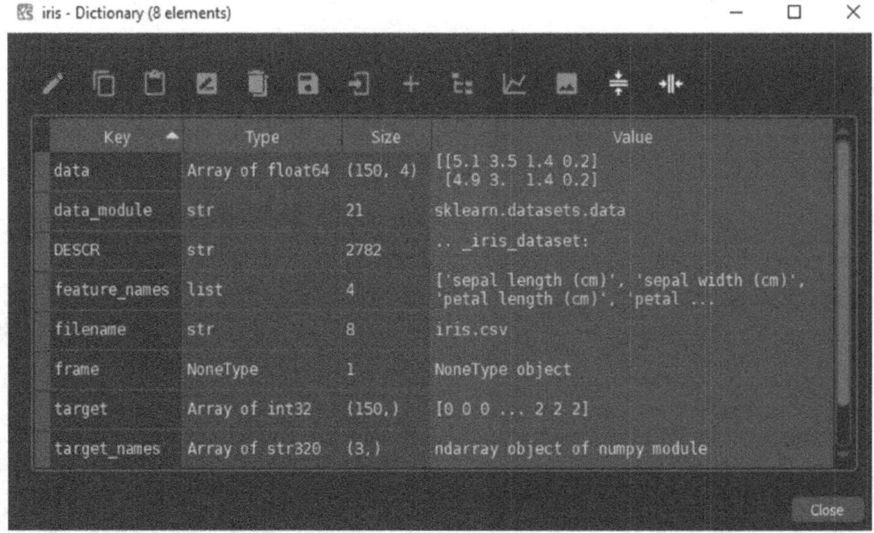

iris - Dictionary (8 elements)

Key	Type	Size	Value
data	Array of float64	(150, 4)	[[5.1 3.5 1.4 0.2] [4.9 3. 1.4 0.2]
data_module	str	21	sklearn.datasets.data
DESCR	str	2782	.. _iris_dataset:
feature_names	list	4	['sepal length (cm)', 'sepal width (cm)', 'petal length (cm)', 'petal ...
filename	str	8	iris.csv
frame	NoneType	1	NoneType object
target	Array of int32	(150,)	[0 0 0 ... 2 2 2]
target_names	Array of str320	(3,)	ndarray object of numpy module

Figure 2.17 – Iris dataset object view

The data contains the measurements of four variables in cm. These measurements are defined in `feature_names`. Here, we have sepal length (cm), sepal width (cm), petal length (cm), and petal width (cm). Another important variable in the Iris dataset is the `target` variable, which represents the Iris species.

Exploring the data and the associated variable names

As shown in the following screenshot, `feature_names` is a list of four strings (text) that associate the variables (rows, sepal length (cm), sepal width (cm), petal length (cm), and petal width (cm)) and their associated measurements, which are shown as numbers with one decimal place (floats):

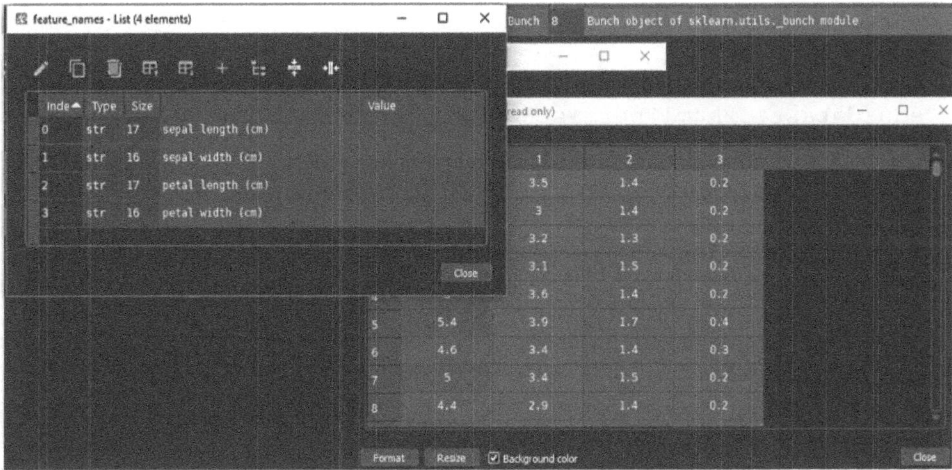
Figure 2.18 – Iris dataset

Now, have a look at the following interface (you can access it by double-clicking the **target** object in the **Variable Explorer** area):

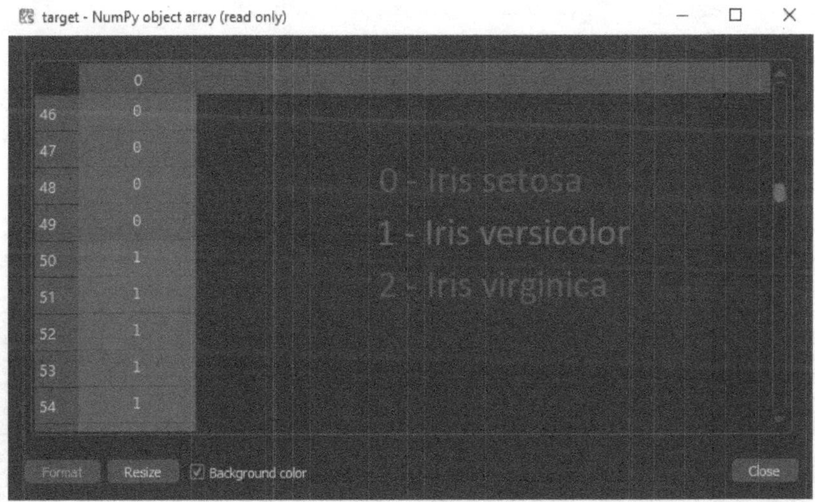

Figure 2.19 – Iris flower species in the dataset

The target variable is composed of integers 0, 1, and 2, which correspond to different Iris species. 0 is associated with Iris setosa, 1 with Iris versicolor, and 2 with Iris virginica.

In Python, the most frequently used type of a data object is called a DataFrame. DataFrames can contain multiple types of variables. They can contain strings, integers, and floats, all merged into one table. This is one of the reasons why it's very effective to load the data directly as a DataFrame.

Here's how you can do it:

1. First, download the Iris dataset from Harvard Dataverse at `https://dataverse.harvard.edu/dataset.xhtml?persistentId=doi:10.7910/DVN/R2RGXR`.
2. Once you've downloaded the file, place it in an easy-to-access place, such as your desktop or some dedicated folder.

For simplicity, we'll use the **Downloads** location to show you how to load the downloaded Iris dataset into your Python session.

If you check the properties of the downloaded file, you'll see that it's an Excel file with the `.xls` extension. This means that the pandas Excel loading function will be needed to import the dataset. The `pandas` package contains several functions that can be used to read Excel or text files. In this case, the data is in a Microsoft Excel file, so a specific function to load `.xls` or `.xlsx` files will be needed to load the data.

You can use the `pd.read_excel()` function to load Excel files into Python, like so:

```
#Load Iris dataset as an excel file
#in case of xrld no package error(use pip install xlrd)
data=pd.read_excel(r'C:\Users\MEDIN\Downloads\Iris.xls',
     sheet_name='Data')
```

You can check the loaded DataFrame in the **Variable Explorer** area. The Python object that's loaded is a DataFrame. This is the default type of Python object the `pd.read_csv()` function will create.

Now, let's explore the loaded Iris DataFrame by running the `read_excel()` function in **Variable Explorer**:

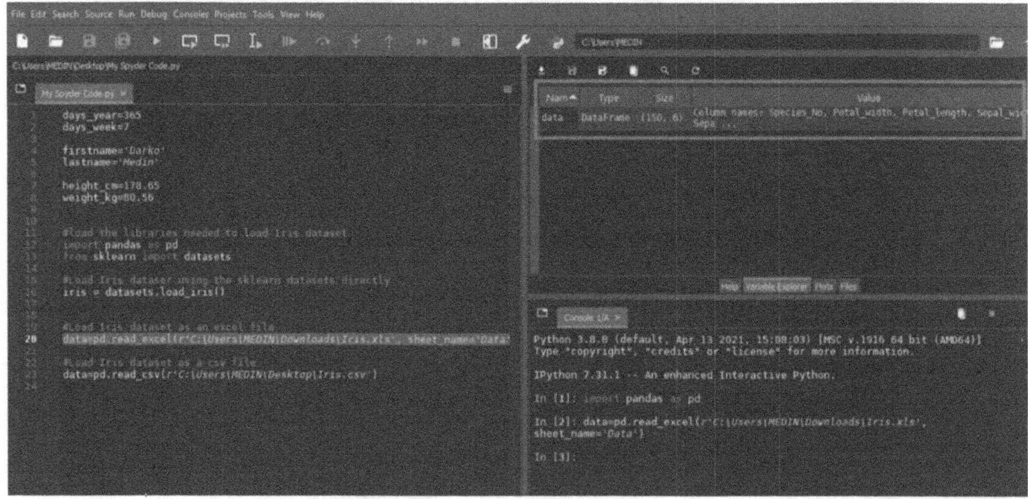

Figure 2.20 – Using the read_excel function to load the data

Here, we're loading a DataFrame whose size is (150,6). This means the DataFrame consists of 150 observations (rows) and 6 variables (columns).

Double-click anywhere on the DataFrame in the **Variable Explorer** area. The DataFrame viewer will open. A new window will open where you can view the DataFrame as a table of rows and columns. However, this time, the names of the columns have been added as variable names, and the target variable – Iris species – has also been added:

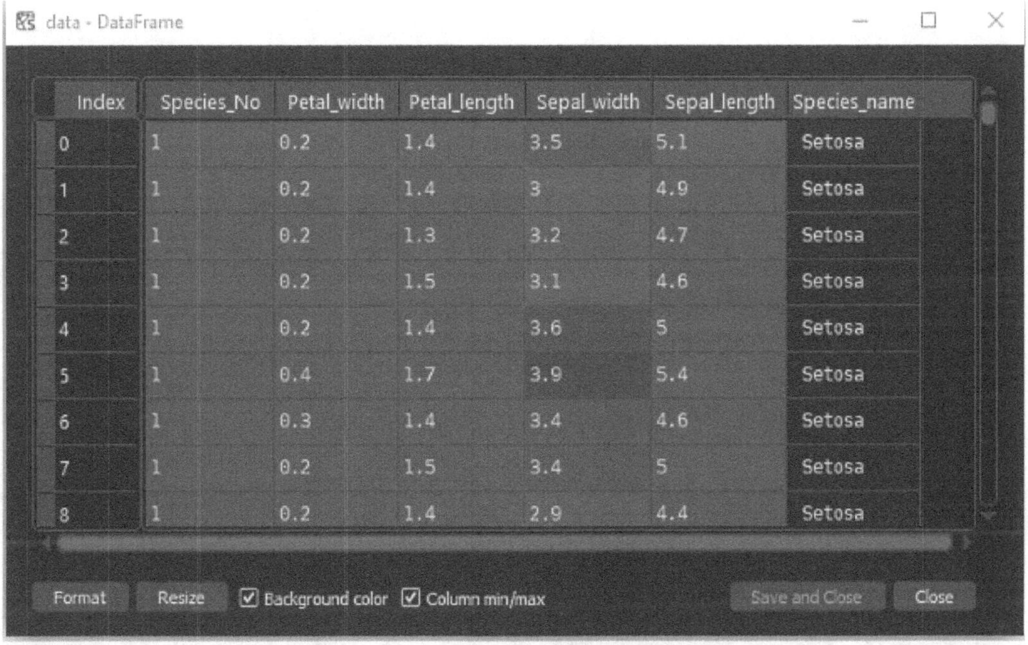

data - DataFrame

Index	Species_No	Petal_width	Petal_length	Sepal_width	Sepal_length	Species_name
0	1	0.2	1.4	3.5	5.1	Setosa
1	1	0.2	1.4	3	4.9	Setosa
2	1	0.2	1.3	3.2	4.7	Setosa
3	1	0.2	1.5	3.1	4.6	Setosa
4	1	0.2	1.4	3.6	5	Setosa
5	1	0.4	1.7	3.9	5.4	Setosa
6	1	0.3	1.4	3.4	4.6	Setosa
7	1	0.2	1.5	3.4	5	Setosa
8	1	0.2	1.4	2.9	4.4	Setosa

Format Resize ☑ Background color ☑ Column min/max Save and Close Close

Figure 2.21 – Exploring the Iris dataset

Here, `Sepal_length` (cm), `Sepal_width` (cm), `Petal_length` (cm), and `Petal_width` (cm) are floating-point variables, while `Species_name` is a string or category.

Another frequently used file format in biostatistics is the text format.

One of the standard ways to store data is as a CSV or comma-separated text file. To access `Iris.csv` with ease, open the previous Excel file. Then, choose **File| Save as** and set the filename as `Iris`. Once you've done this, set `.csv` in the file type drop-down menu. This will create another file in the downloads folder but this time as a CSV file called `Iris.csv`.

So, how can you load `Iris.csv` into Python? You can do this using the `pd.read_csv()` function.

Here's the code:

```
#Load Iris dataset as a csv file
data=pd.read_csv(r'C:\Users\MEDIN\Desktop\Iris.csv')
```

The `iris.xls` file contains two sheets. Note that the preceding code doesn't create the CSV file directly. Instead, it asks us to create the first sheet of the CSV file. Choose this option or use the following code to do so:

```
data=pd.read_excel(r'C:\Users\MEDIN\Downloads\Iris.xls',
    sheet_name='Data')
```

Let's explore the DataFrame in the **Variable Explorer** area:

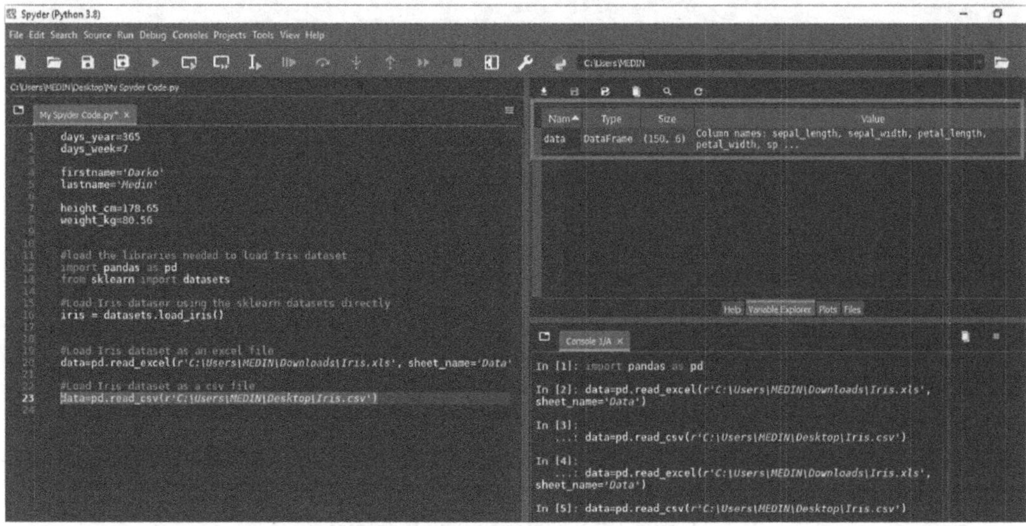

Figure 2.22 – Using the read_csv function to load the data

Note that the `pd.read_csv()` function also loaded a DataFrame consisting of 150 observations and 6 columns. The loading process is very similar to the one for Excel files.

Now, double-click on the DataFrame and explore its content further. Here, we're exploring the content of the Iris DataFrame:

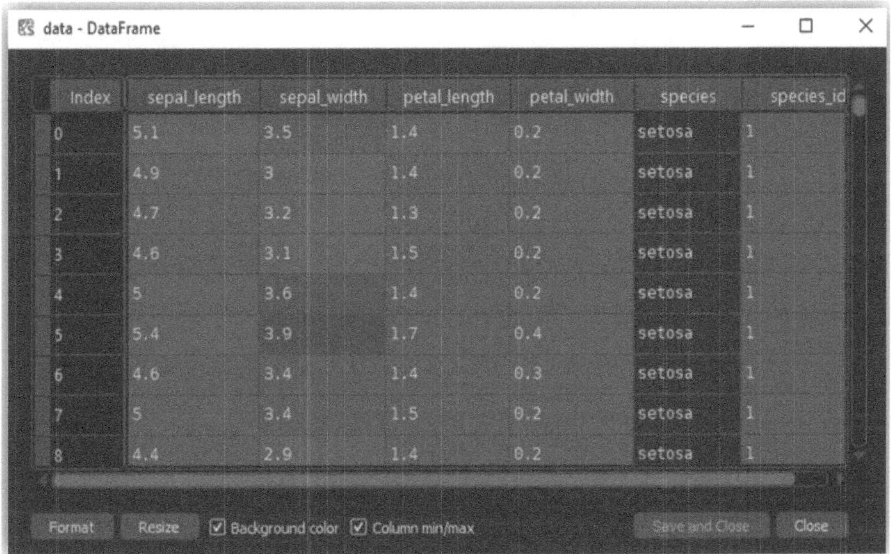

data - DataFrame

Index	sepal_length	sepal_width	petal_length	petal_width	species	species_id
0	5.1	3.5	1.4	0.2	setosa	1
1	4.9	3	1.4	0.2	setosa	1
2	4.7	3.2	1.3	0.2	setosa	1
3	4.6	3.1	1.5	0.2	setosa	1
4	5	3.6	1.4	0.2	setosa	1
5	5.4	3.9	1.7	0.4	setosa	1
6	4.6	3.4	1.4	0.3	setosa	1
7	5	3.4	1.5	0.2	setosa	1
8	4.4	2.9	1.4	0.2	setosa	1

Format Resize Background color Column min/max Save and Close Close

Figure 2.23 – Making sure the data has been loaded correctly

As in the previous loading process, petal length (cm), petal width (cm), sepal length (cm), and sepal width (cm) are floating-point variables and the species names are strings or categories that are loaded and ready for further analysis.

The data loading process is one of the most important to follow in biostatistics. Making sure the data has been loaded and checking all the variables in a DataFrame is one of the essential parts of data analysis.

Summary

In this chapter, we learned how to install Python through Anaconda Navigator. We also learned how to navigate its interfaces and use its functionalities. After, we learned how to install and use Jupyter Notebook and the Spyder IDE, as well as how to install basic Python packages such as `numpy` and `pandas`.

Finally, we learned how to load data directly from packages by using Excel or CSV files while using the Iris dataset as an example.

The next chapter will teach us how to clean and describe data using Python.

3

Exercise 1 – Cleaning and Describing Data Using Python

Welcome to *Chapter 3*! In this chapter, you'll learn how to clean and describe data using Python. Cleaning and describing data also include aspects such as validating data and exploring it, a process known as **exploratory data analysis** (**EDA**). This chapter will introduce you to loading, validating, and exploring such data.

In this chapter, we're going to cover the following main topics:

- Data types
- Loading the *Exercise 1* data using Python
- Cleaning missing values and invalid data
- Finding NaN values and invalid data types and addressing them
- Performing descriptive statistics analysis using EDA in Python

Technical requirements

Before we proceed, make sure you follow these steps to prepare your Python environment:

1. First, make sure you've installed Anaconda Navigator, Python, and the Spyder IDE, as shown in *Chapter 2, Getting Started with Python for Biostatistics.*
2. Install the necessary packages for this exercise by navigating to Anaconda Prompt (Windows) or your Terminal (macOS/Linux):

   ```
   pip install matplotlib
   pip install seaborn
   ```

   The `pandas` and `numpy` packages will be installed automatically.

3. You can install any packages in the future by using `pip install` or `conda install` (through Anaconda Prompt). The code to do so can be found at `pypi.com` (for example, you can type `matplotlib pypi` in any browser and find the installation instructions on the pypi website.

Data types

There are various types of data. One of the first aspects of data to understand before any meaningful work can be done with it is which data types are present in the dataset and what these types of data represent. Imagine collecting data from lab blood tests – we would usually see some numbers and reference ranges or numerical data. There are also types of data that aren't numerical, such as observations of different eye colors or bird species. These types of data are categorical. Note that other types of data exist.

Let's understand the main data types we can use:

- **Quantitative data**: This type of data can be quantified using different units (for example, cm and ml/L). Because this type of data is quantified using numbers, which specify how many of the units are present in a quantitative value, this type of data is often called **numeric data**. Some quantitative data types can be unitless, such as ratios, where two numbers are divided.
- **Qualitative data**: This type of data can't be quantified using measurement units, but it can be observed and placed in a specific category (for example, species of plants). Because of this, qualitative data is often also called **categorical data**.

Both quantitative and qualitative data can be further separated into data subtypes.

The main subtypes of quantitative data are continuous and discrete, whereas the main subtypes of qualitative data are nominal and ordinal. These can be seen in the following figure:

Figure 3.1 – Data types in biostatistics

Let's take a closer look at these subtypes, starting with quantitative types:

- **Continuous**: Examples of continuous data include leaf length (13.23 mm), height (170.45 cm), and glucose level (7.5 mmol/L).

  What makes this data continuous? This type of data can be separated into smaller and smaller units infinitely. If you were to use a more accurate instrument of measurement, leaf length could be 13.234563413 and so on to infinity.

- **Discrete**: Discrete data can't be divided into a limitless number of subunits of measurements like continuous data can. The most basic type of discrete data is a count. One such example is cards. You can't have 13.5 cards; the number must be whole. This is what makes the data discrete. Some examples are the number of subjects in a study, the population of people, the number of plants in a forest, and so on.

Now, let's explore the two subtypes of qualitative data – that is, ordinal and nominal:

- **Nominal**: Nominal data belongs to the categorical data type. This means it represents the categories in our data. One such example is the flowers of the Iris species. In the previous chapter, we mentioned three Iris species – **Iris setosa**, **Iris versicolor**, and **Iris virginica**. Each is a category of Iris flower. They can't be quantified but rather only categorized. This categorization of Iris species doesn't represent a scale but rather the species belonging to a specific category that isn't quantitative but qualitative. Other examples of nominal data include eye color (blue, green, brown, and so on) and blood type (A, B, AB, O, and so on).
- **Ordinal**: The Iris species can be categorized into three areas based on the size of their leaves – small, medium, and large. These three categories represent a certain order from smaller to larger leaf sizes. For this reason, this categorization is called ordinal categorization and is another form of qualitative data. Another example of ordinal data is the therapy effect scale – that is, no efficacy, low efficacy, medium efficacy, and high efficacy.

Now, let's explore some terms and metrics that are used in **EDA**.

Terms and metrics in EDA

To be able to understand statistics in detail, we need to learn about different terms and metrics that are used to describe data:

- **Count**: This is the number of observations or subjects in a study. The count of observations is often marked as n (for sample) or N (for full population) in biostatistics, though the term *count* can also be used. Each of the variables – that is, `Petal_width`, `Petal_length`, `Sepal_width`, and `Sepal_length` – has a count of 144 observations.

- **Mean**: Mean is another term for the average of the values. It's calculated as follows:

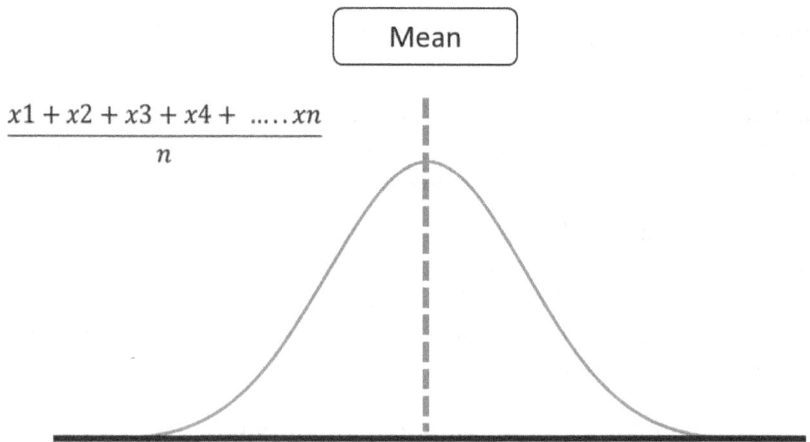

Figure 3.2 – Formula for calculating mean values

As we can see, it's calculated by adding all the observation values and dividing them by the count (n).

- **Standard deviation (Std)**: This is a standard deviation of values. This metric tells us how much the data is dispersed around the mean. In that sense, the mean is a central tendency value, and standard deviation tells us how much the data deviates from the mean. Standard deviation can be used to describe the amount of variability in the data. The larger the standard deviation, the larger the variability. This is calculated as follows:

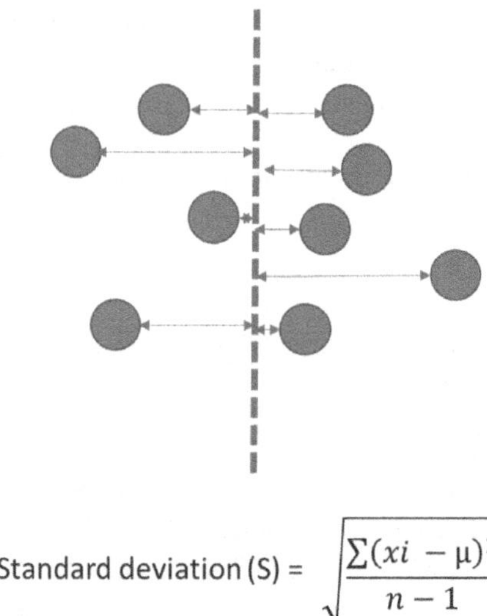

$$\text{Standard deviation (S)} = \sqrt{\frac{\Sigma(xi - \mu)^2}{n - 1}}$$

Figure 3.3 – Formula for calculating standard deviation

This formula indicates that σ (sigma) is the square root of the sum of all squared differences of values relative to the mean, divided by the number of observations.

- **Minimum (min)**: This is the minimum of the values. This metric outputs the minimal value for each of the variables.
- **Q1 (25%)**: This metric is also called **quartile 1 (Q1)**. The reason why there's a 25% mark is that Q1 marks the value that acts as the threshold for the following situation: 25% of the values are below Q1 and 75% are above Q1.
- **Q2 (50%)**: This metric is also called **quartile 2 (Q2)**. The 50% mark means that Q2 marks the value that acts as the threshold for the following situation: 50% of the values are below Q2 and 50% of the values are above Q2.
- Since Q2 separates the data into two segments containing 50% of the values, it's considered the central point and is called the **median**. Along with mean values, medians comprise the central tendency values.
- **Q3 (75%)**: **Quartile 3** (**Q3**) marks the threshold for 75% of the values, so only 25% of the values are above the Q3.
- **Maximum (max)**: The maximum value for each of the variables.

In the next section, we'll learn how to load the data for our exercise.

Loading the Exercise 1 data using Python

For this exercise, use the same `iris.csv` file from the previous chapter and open it using any spreadsheet software. We'll be using Google Sheets in this book to show you how to edit spreadsheet files. In this chapter, you'll learn how to deal with missing values. The Iris dataset is preprocessed in its original form, so it doesn't have any missing or invalid values. For this exercise, missing values will be artificially introduced.

Follow these steps to complete this exercise:

1. Open the `iris.csv` file.
2. Randomly delete (as shown in *Figure 3.4*) some of the cells (6 for this example) for the `sepal_length` row or other columns.
3. Randomly delete some of the cells for the `petal_length` row.
4. In the `petal_length` row, add `Nan` as a term.

5. Save the file as `Iris_m.csv`. You'll see the output on the next page:

Iris_m

File Edit View Insert Format Data Tools Extensions Help Last edit was seconds ago

100% $ % .0 .00 123 Default (Ari... 10 B I S A

I5 fx

	A	B	C	D	E	F	G
1	sepal_length	sepal_width	petal_length	petal_width	species	species_id	
2	5.1	3.5	1.4	0.2	setosa	1	
3	4.9	3	1.4	0.2	setosa	1	
4	4.7	3.2	1.3	0.2	setosa	1	
5		3.1	1.5	0.2	setosa	1	
6	5	3.6	1.4	0.2	setosa	1	
7		3.9	1.7	0.4	setosa	1	
8	4.6	3.4		.3	setosa	1	
9	5	3.4	Nan	.2	setosa	1	
10		2.9	1.4	0.2	setosa	1	
11	4.9	3.1	1.5	0.1	setosa	1	
12	5.4	3.7	1.5	0.2	setosa	1	
13	4.8	3.4	Nan	0.2	setosa	1	
14		3	1.4	0.1	setosa	1	
15	4.3	3	1.1	0.1	setosa	1	
16	5.8	4	1.2	0.2	setosa	1	
17	5.7	4.4	1.5	0.4	setosa	1	
18	5.4	3.9	Nan	0.4	setosa	1	
19	5.1	3.5	1.4	0.3	setosa	1	

Empty cell

Nan value

Figure 3.4 – The Iris dataset as a spreadsheet

You can use the same code from the previous chapter to load this new dataset with missing values, with a few small changes. So, let's implement those changes:

1. First, load the necessary libraries:

```
#load the libraries needed to perform exercise 1
import pandas as pd
#Load the data:
data=pd.read_csv(r'C:\Path of your file\Iris_m.csv')
```

2. This will create a `pandas` DataFrame in the variable environment. It's always a good idea to check whether the DataFrame has been created in the **Variable Explorer** area:

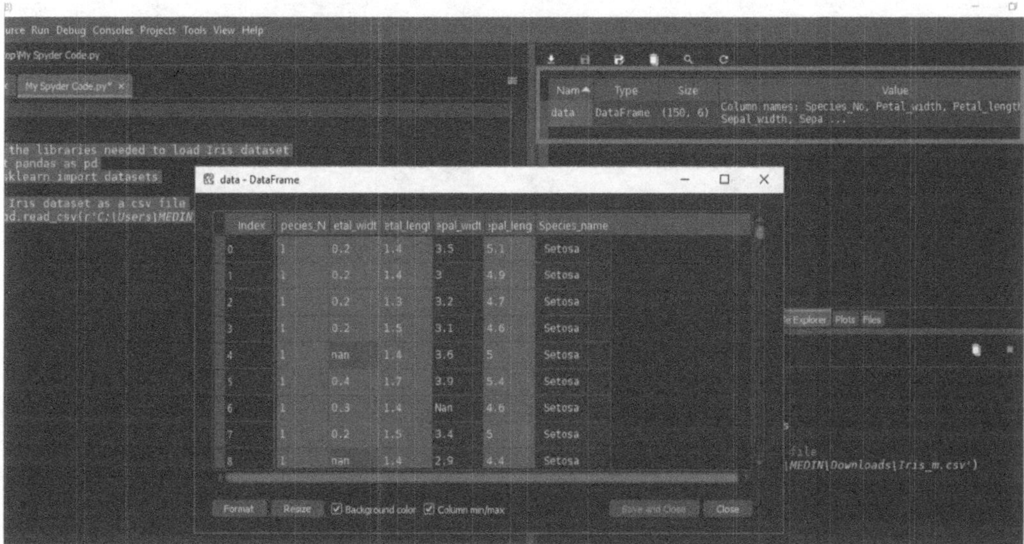

Figure 3.5 – Iris DataFrame in Variable Explorer

3. Click on the **data** object in the **Variable Explorer** area. The **View** pane will open, where you'll see the rows and columns of the data. The rows correspond to the samples, while the columns correspond to the variables of the Iris dataset. In this case, we have a new observation: missing and invalid values, **nan** and **Nan**, are present.

In the Spyder viewer, the table column can be expanded using the available sidebars. You can use these to adjust the width of the columns so that all the names are presented in full. Notice that the **Background color** box is checked, which means the numeric values will be colored, while non-numeric values won't be (they'll just have a background color):

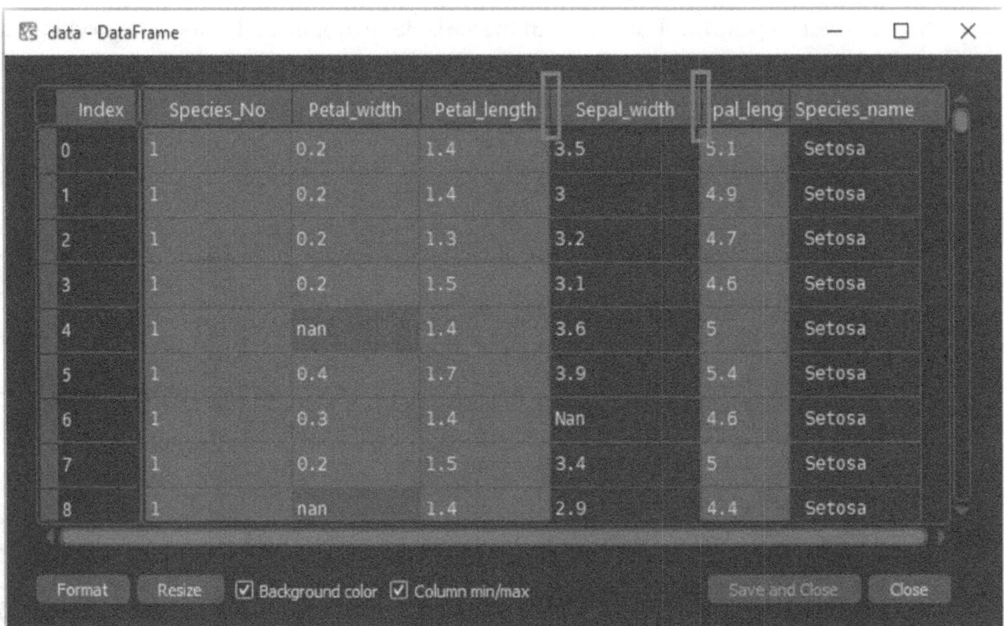

Index	Species_No	Petal_width	Petal_length	Sepal_width	pal_leng	Species_name
0	1	0.2	1.4	3.5	5.1	Setosa
1	1	0.2	1.4	3	4.9	Setosa
2	1	0.2	1.3	3.2	4.7	Setosa
3	1	0.2	1.5	3.1	4.6	Setosa
4	1	nan	1.4	3.6	5	Setosa
5	1	0.4	1.7	3.9	5.4	Setosa
6	1	0.3	1.4	Nan	4.6	Setosa
7	1	0.2	1.5	3.4	5	Setosa
8	1	nan	1.4	2.9	4.4	Setosa

Figure 3.6 – Examining the Iris DataFrame

This means we can visually identify and separate numeric and non-numeric columns. Further using this coloring system, **nan** or missing values can be identified. Indexes are used to mark the rows (by standard samples) of the DataFrame; they can be numbers, as is the case here, or other non-unique types of data, such as strings. Their unique nature enables a tabular system where each row is assigned an identification number or string.

By examining the first three variables – that is, **Species_No**, **Petal width**, and **Petal_length**, we can see that they're all numeric because they're in red. Here, **Petal width** has some **nan** or missing values. These can be identified as samples 4 and 8 in *Figure 3.6*.

Another important observation is that **Sepal_width**, which contains **Nan** (uppercase N) is read by Python as a whole column that's non-numeric.

Cleaning missing values and invalid data

By default, the pandas `read_csv()` function will read a variable as if it's non-numeric (string) if it contains at least one string (text). So, what's the difference between **nan** instances in the **Petal_width** column and the **Sepal_width** column? Python will convert empty cells into **nan** values but will keep the numeric nature of the variable, as is the case for the **Petal_length** variable.

In biostatistics, experimenters might use different words to mark a missing value, such as `Nan` or `NA` (short for *not applicable*), or even whole words such as `missing` or `not applicable`. Remember that `Nan` and `NA` are still strings, so if there's an empty cell, Python will read it as a string and coerce the whole variable into a string variable. This wasn't the case for **Petal_width** since Python read empty cells as **nan** and didn't coerce the variable into a string, instead keeping it numeric. In this case, Python read **nan** as the valid missing value, while **Nan** was just a string. To resolve this problem, all non-numeric values should be converted into one type of missing data – the `nan` value. For numeric variables such as those in the Iris dataset, the `pd.to_numeric` function can be used to coerce all the variables so that they're numeric. If the coerced variables contain any non-numeric values such as **NA**, **Nan**, or other strings, they will all be considered as empty blocks and outputted as the main missing value indicator for Python, **nan**. On the other hand, the variables will be converted into the correct data type – that is, numeric.

Here's the next block of code:

```
#len() function can tell how many samples are in the dataset (rows)
len(data)

#is.na() function can identify empty cells in the data
data.isna()

#adding the .sum() after the isna() will output the number of empty
cells
data.isna().sum()
```

Here, the `isna()` function identifies whether elements in a variable are missing. So, if the missing value is present, it will be identified as `TRUE = 1`, while if it isn't present, it will be identified as `FALSE = 0`. In this case, we can add `.sum()` to it, which will calculate the sum of missing values in each variable.

Here's the output:

```
Out[2]:
Species_No        0
Petal_width       5
Petal_length      0
Sepal_width       3
Sepal_length      0
Species_name      0
dtype: int64
```

As we can see, the `Petal_width` variable has 5 missing values, while `Sepal_width` has 3 missing values. Keep in mind that the `Sepal_width` variable also has `Nan`, with an uppercase `N`, so it won't be counted as a missing value. Instead, the whole column will be read as non-numeric. However, it also contains three `nan` values that are read by the `isna()` function as missing values.

To observe the missing values, click on the `data` DataFrame in the **Variable Explorer** area again. The **Python Viewer** area will open:

```
#Coericng the data to numeric will also convert all invalid data to
empty cells
data['Sepal_width']=data['Sepal_width'].apply(
    pd.to_numeric, errors='coerce')
```

> **Note**
>
> In Python programming, both single (' ') and double (" ") quotes can be used to create string definitions or definitions of textual, word-based variables. They're functionally identical, but it's good practice to use one of them consistently. There are some minor differences in some situations, but that's outside the scope of this book.

Here's the output:

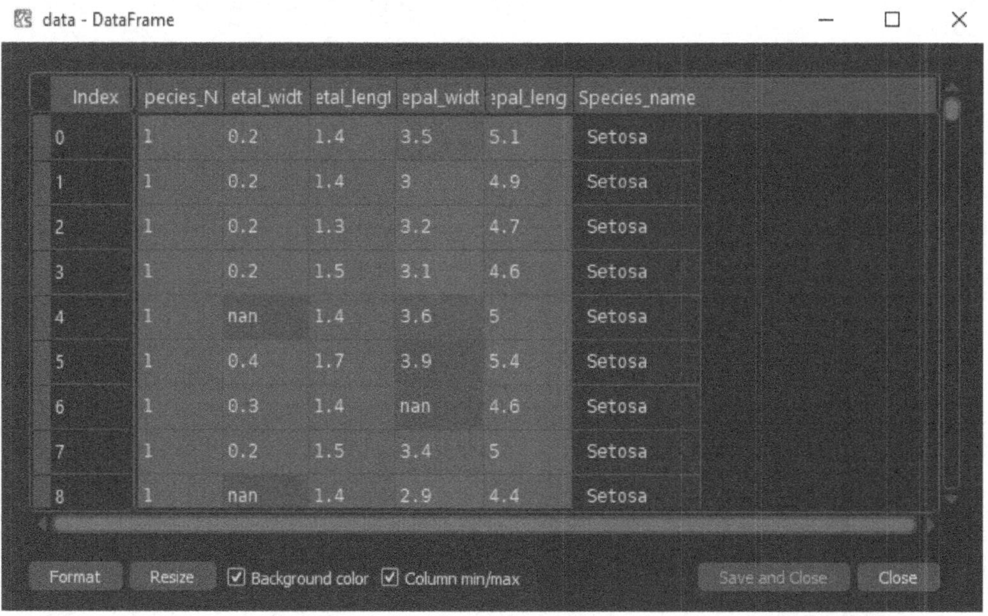

Index	pecies_N	etal_widt	etal_lengl	epal_widt	pal_leng	Species_name
0	1	0.2	1.4	3.5	5.1	Setosa
1	1	0.2	1.4	3	4.9	Setosa
2	1	0.2	1.3	3.2	4.7	Setosa
3	1	0.2	1.5	3.1	4.6	Setosa
4	1	nan	1.4	3.6	5	Setosa
5	1	0.4	1.7	3.9	5.4	Setosa
6	1	0.3	1.4	nan	4.6	Setosa
7	1	0.2	1.5	3.4	5	Setosa
8	1	nan	1.4	2.9	4.4	Setosa

Figure 3.7 – Variable Explorer

Notice that the background color of all the variables has been restored since **Sepal_width** hasn't been converted into a numeric variable. Also, note that Python coerced **Nan** strings into Python **nan** values.

Now, let's learn how to find all the nan values and how to address them.

Finding NaN values and invalid data types and addressing them

At this point, all the missing values are uniform nan value. So, the next step – that is, removing the nan instances – can be initiated using the data.dropna() function. Let's learn how to implement data.dropna().

First, let's perform EDA:

```
#Define the variables to consider in EDA
variables=['Petal_width', 'Petal_length', 'Sepal_width',
    'Sepal_length']


#Drop the missing or invalid data
cleandf=data.dropna(subset=variables)

#Create a new object to store the data, but this time with shorter
name
df=cleandf
```

Now, open **Variable Explorer** in your Spyder interface (the lower right part of the interface).

You'll see the following output:

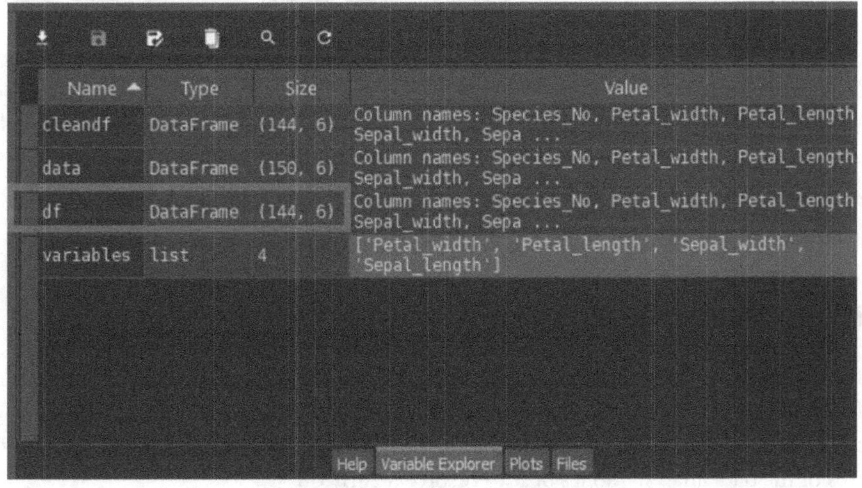

Figure 3.8 – The df data object

Once the `dropna()` function is used, all the rows containing **nan** or the missing numeric value will be removed. Since six rows contain the **nan** values, the total length of the clean dataset will be 150 minus 6 (144).

Now, open the **df** DataFrame and check whether the **nan** values have been removed:

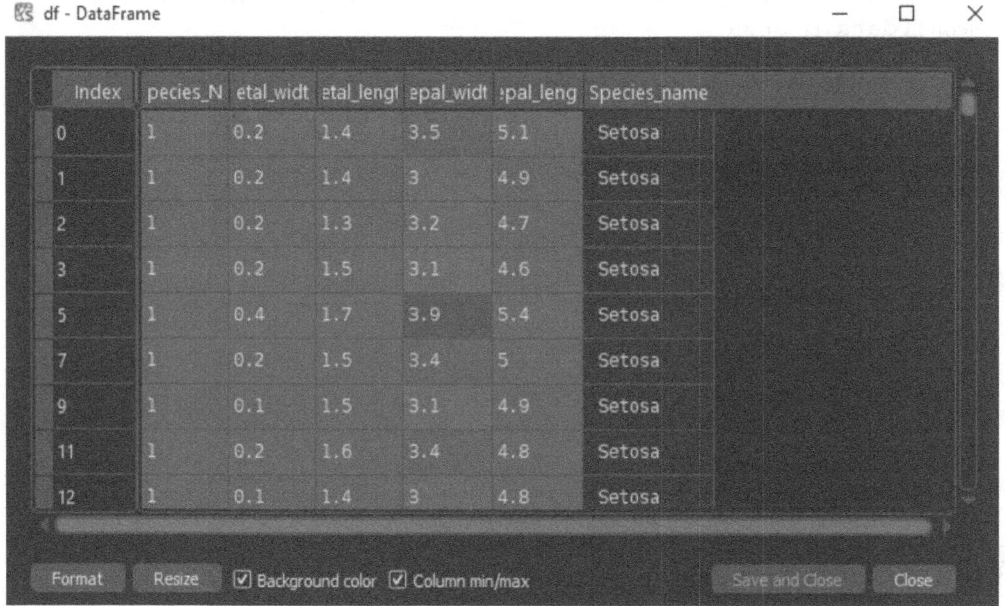

Figure 3.9 – Clean dataset

Use the scroll pane on the right to examine the whole dataset, this time checking the names of the Iris species. As we can see, each of the variables (`Petal_width`, `Petal_length`, `Sepal_width`, and `Sepal_length`) has a count of 144 observations.

Currently, the data contains one wrong Iris species name. Let's check it out.

Identifying the wrong species name

Upon scrolling down, we'll see that one of the species names contains a typo. Instead of Iris Virginica, the species is named Iris Verginica. This problem must also be addressed in the data cleaning phase. If it's not addressed at this point, the species name might show up as invalid in the plots and tables in later phases of the data analysis, so it's best to resolve it early on.

Note that the first variable, **Species_No**, is a numerical representation of Iris species. It won't be needed in future analyses, so dropping this column is also advised. Please refer to the following screenshot:

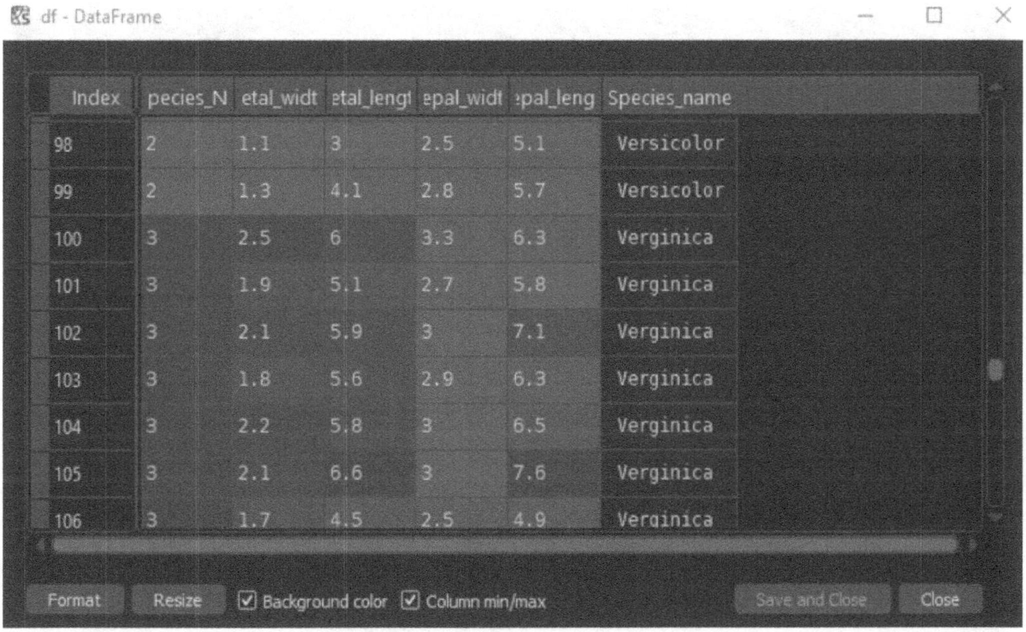
df - DataFrame
Index
pecies_N
etal_widt
etal_lengt
epal_widt
epal_leng
Species_name
98 2 1.1 3 2.5 5.1 Versicolor
99 2 1.3 4.1 2.8 5.7 Versicolor
100 3 2.5 6 3.3 6.3 Verginica
101 3 1.9 5.1 2.7 5.8 Verginica
102 3 2.1 5.9 3 7.1 Verginica
103 3 1.8 5.6 2.9 6.3 Verginica
104 3 2.2 5.8 3 6.5 Verginica
105 3 2.1 6.6 3 7.6 Verginica
106 3 1.7 4.5 2.5 4.9 Verginica
Format
Resize
Background color
Column min/max
Save and Close
Close

Figure 3.10 – Wrong species names

To resolve the invalid name, we can use `df.replace`. Note that to remove the **Species_No** column, we'll require the `df.drop()` function. These functions can be seen in the following code:

```
#Correct the invalid Species names
df = df.replace('Verginica','Virginica ', regex=True)
df=df.drop('Species_No', axis=1)
```

This chunk of code will correct **Iris Verginica** to the actual name, Iris Virginica, and remove the first column, which isn't needed.

Here's the output DataFrame:

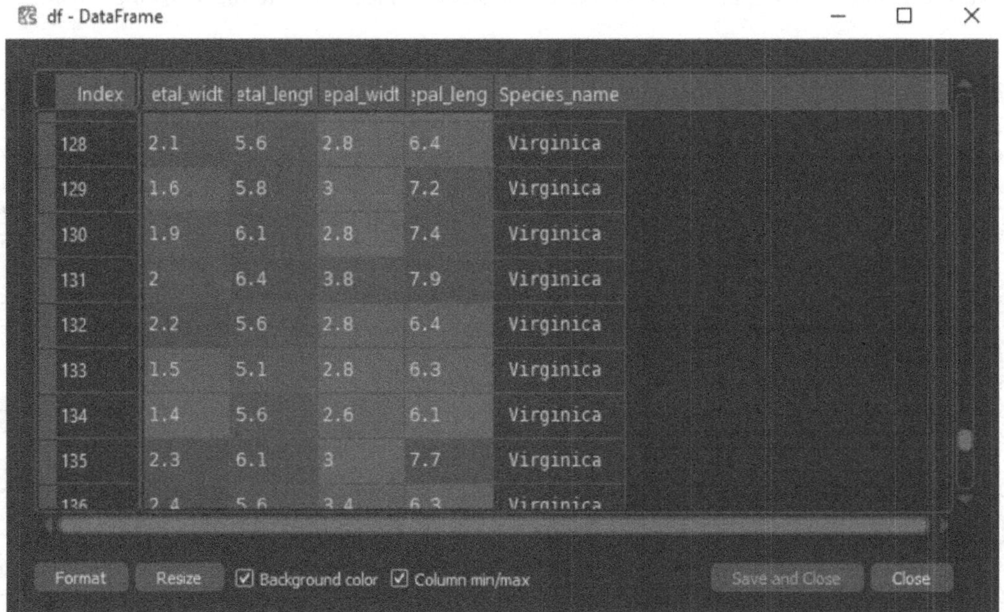

Index	etal_widt	etal_lengt	epal_widt	epal_leng	Species_name
128	2.1	5.6	2.8	6.4	Virginica
129	1.6	5.8	3	7.2	Virginica
130	1.9	6.1	2.8	7.4	Virginica
131	2	6.4	3.8	7.9	Virginica
132	2.2	5.6	2.8	6.4	Virginica
133	1.5	5.1	2.8	6.3	Virginica
134	1.4	5.6	2.6	6.1	Virginica
135	2.3	6.1	3	7.7	Virginica
136	2.4	5.6	3.4	6.3	Virginica

Figure 3.11 – Corrected species names

Now that we've cleaned the missing and invalid values, we can be sure that the data has been prepared for the next step, which is exploring it. Doing this means describing the data using descriptive statistics and visualizing different aspects of the data.

Performing descriptive statistics analysis in Python

As shown in *Figure 3.11*, the missing values have the removed, the species name has been corrected, and the first row has been removed. This means that the **df** object is now clean of any invalid data and can be used for further analysis.

The next step is essential and is called EDA:

```
#Perform the descriptive statistics in Iris dataset
df.describe()
```

Here's the output:

```
Out[2]:
       Petal_width  Petal_length  Sepal_width  Sepal_length
count   144.000000    144.000000   144.000000    144.000000
mean      1.239583      3.854167     3.045139      5.881944
std       0.751336      1.736078     0.436424      0.819999
min       0.100000      1.000000     2.000000      4.300000
25%       0.400000      1.600000     2.800000      5.175000
50%       1.350000      4.400000     3.000000      5.800000
75%       1.800000      5.100000     3.300000      6.400000
max       2.500000      6.900000     4.400000      7.900000
```

Here, there are four variables of interest: `Petal_width`, `Petal_length`, `Sepal_width`, and `Sepal_length`. They all contain 144 observations (count). The variable with the lowest mean value is `Petal_width` (`1.239` mm), while the variable with the largest mean value is `Sepal_length` (`5.881` mm).

This situation is very similar when median values are used. These values show central points in the data. However, we can use standard deviation and the Q1-Q3 (interquartile range) values to evaluate the dispersion of the data. Here, `Petal_length` has the highest level of variability and dispersion with a standard deviation of `1.736`, while the lowest variability variable is `Sepal_width`.

The mean can be combined with the standard deviation so that `Petal_length` is `3.85 mm` (±1.73 is the standard deviation and is relative to the mean). Using this format, the variable can be described in terms of its central point (± dispersion) or mean (± standard deviation).

Why do we use the ± sign? Because the dispersed data is generally contained both below and above the mean.

Now, let's learn how to view the descriptive statistics in each of the Iris species:

```
#Perform the descriptive statistics for each of the Iris species
dtable=df.groupby(['Species_name']).describe()
```

Here's the output:

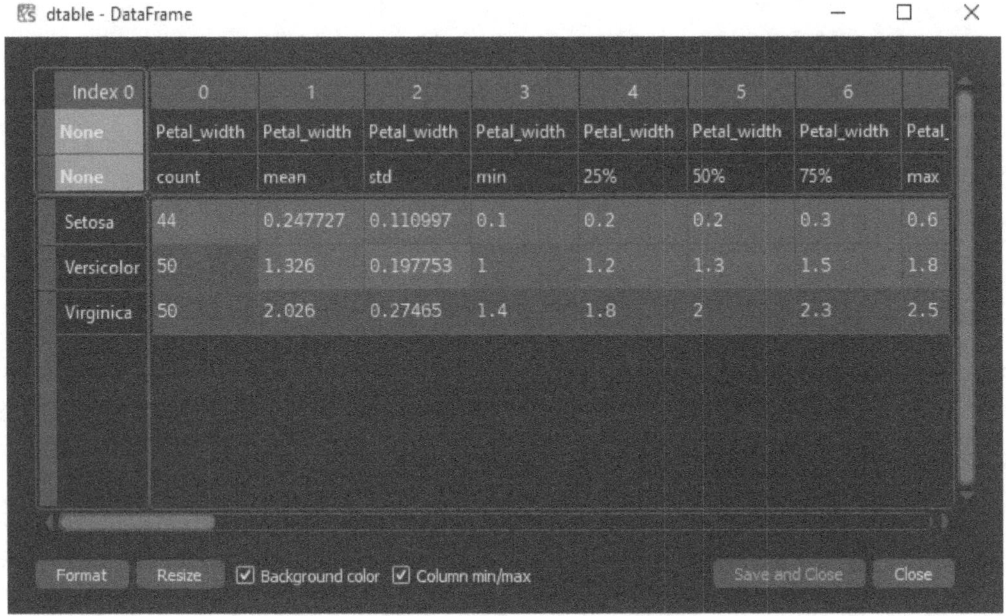

Index 0	0	1	2	3	4	5	6	
None	Petal_width	Petal_width	Petal_width	Petal_width	Petal_width	Petal_width	Petal_width	Petal_
None	count	mean	std	min	25%	50%	75%	max
Setosa	44	0.247727	0.110997	0.1	0.2	0.2	0.3	0.6
Versicolor	50	1.326	0.197753	1	1.2	1.3	1.5	1.8
Virginica	50	2.026	0.27465	1.4	1.8	2	2.3	2.5

Figure 3.12 – Descriptive statistics non-transposed

Now, let's learn how to view the descriptive statistics results in a transposed view:

```
#Transpose the descriptive table
dtable=dtable.transpose()
```

For the descriptive statistics to be more intuitive, it's always better to transform the structure of the table so that it's easier to read. In this case, we used the `transpose()` function to transpose the table and observe all the metrics vertically. To explore the output, simply click on the `dtable` object in the **Variable Explorer** area:

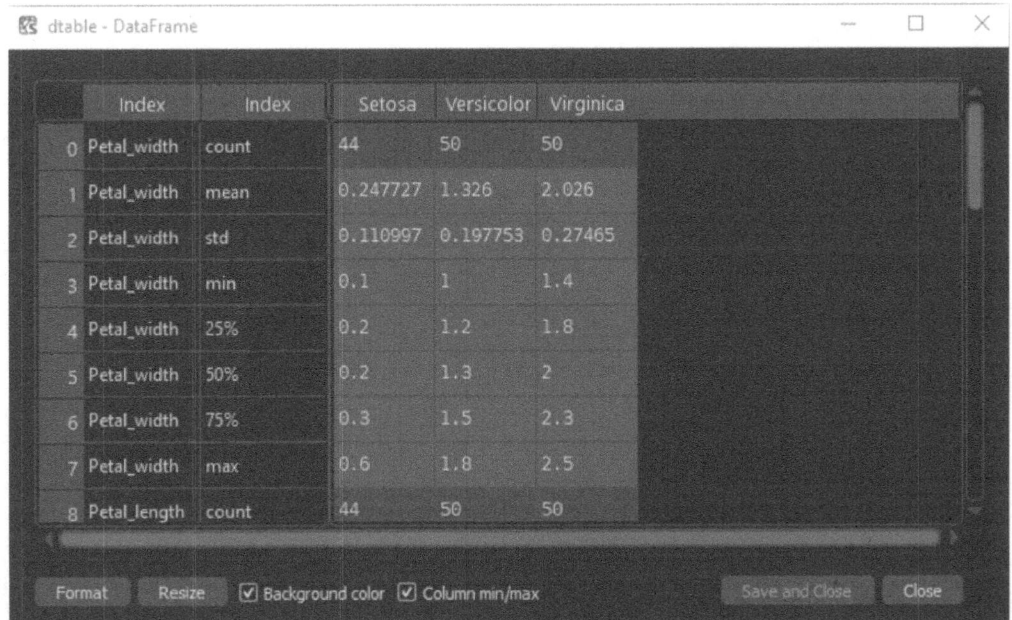

dtable - DataFrame

	Index	Index	Setosa	Versicolor	Virginica
0	Petal_width	count	44	50	50
1	Petal_width	mean	0.247727	1.326	2.026
2	Petal_width	std	0.110997	0.197753	0.27465
3	Petal_width	min	0.1	1	1.4
4	Petal_width	25%	0.2	1.2	1.8
5	Petal_width	50%	0.2	1.3	2
6	Petal_width	75%	0.3	1.5	2.3
7	Petal_width	max	0.6	1.8	2.5
8	Petal_length	count	44	50	50

Figure 3.13 – Descriptive statistics transposed

EDA starts with the most basic descriptive characteristic of the data: its distribution. One plot that can show how the data is distributed is the histogram plot. This plot shows samples on the *X*-axis (horizontal) and the values of variables on the *Y*-axis (vertical).

In this section, we learned how to describe the Iris data using metrics such as the mean, median, and standard deviation. However, one of the best ways to describe the data is to describe its distribution – that is, we can describe how the data points are visually distributed.

Continuous and discrete distributions

Normal distribution, also known as Gaussian distribution, is a data probability distribution that's used for a set of continuous observations of a variable. Why is this important? The Iris data contains four continuous variables, and it's expected for them to have a normal distribution. A typical normal distribution is bell-shaped. Here's what it looks like:

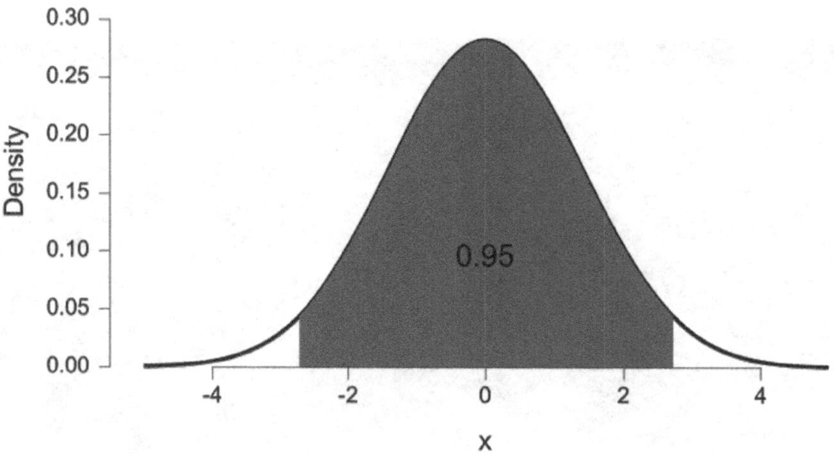

Figure 3.14 – Normal probability distribution

From this plot, we can see where the majority of observations are expected to lie. In the preceding example, the blue color signifies where 95% of the values lie in a normal distribution. The rest represent tails. As we can see, there are two tails in a normal distribution, each consisting of 2.5% of the normal distribution.

Contrary to continuous distributions such as Gaussian, discrete distributions often have heavily skewed shapes and discrete values. The following is an example of a discrete distribution called a **negative binomial distribution**. It's frequently used in biological data such as RNA gene expression data:

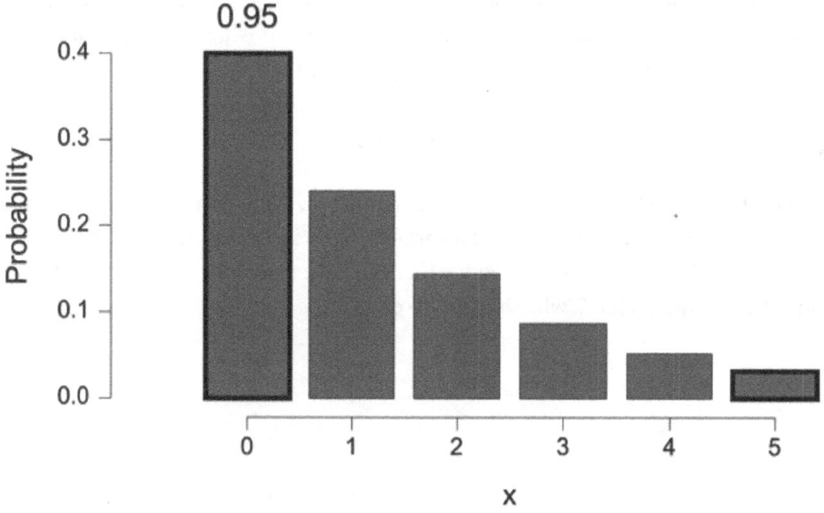

Figure 3.15 – Negative binomial distribution

The preceding figure shows what discrete count data often looks like. Here, the mean isn't the central point, and the data is heavily skewed so that it favors the lower values compared to the higher ones. In other words, in this case, most of the data is focused close to 0 and as the values increase, there are fewer observations present with those values. This tells us that the mean isn't always the best representative of the whole data, only a specific central location. This is sometimes useful to describe the data, but not completely. In negative binomial distribution, most of the values focus around 0. This helps us understand that data may have different types of distributions and different data methods may be needed to assess it.

Now, let's see what a real dataset visualization looks like by observing the Iris dataset.

Visualizing the Iris data

Let's produce some histograms and see whether the data follows a normal distribution:

```
#Perform univariate visual EDA on 4 Iris numeric variables
import matplotlib
import matplotlib.pyplot as plot
import seaborn as sns
df[variables].hist()
```

Here's what the histograms look like (navigate to **Plots** in the same panel where **Variable Explorer** is, at the bottom right of the Spyder interface):

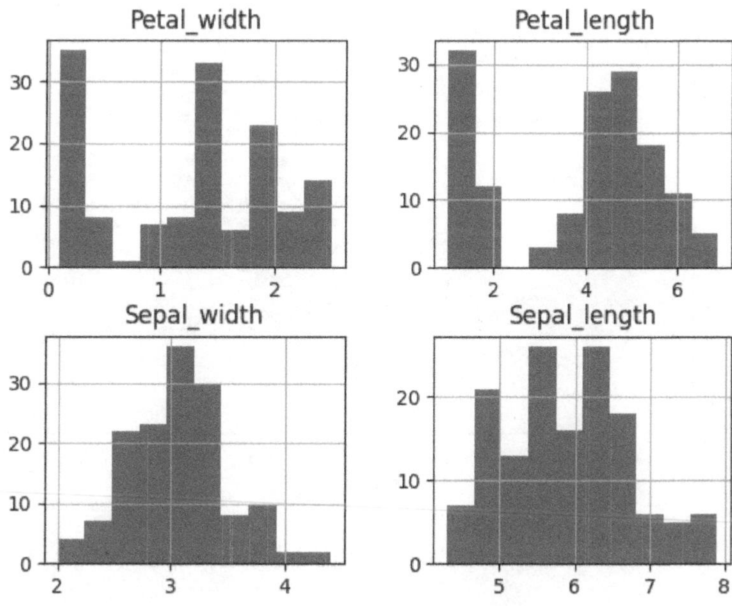

Figure 3.16 – Iris data visualizations

By using the `hist ()` function from `matplotlib`, we can plot our histograms. They are mainly used to show the data's distribution. Overall, four histograms have been created for the `Petal_width`, `Petal_length`, `Sepal_width`, and `Sepal_length` variables. All four variables have unclear distributions, with gaps that are more or less distorted. This is because three different species have been included in the dataset and they have their own distributions. This means that EDA should be continued on each of the species separately so that we can make better conclusions.

To continue our EDA based on each of the species, visualize the data using a scatter plot. This is a simple plot where each point is plotted against the *X* and *Y* axes. This scatter plot will provide insights into the differences between species in the data. Generally, scatter plots have two dimensions, so two variables need to be inputted to create them. In this case, we'll use `Sepal_length` and `Sepal_width` and the observations will be colored according to the species name. Please refer to the following code:

```
#Perform bivariate visual EDA using Sepal_length and Sepal_width
sns.scatterplot(x=df["Sepal_length"],
                y= df["Sepal_width"],
                hue=df["Species_name"])
```

We'll get the following output:

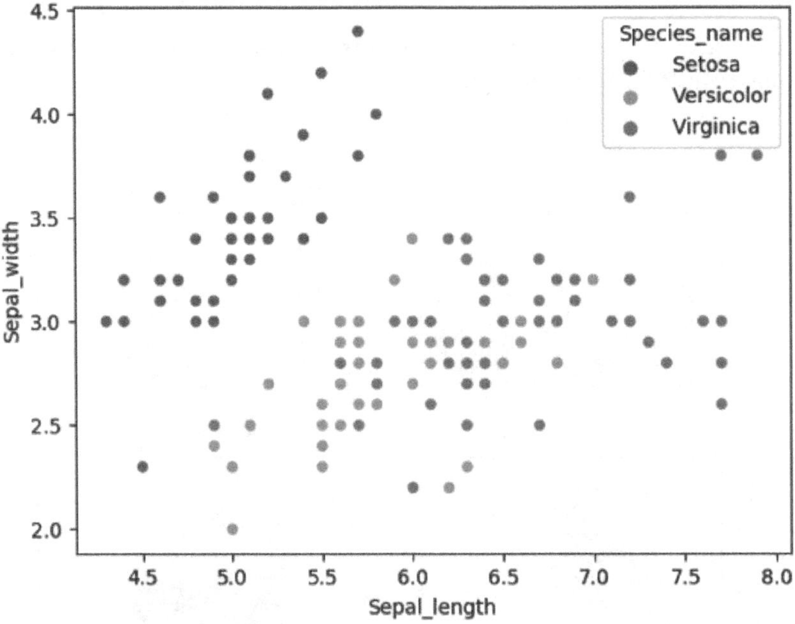

Figure 3.17 – Iris dataset – sepal leaf scatter plot

As we can see, Iris Versicolor and Iris Virginica are similar, while Iris Setosa is quite different and forms a separate cluster of observations. To use the scatter plot even further, combine it with the distribution plot, like so:

```
#Perform bivariate EDA using both scatter plots and histograms
sns.jointplot(x=df["Sepal_length"],
              y= df["Sepal_width"],
              hue=df["Species_name"])
```

As you can see, we used `sns.jointplot()` to plot a scatter plot and density plot at the same time. The `hue` argument is used to define which variable (in this case, `Species_name`) will be used to color the points based on the species.

Here's the output:

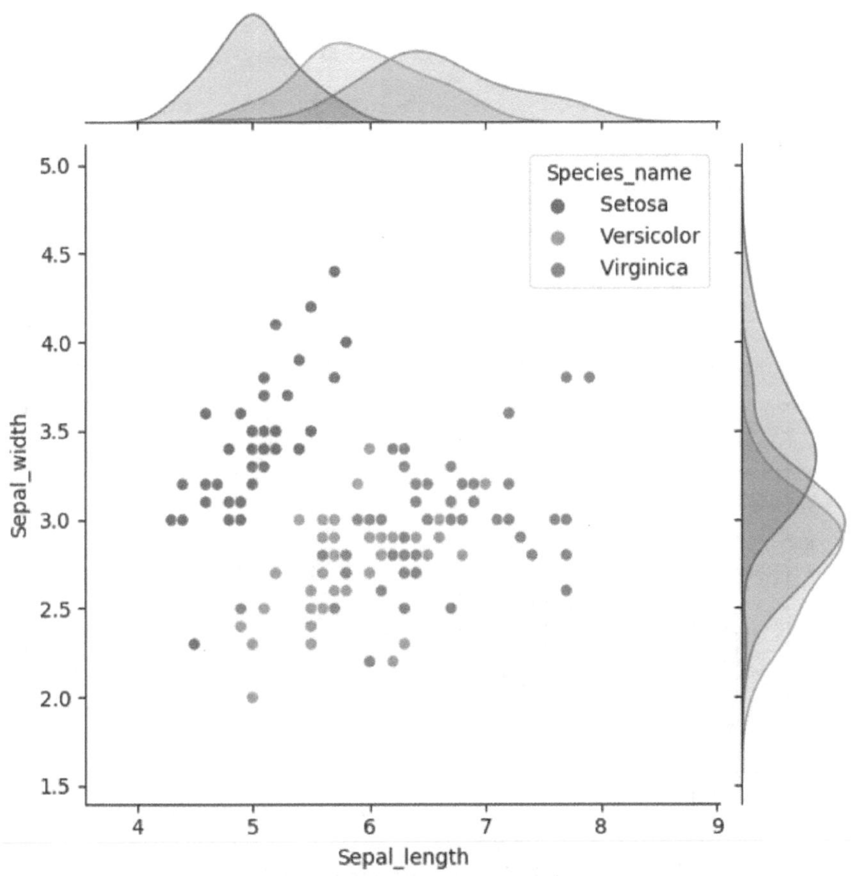

Figure 3.18 – Combined probability density and scatter plot

As we can see, using the X-axis (**Sepal_length**) and the legend, all three species have a distribution that's very similar to the normal bell-shaped curve. The Iris Setosa distribution is on the left compared to other species, which means that on average, it has the lowest value of **Sepal_length**. Iris Virginica has the highest value of **Sepal_length**, while the Iris Versicolor values are in the middle. Upon analyzing **Sepal_width**, we'll see that all three species have values that resemble the normal distribution. However, on the vertical axis, Iris Virginica and Iris Versicolor have similar values, while the blue distribution (Setosa) is still lower. From this, we can conclude that, with high probability, the variables of interest follow a normal distribution. This means that averages are good central tendency parameters of the data. Central tendency is a term that's used to describe parameters in the data that are central values and explain most data points if the data distribution is bell-shaped.

Now, let's use the medians and quartiles to evaluate the data further. These can help us identify the specific threshold values under where 25%, 50%, and 75% of all data points are. The most efficient way to do this is by creating a boxplot, like so:

```
#Create the box plot for each of the variables of interest
plot.boxplot(df[variables])
```

Here's the output:

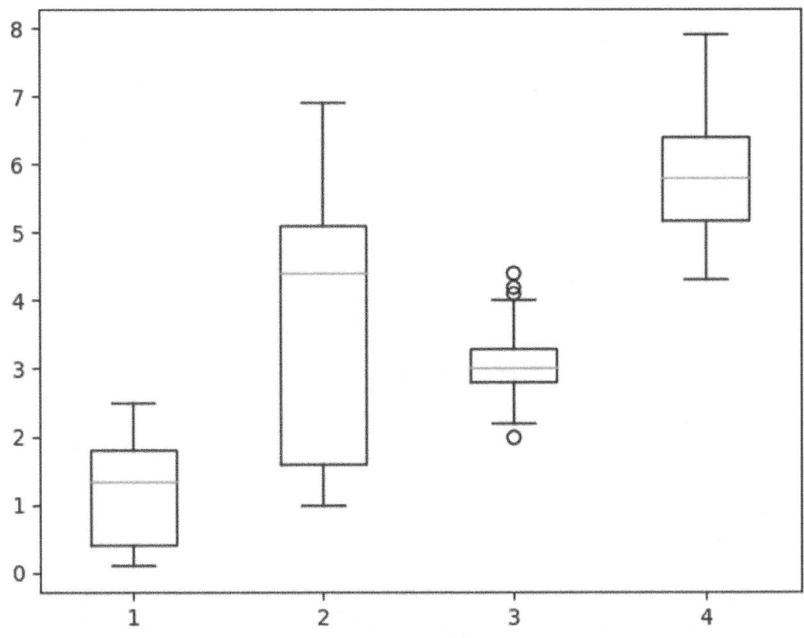

Figure 3.19 – Boxplot for the variables

Marks **1**, **2**, **3**, and **4** correspond to the **Petal_width**, **Petal_length**, **Sepal_width**, and **Sepal_length** variables; this is how they were ordered in the dataset. The orange lines are medians. From this, we can confirm that **Sepal_length** has the lowest median values, while **Sepal_length** has the highest median values. How can we interpret this further? Here, the lower and upper borders of the boxes are Q1 and Q3.

As mentioned previously, we can't make any conclusions until the data has been separated based on the species category.

We can use the following code to make a boxplot for each of the species:

```
#Sepal length EDA based on each of the species separately
import seaborn as sns
sns.boxplot(x=df["Species_name"], y=df["Sepal_length"] )
```

Here's the resulting boxplot, which has boxes and whiskers for different Iris species:

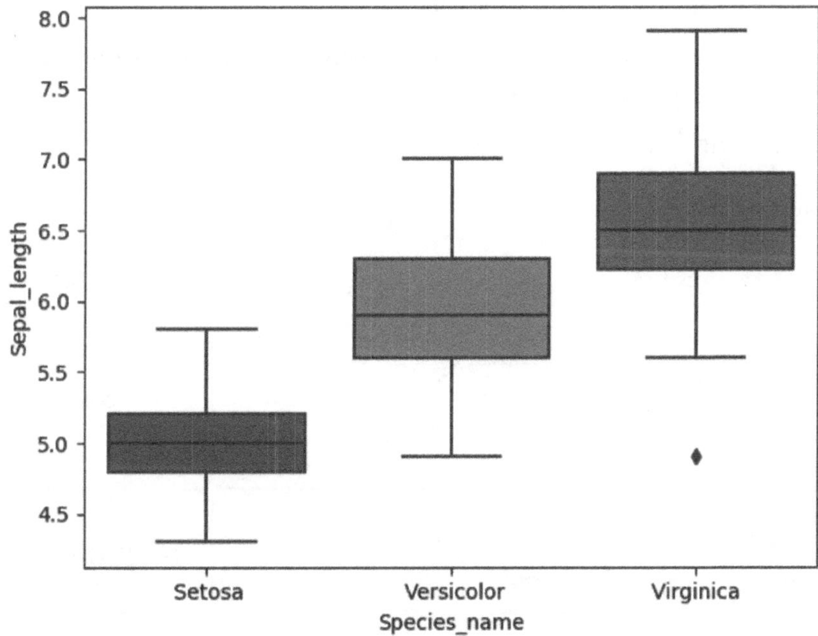

Figure 3.20 – Colored boxplot

Now, we can see the medians and interquartile ranges (boxes) for each of the species. But this time, we can see one outlier (diamond mark for the Virginica species) that wasn't present in the previous plot for **Sepal_length**. This is good evidence of how separating the data based on species can serve to make a more detailed EDA.

Finally, we can see that the Virginica species has the largest median (central line in the box), whereas Versicolor and Setosa have lower medians in descending order. We can also see that the lower and upper edges of the boxes, which represent the thresholds for 25% and 75% of the data, also follow the same pattern. From this, we can conclude that the sepal length has a median value for species that decreases from Virginica to Versicolor and Setosa.

One of the final questions we must ask regarding EDA is what we do with the outliers.

First, we must understand what outliers are. Imagine that we have a set of measurements that are in the expected range and another observation that's very distant from them.

In statistics, an outlier is a data point that differs significantly from other values. Outliers can be present due to variability in the measurement (normal occurrence), though they may be the result of experimental error or data being invalid for some other reason. In biostatistics and other life sciences, outliers should be identified but not removed if the data is valid. Being an outlier doesn't mean the data isn't accurate or invalid. The Iris dataset has been validated and all the observations are valid, so no further action is needed for outliers to be removed.

How can we add all combinations of variables for each of the species on one plot and have a complete EDA visualization? By making a paired scatter plot:

```
sns.pairplot(df, hue='Species_name', palette='OrRd' )
```

Here's the output:

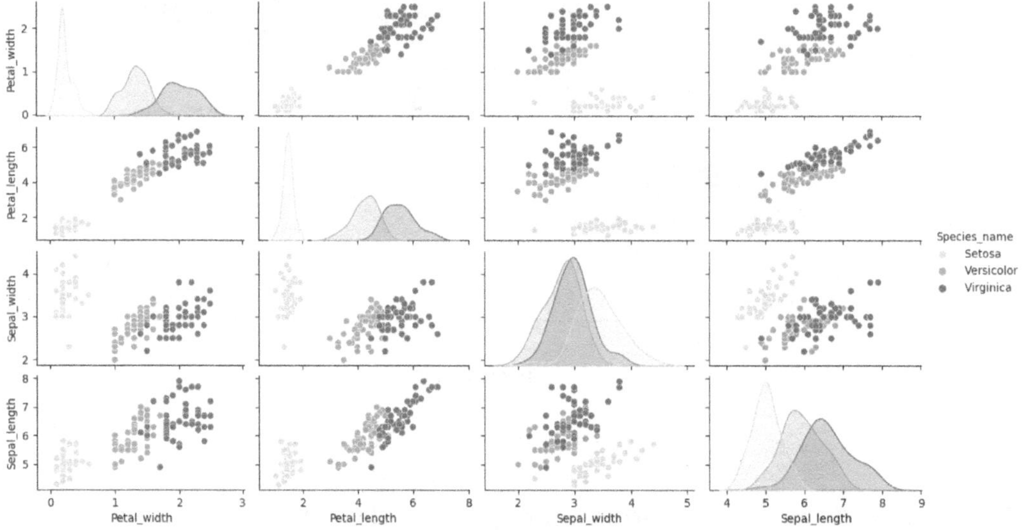

Figure 3.21 – Pairplot

As we can see, each combination of variables is now plotted and all of their distributions have been compared. By comparing the *X* and *Y* axes, we can see the scatter plots and distribution plots for each pair of variables. As an example, take a look at the lower left graph; its sepal length on the *Y*-axis and petal width on the *X*-axis are shown as a scatter plot. However, at the top left, we have the density plots for distributions of classes in the data. So, by pairing the *X* and *Y* axes' variable labels, we can observe both scatter plots and the distributions of the classes for all variable pairs.

Three important aspects can now be described for the variables and species:

- Difference in distributions
- Clustered observations
- Gaps in the data

Let's interpret this this:

- The difference in distributions is largest in the **Petal_width** and **Petal_Length** variables. Iris Setosa generally has lower measurement values (except for **Sepal_width**).
- The Setosa species are a separate cluster, while the Virginica and Versicolor species are more similar and less separated.
- Gaps in the data are present between the Iris Setosa species and others.

This concludes *Exercise 1*! Great work – you are now ready for the first exemplar project, where you'll learn how to load, clean, and describe diabetes data in Python.

In this exercise, you learned how to load, clean, describe, and visualize data using the Iris dataset. Now, let's summarize this chapter.

Summary

In this chapter, we learned how to load data using Python. We also learned how to clean the data and address the missing values and invalid data. After, we learned about one of the most important aspects of biostatistics, which is how to describe data. We did this using the exemplar Iris dataset.

In the next chapter, you'll learn how to load, clean, and describe a more complex biological dataset in Python – the diabetes dataset.

4

Part 1 Exemplar Project – Load, Clean, and Describe Diabetes Data in Python

Welcome to *Chapter 4*! In this chapter, you will learn how to practically apply the code learned in the previous chapter. You will also learn how to interpret and examine the *Diabetes* dataset and create data visualizations as output as a practical end goal.

In this chapter, we're going to cover the following main topics:

- Loading and examining the *Diabetes* dataset
- Validating and describing the *Diabetes* dataset
- Creating the data visualizations and table outputs

Loading and examining the Diabetes dataset

For this exemplar project, principles similar to that of loading the data are used as before: loading the `.csv` file in the Python environment (Spyder IDE). Before starting with loading the data or any other further procedures, all the required libraries are loaded.

We will be using the `pandas` library to load, process, and describe the data, `sklearn` to get the dataset, and other libraries such as `matplotlib` and `seaborn` to make different data visualizations. Make sure all of them are installed (i.e., run `pip install packagename` in Command Prompt on Microsoft Windows or Terminal on Linux/macOS, as explained in *Chapter 2* and *Chapter 3*); you can visit `pypi.com` for more information on installing the packages.

Here are the required libraries:

```
#load the libraries needed to perform the exemplar project 1
import pandas as pd
from sklearn import datasets
import matplotlib
import matplotlib.pyplot as plot
import seaborn as sns
```

The dataset used for this project is the *Diabetes* dataset found here: `https://data.mendeley.com/datasets/wj9rwkp9c2/1` (*Rashid, Ahlam (2020), "Diabetes Dataset", Mendeley Data, V1, doi: 10.17632/wj9rwkp9c2.1*). The data was collected from the Iraqi society.

The dataset consists of data from healthy subjects (controls), diabetes subjects, and a third segment, predicted diabetes subjects. Their biomedical data and laboratory analyses were collected and included in the dataset.

The variables in the dataset are as follows:

- The `Index` number of the patient
- `Gender`
- `AGE`
- Creatinine ratio (`Cr`)
- Body mass index (`BMI`)
- `Urea`
- Cholesterol (`Chol`)
- Fasting lipid profile total, including `LDL` (short for **low-density lipoprotein**) and `VLDL` (short for **very low-density lipoprotein**)
- Triglycerides (`TG`)
- `HDL` (short for **high-density lipoprotein**)
- `HbA1c` (glycated hemoglobin)

The next step is to load the dataset using `pd.read_csv()`. Make sure you know where you saved the dataset on your computer so you can input the path in the following code. In this case, the `Downloads` folder is used:

```
#Using the pandas read_csv the file is loaded from its path
data=pd.read_csv(r'C:\Users\MEDIN\Downloads\Dataset of Diabetes .csv')
```

Now, use the Variable Explorer and double-click the data object created. Refer to the following screenshot:

data - DataFrame

Index	ID	No_Pation	Gender	AGE	Urea	Cr	HbA1c	Chol	TG	HDL	LDL	VLDL	BMI	CLASS
0	502	17975	F	50	4.7	46	4.9	4.2	0.9	2.4	1.4	0.5	24	N
1	735	34221	M	26	4.5	62	4.9	3.7	1.4	1.1	2.1	0.6	23	N
2	420	47975	F	50	4.7	46	4.9	4.2	0.9	2.4	1.4	0.5	24	N
3	680	87656	F	50	4.7	46	4.9	4.2	0.9	2.4	1.4	0.5	24	N
4	504	34223	M	33	7.1	46	4.9	4.9	1	0.8	2	0.4	21	N
5	634	34224	F	45	2.3	24	4	2.9	1	1	1.5	0.4	21	N
6	721	34225	F	50	2	50	4	3.6	1.3	0.9	2.1	0.6	24	N
7	421	34227	M	48	4.7	47	4	2.9	0.8	0.9	1.6	0.4	24	N
8	670	34229	M	43	2.6	67	4	3.8	0.9	2.4	3.7	1	21	N
9	759	34230	F	32	3.6	28	4	3.8	2	2.4	3.8	1	24	N
10	636	34231	F	31	4.4	55	4.2	3.6	0.7	1.7	1.6	0.3	23	N
11	788	34232	F	33	3.3	53	4	4	1.1	0.9	2.7	1	21	N
12	82	46815	F	30	3	42	4.1	4.9	1.3	1.2	3.2	0.5	22	N
13	132	34234	F	45	4.6	54	5.1	4.2	1.7	1.2	2.2	0.8	23	N

Format | Resize | Background color | Column min/max | Save and Close | Close

Figure 4.1 – Dataset in the Variable Explorer

When the viewer is opened, you will be able to see all the variables as column names and samples as rows. Next, we use the `len` function to see how many rows (samples) are present in the dataset:

```
#len() function to answer how many samples are in the dataset (rows)
len(data)
```

Here is the output:

```
len(data)
Out[1]: 1000
```

With this, we have successfully loaded and examined the *Diabetes* dataset.

Validating and describing the Diabetes dataset

After we load and examine the dataset, we can take the next step, which is validating and describing the *Diabetes* dataset. This includes several procedures, including checking for missing values (`nan`), simplifying the dataset structure and removing the unnecessary variables, fixing potential wrong names in the classes (the `CLASS` variable) and categories, and making sure the structure of the dataset is as described on the official website.

We will perform each of these procedures here.

First, we check for missing values as follows:

```
#The .sum() after the isna() outputting the number of empty cells
data.isna().sum()
```

As there are no missing values (the output of `data.isna().sum` is `0`), you can proceed to simplify the dataset structure and remove the unnecessary variables. For this project, the `ID` and `No_Pation` variables (which are just unique identifiers for samples) are not needed, so they are removed to have a simpler structure of the dataset with only the relevant data. The `CLASS` variable is of special interest for this project as it contains the clinical information on the subjects' diabetes status:

```
#Define the shorter name for the dataframe - df
df=data

#Remove the unwanted columns
df=df.drop(['ID', 'No_Pation'], axis=1)
df = df[df["CLASS"].str.contains("P") == False]
```

The `P` class represents predicted diabetes cases, so these cases should also be removed. Why? Because working with predicted and not real classes can cause bias, and only diagnostically confirmed cases should be kept in the dataset.

To analyze these classes better, the dataset is grouped by gender and then described, as follows:

```
#Group by gender and transpose to have a better table
dtable1=df.groupby(['Gender']).describe()
dtable1=dtable1.transpose()
```

Note

As can be seen, there are different classes and categories in the data. The most important variable is the `class` variable (the `class` variable is the categorical variable with disease status classes, labeled as `CLASS` in the data). In this data, there are `Gender` and `CLASS`, which represent logically the gender of the subjects (`F` is female, and `M` is male) and the presence and absence of diabetes (`Y` for presence, and `N` for absence).

Here is the result of grouping by gender (splitting the data based on females or males in the dataset):

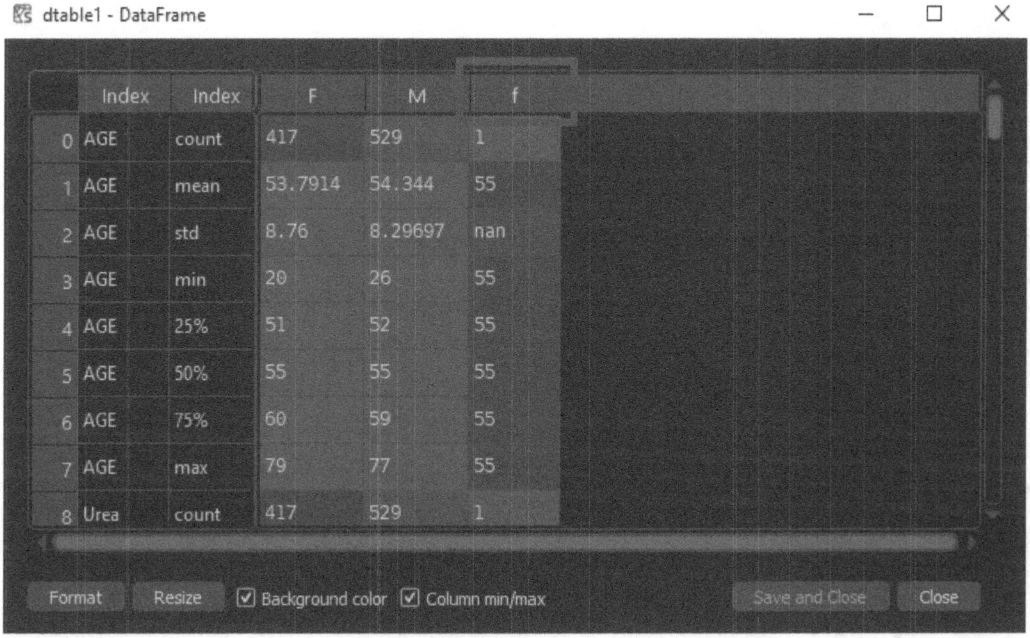

dtable1 - DataFrame

	Index	Index	F	M	f
0	AGE	count	417	529	1
1	AGE	mean	53.7914	54.344	55
2	AGE	std	8.76	8.29697	nan
3	AGE	min	20	26	55
4	AGE	25%	51	52	55
5	AGE	50%	55	55	55
6	AGE	75%	60	59	55
7	AGE	max	79	77	55
8	Urea	count	417	529	1

Format Resize Background color Column min/max Save and Close Close

Figure 4.2 – Grouping by gender and invalid data

Instead of having two gender categories (`F` and `M`), three are present in the data (`F`, `M`, and `f`). This is due to a typo where `F` is written as a lowercase `f` and this should be corrected and validated later.

Here is how you correct the `f` typo using the `replace` function for the `pandas` DataFrame:

```
#Fix the problem with invalid class and repeat the grouping by Gender
df = df.replace('f','F', regex=True)
dtable1=df.groupby(['Gender']).describe()
dtable1=dtable1.transpose()
```

The descriptive table is created again to validate the correction:

dtable1 - DataFrame

	Index	Index	F	M
0	AGE	count	418	529
1	AGE	mean	53.7943	54.344
2	AGE	std	8.74969	8.29697
3	AGE	min	20	26
4	AGE	25%	51	52
5	AGE	50%	55	55
6	AGE	75%	60	59
7	AGE	max	79	77
8	Urea	count	418	529

Format Resize ☑ Background color ☑ Column min/max Save and Close Close

Figure 4.3 – Validated data

As can be seen, there are now just two categories, `F` and `M`. Every variable has eight metrics: `count`, `mean`, `std` (standard deviation), `min` (minimum), Q1 (`25%`), median (`50%`), Q3 (`75%`), and `max` (maximum). Start by evaluating the `count` metric. The count for the `AGE` variable is *F = 418* and *M = 529*. This means that there are 418 observations for the `AGE` variable in the female category and 529 observations for `AGE` in the male category. Since there is no missing data, all the variables will have these numbers of observations, and you can confirm this by scrolling down.

From now on, you can use any count of any variable to confirm the number of males and females or other categories in the data. In biostatistics, the number of subjects is often denoted as *N*. So then we have *N (females) = 418* and *N (males) = 529*.

Then, we evaluate the means or averages for the ages of females and males. It can be seen that the mean age for females is *53.79* and *54.34* for males. This shows that age is not that different between the genders in the data. Further, the median for both groups, females and males, is `55`, which is another indicator that females and males have similar ages in the data.

Now, scroll down and evaluate other variables in the descriptive statistics, as shown here:

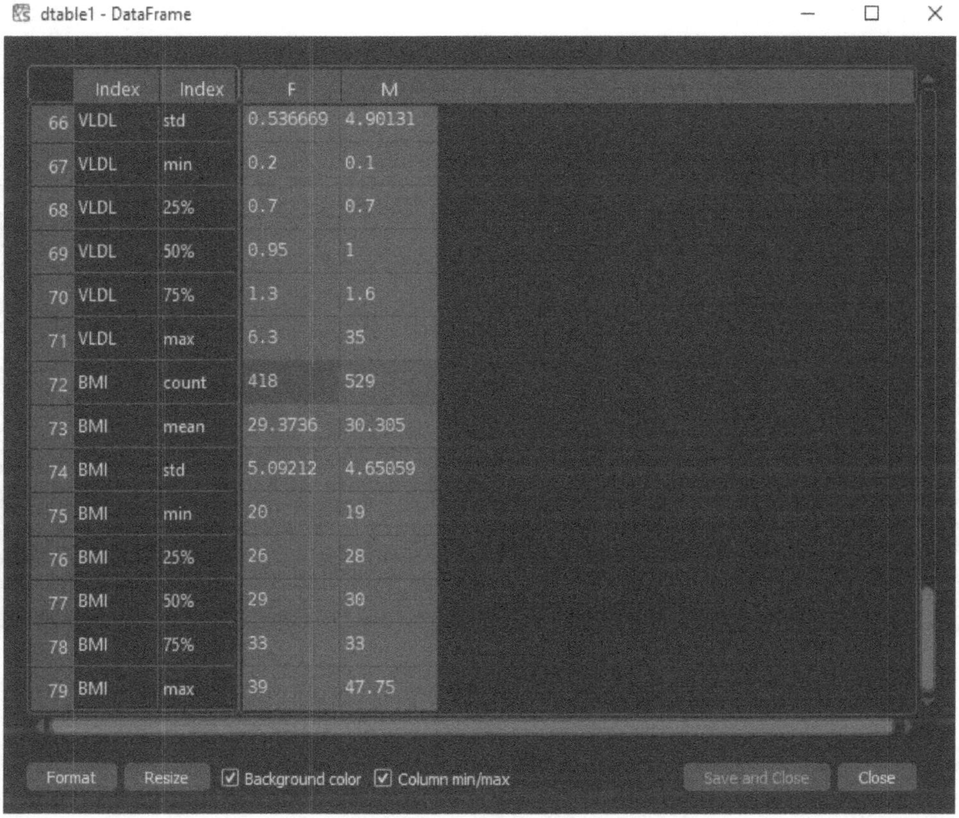

dtable1 - DataFrame

	Index	Index	F	M
66	VLDL	std	0.536669	4.90131
67	VLDL	min	0.2	0.1
68	VLDL	25%	0.7	0.7
69	VLDL	50%	0.95	1
70	VLDL	75%	1.3	1.6
71	VLDL	max	6.3	35
72	BMI	count	418	529
73	BMI	mean	29.3736	30.305
74	BMI	std	5.09212	4.65059
75	BMI	min	20	19
76	BMI	25%	26	28
77	BMI	50%	29	30
78	BMI	75%	33	33
79	BMI	max	39	47.75

Format Resize Background color Column min/max Save and Close Close

Figure 4.4 – VLDL by gender

It can be seen that `VLDL` and `BMI` are very similar between females and males. You can scroll up and down and evaluate other variables between the genders. This dataset contains two very distinct groups and you can explore these using the `CLASS` variable: `Y` are those with diabetes and `N` are those without diabetes or the control group. This comparison is the main focus, although it's important to take into consideration the possible gender-based differences in the data.

Here is how to group them by the `CLASS` diabetes variable:

```
#Group by class and see if there are any invalid classes in the CLASS
dtable2=df.groupby(['CLASS']).describe()
dtable2=dtable2.transpose()
```

This is how it looks:

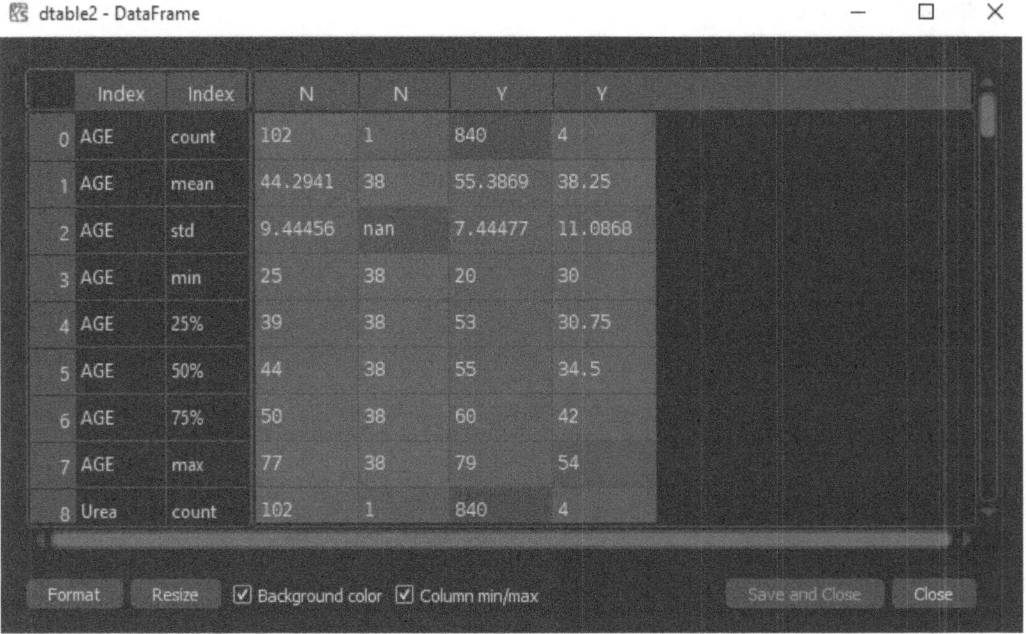
dtable2 - DataFrame

	Index	Index	N	N	Y	Y
0	AGE	count	102	1	840	4
1	AGE	mean	44.2941	38	55.3869	38.25
2	AGE	std	9.44456	nan	7.44477	11.0868
3	AGE	min	25	38	20	30
4	AGE	25%	39	38	53	30.75
5	AGE	50%	44	38	55	34.5
6	AGE	75%	50	38	60	42
7	AGE	max	77	38	79	54
8	Urea	count	102	1	840	4

Figure 4.5 – Grouping by class and invalid classes

There are invalid spaces in the CLASS variable, which is the reason why it outputs four instead of two classes of Y and N. The next step is to fix this by removing the spaces (using the strip() function) around the Y and N letters, which will correct the invalid data in the CLASS variable.

Here is the code for it:

```
#Fix the invalid spaces in the CLASS variable and repeat the grouping
df['CLASS']=df['CLASS'].str.strip()
dtable2=df.groupby(['CLASS']).describe()
dtable2=dtable2.transpose()
```

Now, open the dtable2 file in the **Variable Explorer** and you will see the following:

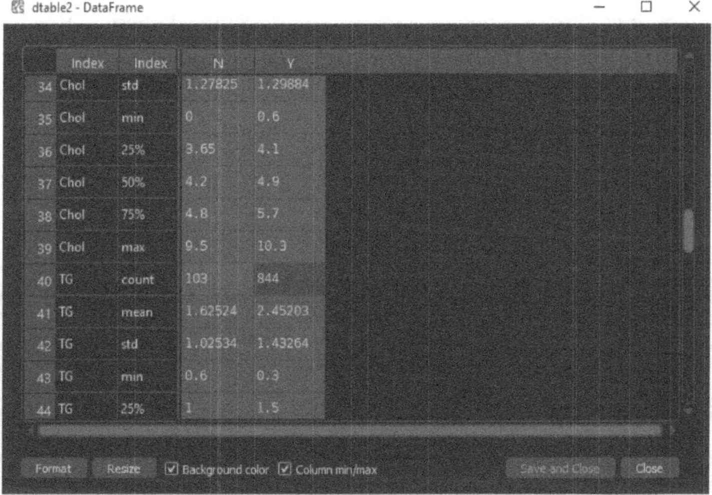

	Index	Index	N	Y
34	Chol	std	1.27825	1.29884
35	Chol	min	0	0.6
36	Chol	25%	3.65	4.1
37	Chol	50%	4.2	4.9
38	Chol	75%	4.8	5.7
39	Chol	max	9.5	10.3
40	TG	count	103	844
41	TG	mean	1.62524	2.45203
42	TG	std	1.02534	1.43264
43	TG	min	0.6	0.3
44	TG	25%	1	1.5

Figure 4.6 – Corrected classes

The `Y` and `N` classes are now fixed and further analysis can be performed. From the preceding screenshot, it can be seen that diabetics had a much higher average `TG` (triglycerides) value (*2.45*) compared to the control group (non-diabetics), which had *1.62*. Now, let us see how this can be grouped.

A more detailed grouping for descriptive statistics

As can be seen, the table (using the dataset values) is quite large and contains most of the descriptive information about control (non-diabetics) and diabetes subjects. Scroll down even further and you can see that `BMI` is also an interesting parameter in relation to diabetes, as shown in the following screenshot:

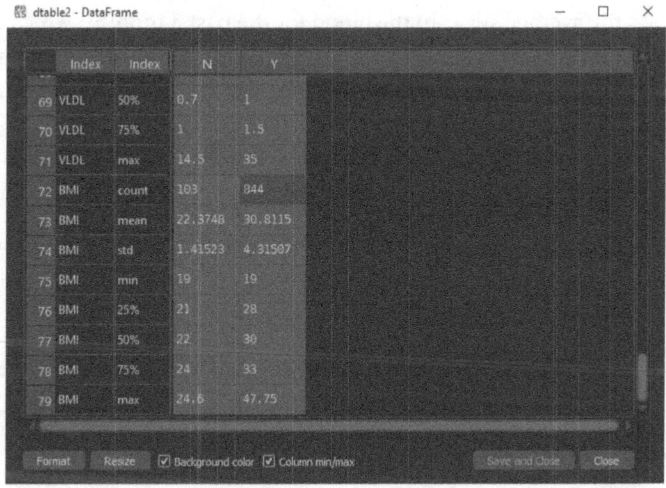

	Index	Index	N	Y
69	VLDL	50%	0.7	1
70	VLDL	75%	1	1.5
71	VLDL	max	14.5	35
72	BMI	count	103	844
73	BMI	mean	22.3748	30.8115
74	BMI	std	1.41523	4.31507
75	BMI	min	19	19
76	BMI	25%	21	28
77	BMI	50%	22	30
78	BMI	75%	24	33
79	BMI	max	24.6	47.75

Figure 4.7 – Clean class dataset descriptives

The average `BMI` count is *22.37* for control subjects and *30.81* for diabetics. This is a big difference and shows that obesity is related to diabetes (both casually and as a risk factor), which is another fact referenced scientifically in the past. Further, the `std` metric is around three times larger for diabetes subjects, meaning increased variability of `BMI` for this group.

Now, scroll up and find one of the most important parameters to evaluate diabetes: glycated hemoglobin (`HbA1c`). Here is the table for it:

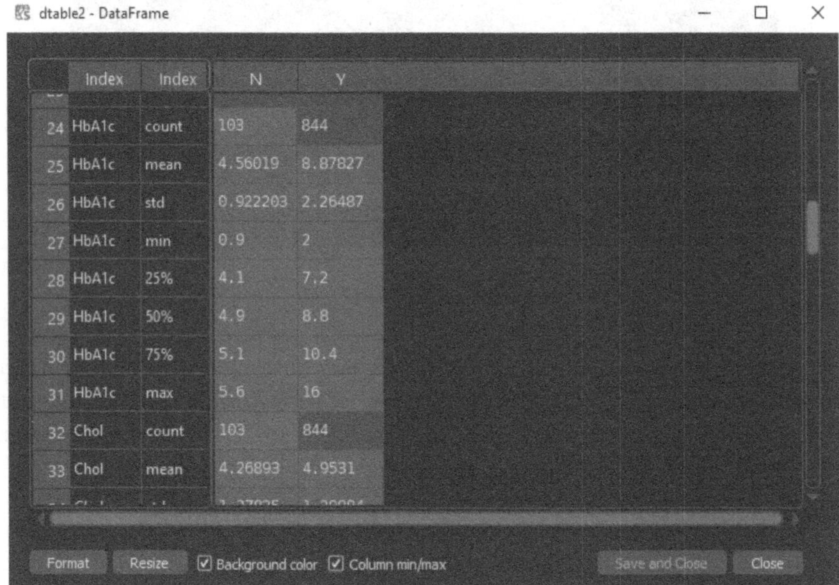

dtable2 - DataFrame

	Index	Index	N	Y
24	HbA1c	count	103	844
25	HbA1c	mean	4.56019	8.87827
26	HbA1c	std	0.922203	2.26487
27	HbA1c	min	0.9	2
28	HbA1c	25%	4.1	7.2
29	HbA1c	50%	4.9	8.8
30	HbA1c	75%	5.1	10.4
31	HbA1c	max	5.6	16
32	Chol	count	103	844
33	Chol	mean	4.26893	4.9531

Figure 4.8 – HbA1c by class

This parameter shows the average sugar in the blood for the past 3 months, which makes it essential to evaluate. Notice how its average is almost twice as large in diabetes subjects compared to controls. The central point, the median (50% of the data), is also almost twice as large. This makes HbA1c one of the most important diabetes biomarkers explored in the data so far and will be evaluated further in this chapter along with other interesting parameters. In biostatistics, it is often needed to group the data based on not just one variable but multiple different variables.

The best angle for analysis is when grouping on both `Gender` and `CLASS` variables (usually in biomedical research, it's important to groups based on gender, sometimes age group, and, of course, the target variables of interest and biomarkers).

Now, let's try to group the data on multiple variables (`Gender` and `CLASS`):

```
table_final=df.groupby(['Gender','CLASS']).describe()
table_final=table_final.transpose()
```

The final descriptive table is now created. Here is how it looks:

	Index 0	Index 1	0	1	2	3
Index 0		Gender	F	F	M	M
Index 1	None	CLASS	N	Y	N	Y
0	AGE	count	64	354	39	490
1	AGE	mean	43.9844	55.5678	44.641	55.1163
2	AGE	std	8.9894	7.44401	10.1916	7.62575
3	AGE	min	25	20	26	26
4	AGE	25%	38.75	53	39	53.25
5	AGE	50%	44.5	55	44	55
6	AGE	75%	50	60	50	60
7	AGE	max	77	79	77	77
8	Urea	count	64	354	39	490
9	Urea	mean	4.38438	4.84068	5.16667	5.48968
10	Urea	std	1.94416	2.7889	3.21831	3.15504
11	Urea	min	2	1.2	2	0.5
12	Urea	25%	3.15	3.3	3.8	4
13	Urea	50%	4	4.3	4.7	4.9

Figure 4.9 – Grouping by Gender and CLASS

The first two columns are controls versus diabetics in the female group and the second two are controls versus diabetics in the male group. This way, the diabetes subjects can be compared in separate gender subgroups and get a much more accurate comparison without the effect of gender on the description.

The first detail that can be noticed is that `std` for the `Urea` variable is much higher in female diabetics compared to male diabetics. This means that female diabetes subjects have a much larger variability of urea values compared to their male counterparts. Furthermore, the `Urea` average is also slightly increased in diabetic subjects, in both females and males, which was not the case with the urea variability metric (`std`). Accordingly, you can review all other variables of the dataset. The next step is to visualize the data. Let's explore the data through different types of visualizations.

Creating the data visualizations and table outputs

Data visualization is one of the most intuitive ways of presenting the data. Visual patterns are much easier to spot and describe compared to text or numbers. This is why the majority of high-quality research includes not just numerical but also visual representations of the data.

In the *Diabetes* dataset, we will be exploring the potential differences or similarities in the data distribution between diabetes and control subjects. We will explore their lipid profile (`LDL`, `HDL`, and `TG`) and `HbA1c`, a clinical measure used to show how much of the hemoglobin is glycated over a period of 3 months (a value that tells us the status of glucose over 3 months).

Before creating the visualizations, to simplify the process, create a new object that contains four variables: `TG`, `LDL`, `HDL`, and `HbA1c`. This object will be used to create data visualizations easily. Here is how we can do it:

```
#Create an object containing only 4 variables to analyze
variables=['TG', 'LDL', 'HDL', 'HbA1c']
```

Using the Python `seaborn` library and the `sns.displot` function, the distribution of any variable can be analyzed. This visualization is especially important when analyzing numeric variables, as in the case of the `TG`, `LDL`, `HDL`, and `HbA1c` variables.

Here is the code to visualize the distribution of `HbA1c` and separate the distribution based on the diabetes `CLASS` instance:

```
#Evaluate the distribution plot for HbA1c for Controls and Diabetics
sns.displot(data=df[variables],x="HbA1c", hue=df["CLASS"], kind="kde")
```

Here is the graph for it:

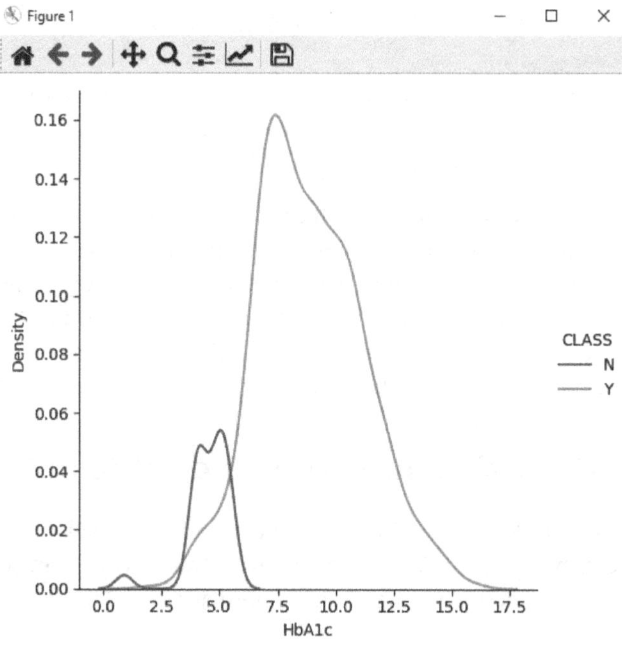

Figure 4.10 – HbA1c distribution density by CLASS

One very important aspect to observe in the previous graph is the horizontal axis (x). The y-axis represents the **KDE** (short for **Kernel Density Estimate**), which is a method for estimating the density of observations. Here, it is defined by the frequency of observations on the y-axis, conditioned on the values of x. It can be seen that the `HbA1c` level is drastically increased in diabetes subjects and most of the distribution (density of observations) for them is above a level of `6`. On the contrary, subjects with no diabetes have most of their `HbA1c` distribution below `6`, and it can be observed that the distribution for controls is to the left compared to diabetes. The peak for the control group distribution is around `5`, while the peak for diabetics is around `7.5`. All of this can be easily explored visually and, as such, is an important part of **exploratory data analysis** (**EDA**).

Now, repeat the process for another variable, triglycerides or `TG`:

```
sns.displot(data=df[variables],x="TG", hue=df["CLASS"], kind="kde")
```

Here is the graph for it:

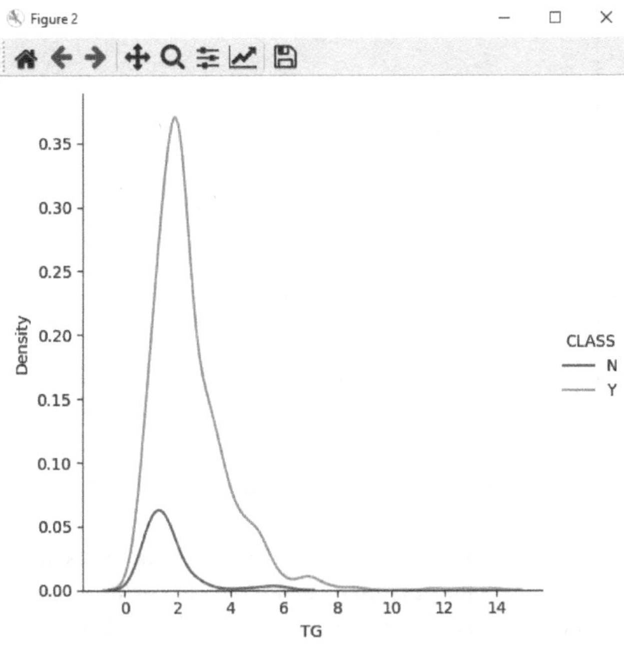

Figure 4.11 – TG distribution density by CLASS

This time, the difference between the control group and diabetics is not as obvious as the peaks are similar, around `2` on the x axis (the `TG` value), but some differences can still be observed. The controls have very little distribution above `3`, which is not the case for diabetics, so a slight increase in `HbA1c` can also be observed for triglycerides (`TG`) in diabetics.

Now, let's create the distribution plot for LDL:

```
sns.displot(data=df[variables],x="LDL", hue=df["CLASS"], kind="kde")
```

Here is the graph for it:

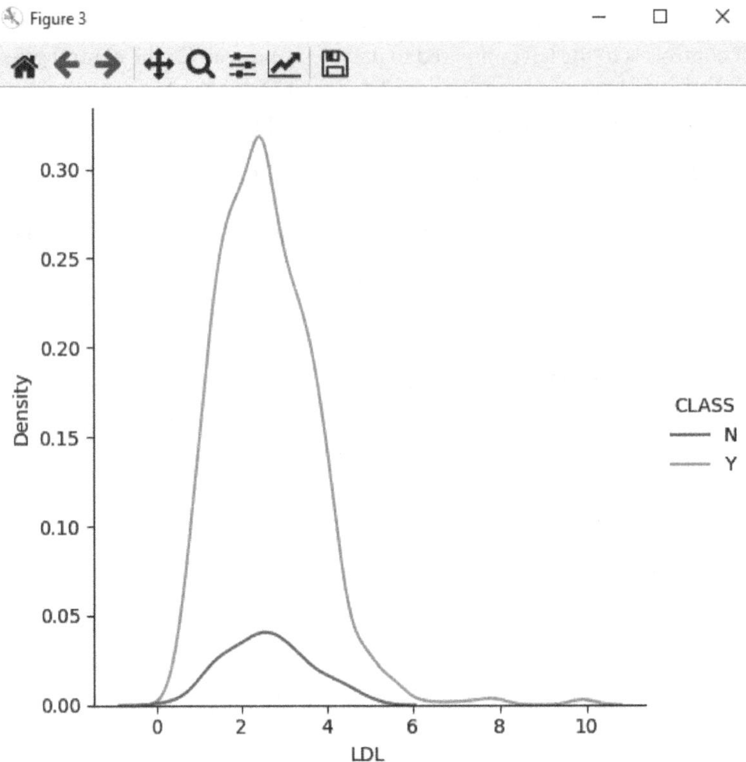

Figure 4.12 – LDL distribution density by CLASS

As for LDL, there is very little difference in terms of the density peak values of HbA1c, (the peaks are in a similar position). Even though the peak for diabetics is higher on the *y* axis, that is not relevant in terms of the differences and mean conditional values on LDL. There is lower variability of LDL in the data for diabetics, but still, the values are mostly a bit above 2, as is the situation with the control group.

The HDL parameter can be beneficial for health and is considered *good* cholesterol. The next step is to evaluate HDL between the control group and diabetics. Use the displot function to evaluate HDL:

```
sns.displot(data=df[variables],x="HDL", hue=df["CLASS"], kind="kde")
```

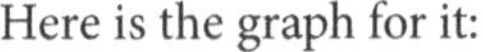

Here is the graph for it:

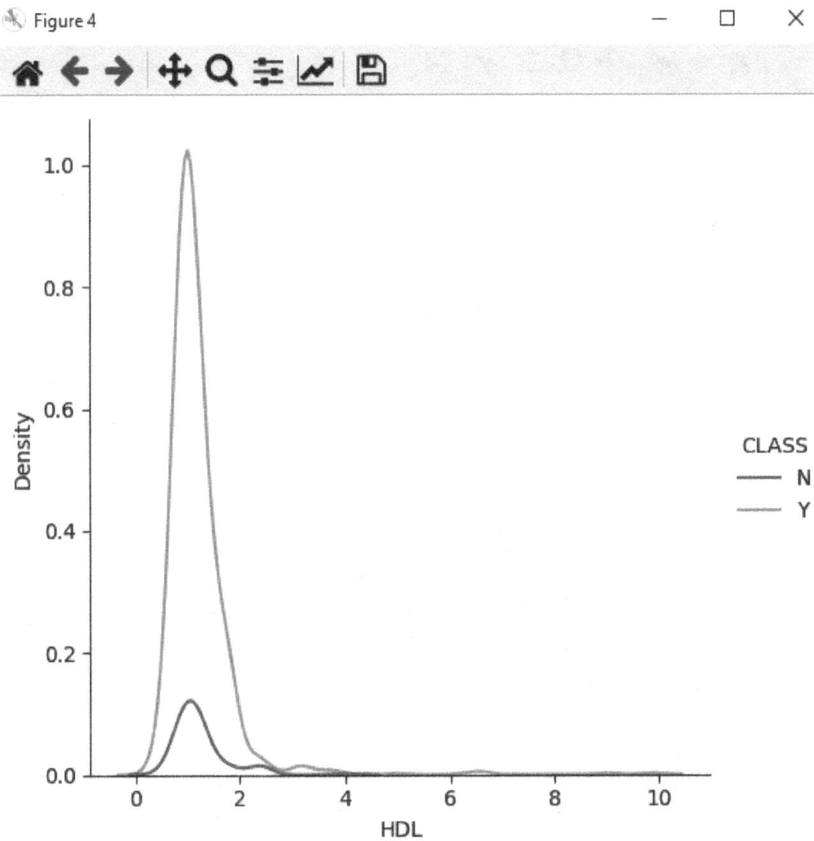

Figure 4.13 – HDL distribution density by CLASS

`HDL` peaks are also similar, as seen on the *x* axis; again, some difference may be present in the peak height, but still not enough to conclude the difference conditions on `HDL`. There is another way to evaluate potential differences in parameters by observing their distribution. Instead of using lines, stacked histograms can be used, by adding `multiple="stack"`, which is the argument for the `sns.distplot()` function, and removing `kind=kde`.

Here is the code for the stacked histogram for `HbA1c`:

```
sns.displot(data=df[variables],x="HbA1c", hue=df["CLASS"],
    multiple="stack")
```

Here is the histogram for it:

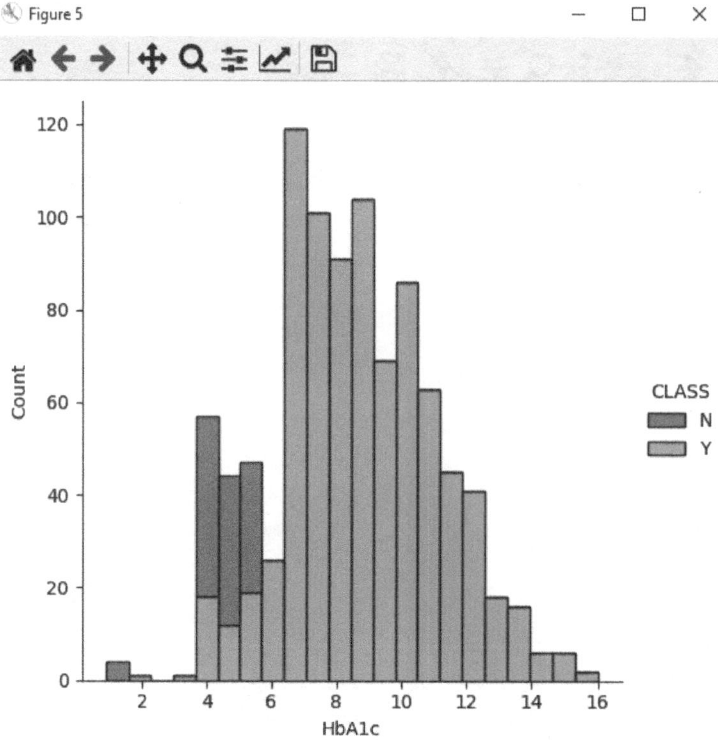

Figure 4.14 – HbA1c histogram by CLASS

The situation is similar to the previous exploration of `HbA1c`. Part of the distribution where the count is larger for controls is between 4 and 6 and below 2. On the other hand, dominant counts of `HbA1c` values are above 6 for diabetes subjects. HbA1c is known to be increased in diabetes subjects and this is confirmed through the EDA of this project.

Repeat the same procedure for `LDL` now:

```
sns.displot(data=df[variables],x="LDL", hue=df["CLASS"],
    multiple="stack")
```

Here is the histogram for LDL:

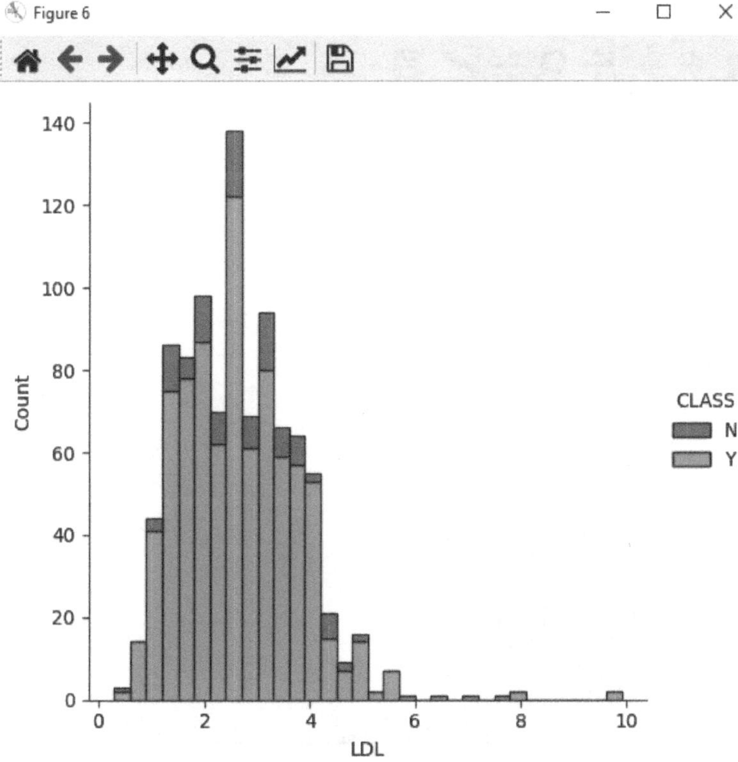

Figure 4.15 – LDL histogram by CLASS

As can be seen, the counts for the N and Y classes can now clearly be observed and the level of homogeneity between differences is notable. Logically, there is more of the dark color (blue if it's an e-book) as there were more diabetes subjects at the beginning of the project, but the difference between N and Y classes is similar across the whole distribution with minor differences in the fact that only the Y class is present above the LDL threshold of 6.

To conclude, the distribution of LDL is similar across both controls and diabetics, with minor differences toward upper values of LDL. Notice how I combined the visual and textual explanations to explore LDL levels across controls and diabetes subjects.

Now, we will explore the TG levels across different subjects:

```
sns.displot(data=df[variables],x="TG", hue=df["CLASS"],
    multiple="stack")
```

Here is the histogram for it:

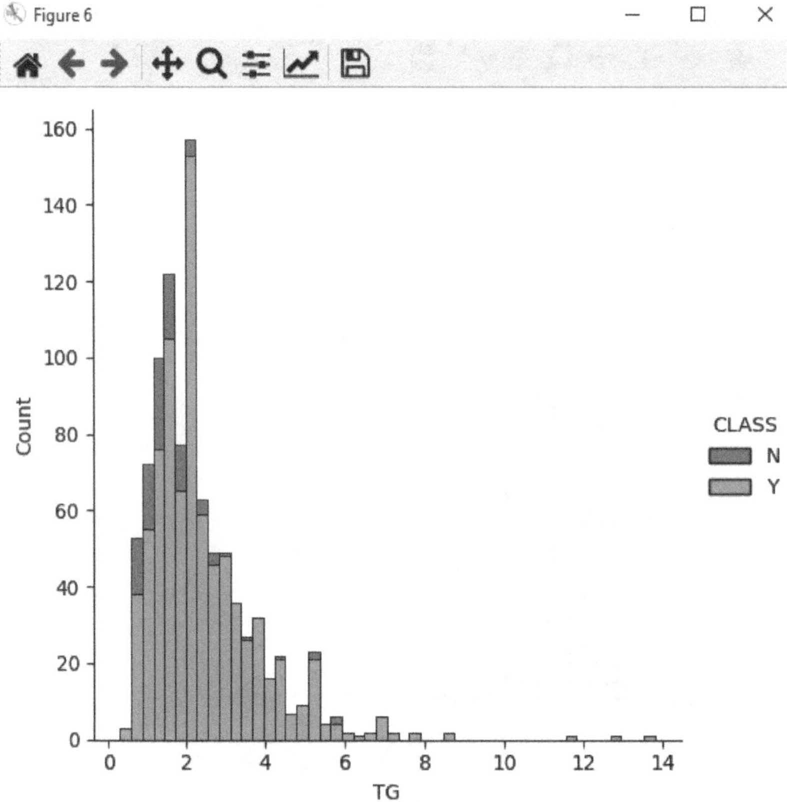

Figure 4.16 – TG histogram by CLASS

The interpretation of this plot is slightly different. It can be seen that the control versus diabetes levels of TG are not the same across the whole plot. The differences increase as the level of TG increases, meaning TG could be another interesting parameter to further explore. Notice how, after a value of 4, there are almost no controls on the plot, while diabetes subjects are still present.

The next variable to explore is the last of the defined variables, which is HDL.

Exploring the HDL levels across different groups (N and Y classes)

Here, we are exploring the HDL level across the different groups as mentioned; a comparison is done between controls and diabetic subjects based on their HDL levels:

```
sns.displot(data=df[variables],x="HDL", hue=df["CLASS"],
    multiple="stack")
```

Here is the plot for it:

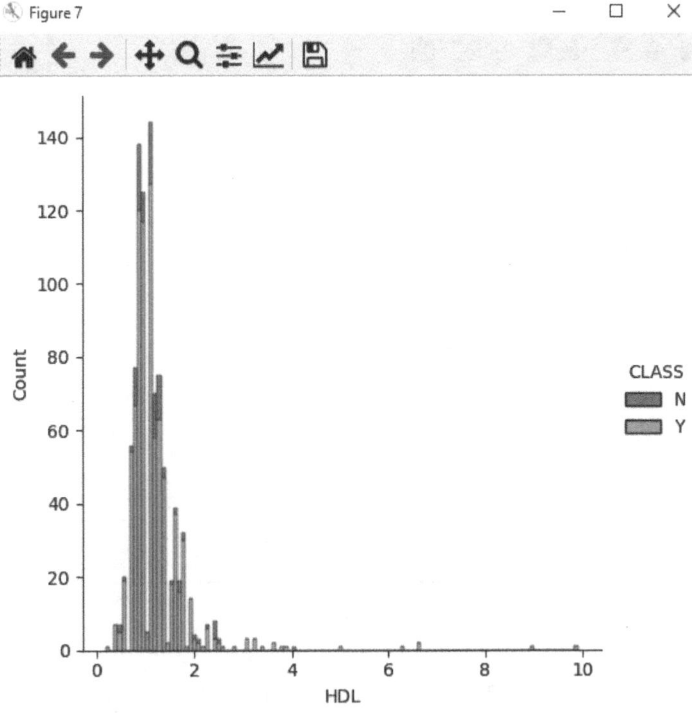

Figure 4.17 – HDL histogram by CLASS

For the `HDL` parameter, no major differences in the distributions of values can be seen across the *x* axis.

To summarize the distribution visual analysis, visual distribution-based EDA has shown that the largest differences between controls and diabetes subjects were identified for the `HbA1c` parameter. As expected, diabetes subjects had much increased glycated hemoglobin. Another different pattern was noted for triglycerides. LDL and HDL were explored and similar distributions were found in both controls and diabetics.

Another type of visualization – Seaborn scatter plot

To have an even more detailed angle on observing each data point individually and also how data points of different classes group together, we can use a scatter plot:

```
#Perform bivariate visual EDA using seaborn scatterplot:
sns.scatterplot(x=df["BMI"],
                y= df["HbA1c"],
                hue=df["CLASS"])
```

Here is the plot for it:

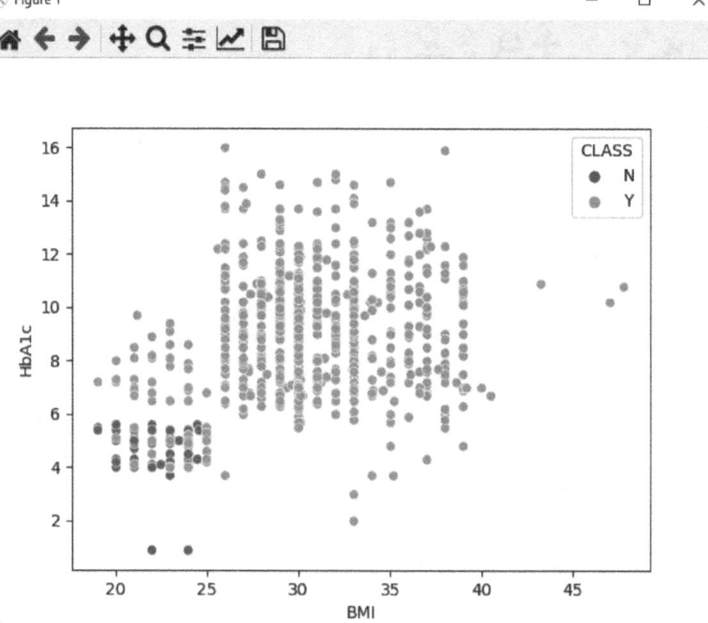

Figure 4.18 – HbA1c and BMI scatter plot by CLASS

A scatter plot includes every data point for different groups plotted based on at least two variables. For this project, the `BMI` and `HbA1c` variables are used. Both the *x* and *y* axes can be used for interpretation. If the *x* axis is observed, it can be seen that `BMI` values for controls (dark dots, or blue dots for the e-book) generally have lower values compared to the majority of points for diabetes subjects (orange dots). A similar observation can be seen for `HbA1c`; as noted before, it is increased for diabetes subjects. Now, we can observe all the points and compare them using two variables (`BMI` and `HbA1c`) simultaneously.

The scatter plot can now be complemented with the distribution plots discussed in the previous visualization and this can make a formidable graph to explore the data.

A scatter plot can show the grouping of all individual data points, and adding the distribution plot we learned about before can show how the data is distributed with lines.

Here is the code for it:

```
#Make the jointplot visualization and separate scatter points by class
sns.jointplot(x=df["BMI"],
              y= df["HbA1c"],
              hue=df["CLASS"])
```

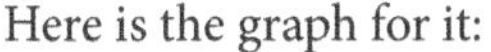

Here is the graph for it:

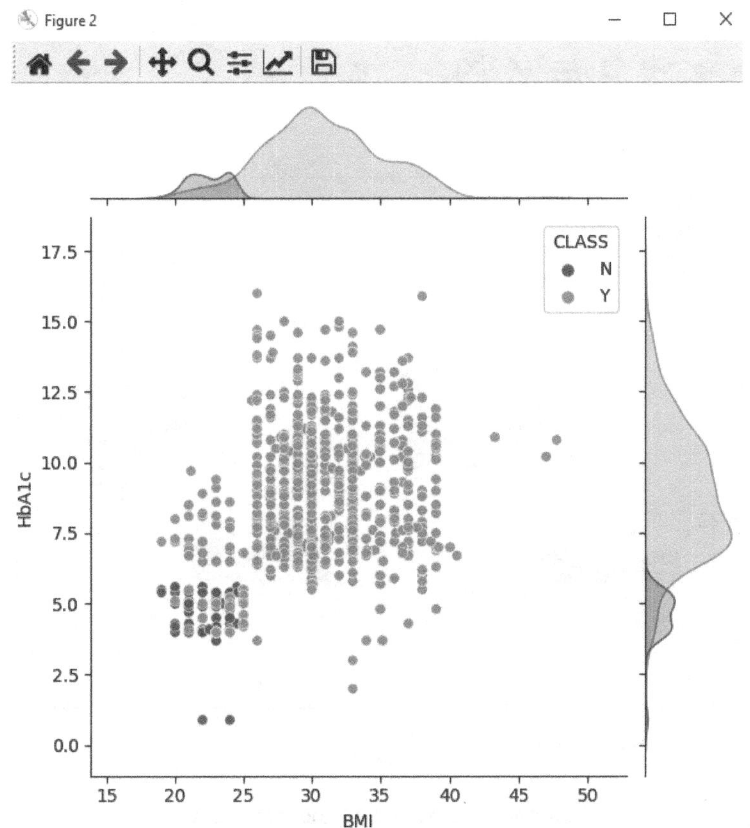

Figure 4.19 – Combined scatter plot and distribution density for HbA1c and BMI

Now, all that was mentioned previously about `BMI` and `HbA1c` can be seen easily: higher `BMI` and `HbA1c` values for diabetes subjects.

Data visualization using boxplots

Since `HbA1c` was the parameter that was identified as important in relation to differences between controls and diabetes subjects, it can be explored further by using the boxplot visualization. Furthermore, the data can be separated based on gender too, so an additional angle of exploration can be added.

Here is the code for it:

```
#Create the boxplot HbA1c vs CLASS
sns.boxplot(x=df["Gender"], y=df["HbA1c"], hue=df["CLASS"] )
```

Here is the boxplot for it:

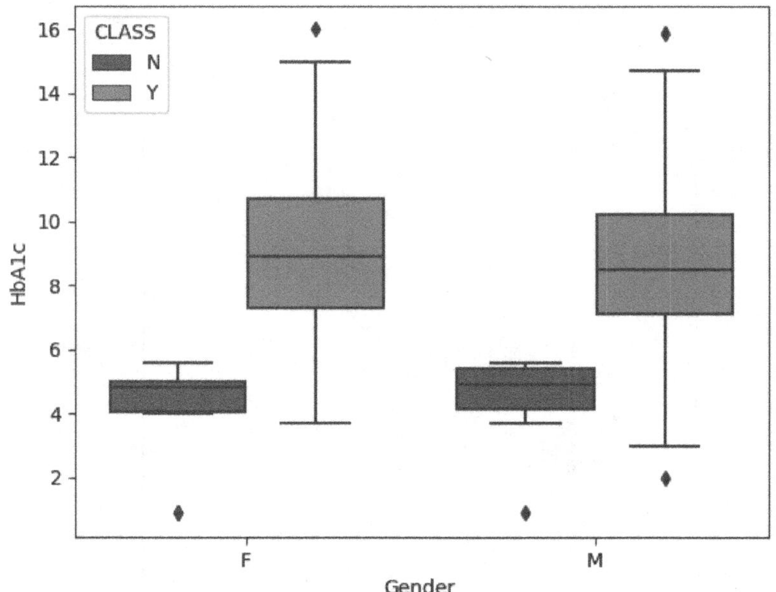

Figure 4.20 – Boxplot by Gender and CLASS

As seen in the plot, the median (central lines of the boxplots), as well as upper and lower ends of the boxes (75% and 25% of the data) are all increased in diabetics compared to controls, which is another confirmation of the previous `HbA1c` data explorations (boxplots for diabetes subjects are on the right side of **F** and **M** on the plot).

Summary

In this chapter, we learned how to practically apply Python functions for loading the data and the *Diabetes* dataset. We also learned how to validate and describe the dataset. Finally, we learned how to create intuitive data visualizations for further describing the data visually.

Congratulations! You have finished the chapter. In the next chapter, the theme is more specialized methods for different biostatistical areas.

Part 2: Introduction to Python for Biostatistics – Methodology and Examples

In *Part 2*, you are introduced to biostatistical concepts and how to apply them using Python. You will learn about hypothesis tests, biostatistical inference, predictive biostatistics, and more. At the end of *part 2*, a practical project using cardiovascular data is included.

This part has the following chapters:

5

Introduction to Python for Biostatistics

Welcome to the chapter! Here, you will learn about many different libraries used for biostatistics in Python. First, you will learn about libraries that can be used for hypothesis tests and practice simple examples. Then, you will learn about values and how these can be used to test different hypotheses and answer different research questions. You will learn how to use different libraries for finding the association between variables using diabetes data, and finally, you will learn specific characteristics of different statistical metrics, how they relate to different hypothesis tests, and when to use them.

In this chapter, we're going to cover the following main topics:

- Libraries for biostatistics hypothesis tests in Python
- Libraries for predictive biostatistics in Python
- Choosing which method to use for answering different scientific or research questions

Libraries for biostatistics hypothesis tests in Python

One of the most important ways to gain insights from biological data is to use hypothesis tests to see if the output we are seeing is statistically relevant or not. This principle is called **testing for statistical significance**. Statistical significance is a term commonly linked to the *p-value*, which represents the probability of obtaining test results that are as extreme as or more extreme than the observed result. So, the lower the p-value, the more credible the result.

The underlying principles of p-values

We can visualize the probability aspects of p-values on a Gaussian curve as follows:

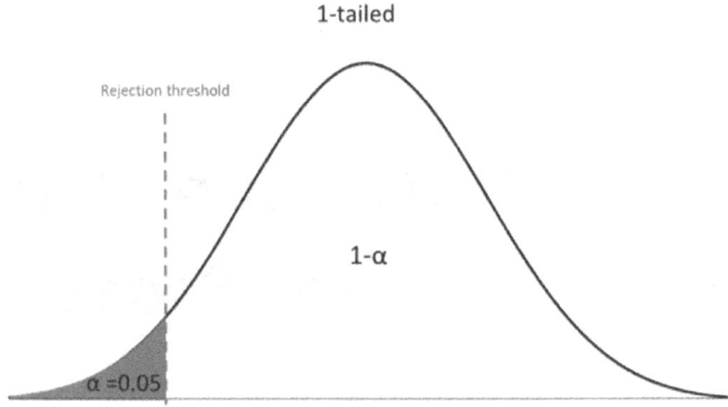

Figure 5.1 – Normal Gaussian curve with one-tailed alpha

The shaded region in the **area under the curve** (**AUC**) represents the probability that the observed results are less extreme than what is shown. This means the blue area is actually an area of uncertainty. If the blue area is less than 5% of the probability, that means the p-value is also less than 0.05. In summary, the p-value provides an indication of the level of credibility regarding the observed result in the output of the analysis.

In a two-tailed p-value representation (the representation where both ends of the distribution are used to evaluate the rejection threshold), the 0.05 threshold is equally distributed on both tails. To understand this, let's again take an example of a 5% probability threshold, which corresponds to a p-value of 0.05:

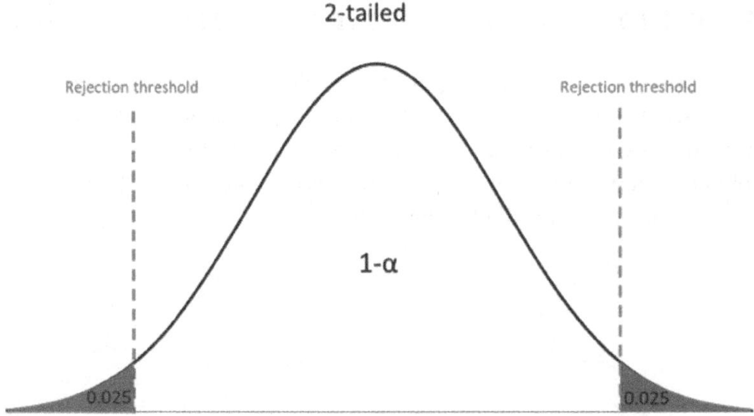

Figure 5.2 – Normal Gaussian curve with rejection threshold

The 0.05 **null hypothesis (N0)** acceptance probability is now composed of two parts: one on the left and one on the right extreme of the distribution, each having a 0.025 probability. These two parts together are called the **alpha value**, and the rest of the probability is called **1-alpha**. The 1-alpha is, again, 0.95 or 95% probability of the result being as extreme as observed.

Each hypothesis test has a null hypothesis (N0), which is generally considered as no statistical difference to what is being tested, and an alternate hypothesis, Na. If the results of a study support N0, this means that the author cannot reject the null hypothesis. Notice how the 'fail to reject' approach is used instead of accepting N0. Usually, hypotheses in studies cannot be formally accepted or rejected; instead, we are generally considering confidence in making conclusions about hypotheses. In a difference scenario, when results support Na, the alternate hypothesis (let's say in a case when the statistical difference between two groups is present), one can consider that there is a level of confidence about that and that Na is the hypothesis favored by the results. Some authors may use the term *accepting Na*, but be careful with this interpretation, as the acceptance of any hypothesis in statistics is generally not absolute.

Generally, when the p-value is less than 0.05, we can accept Na, the alternate hypothesis, meaning the statistically significant difference is present compared to the basic N0 state. Such an example is a comparison of two average values or means. N0 can be set such that there is no significant difference between the two means measured, and Na is the alternate hypothesis: that there is a significant difference between them. If the p-value is less than 0.05, we can accept Na and consider that the difference seen is significant statistically; in other words, at least as extreme as seen in the data. Keep in mind that p-values are not the only aspect to consider; there are other aspects such as effect size, and we will discuss those in the next chapters.

Every metric, such as mean, median, frequency, proportion, or others, is specific and different in terms of the characteristics of the data it represents. For each of these, there are specific hypothesis tests created by statisticians and adapted to specific data types.

Here is a basic scheme where many different hypothesis tests are presented and each assigned to a different metric and different type of research question:

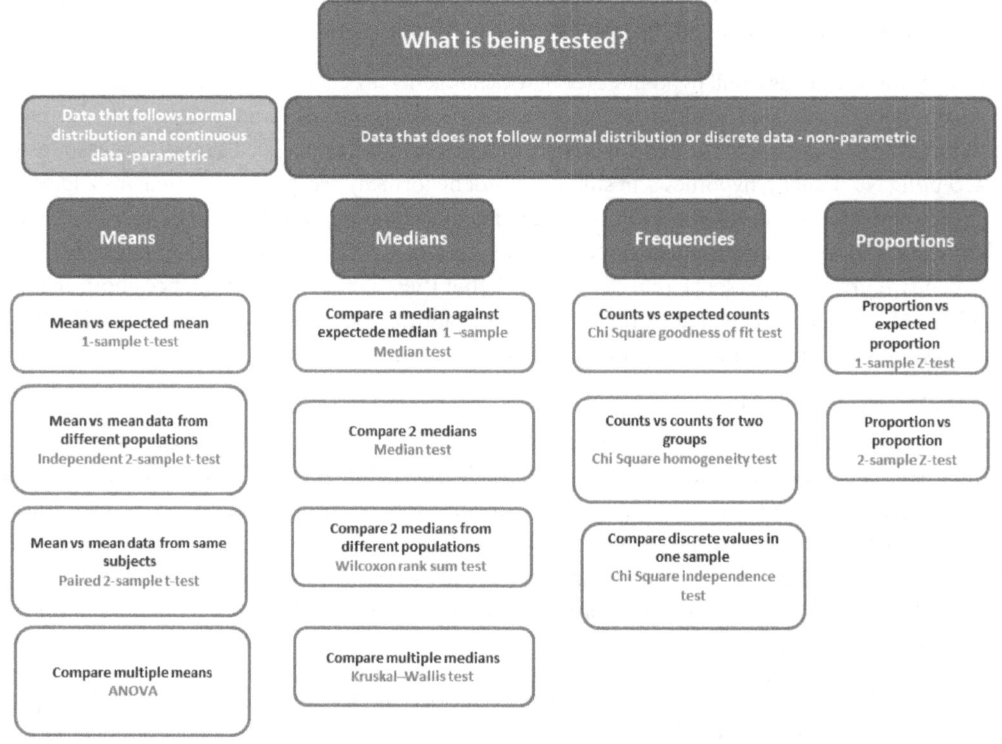

Figure 5.3 – Different statistical tests and what they compare

As can be seen, **means** can be tested using t-tests (tests for comparing means), and these are formally called **student's t-tests**. They can be applied to one sample mean to be compared against an expected hypothetical mean. A **one-sample t-test** is used to compare a single sample against a given value. To compare two measured means, we can use a **two-sample t-test**. If the data comes from different subjects and different populations, a modification called an **independent t-test** is used. If the data comes from the same populations, which is frequently the case in longitudinal studies, following up on the same subjects over different periods of time, then a **paired t-test** is used. Longitudinal studies are frequently used to follow the same groups of subjects over different time periods. This means that if we compare those different periods to each other, they are dependent and paired, and that's why paired tests are used.

This is because the data is paired or bound to the same group of subjects.

Performing tests in Python

Next, let's perform some of these tests practically in Python:

1. Install the following packages in your Anaconda prompt terminal (you can find this in the start menu of your computer once you install Anaconda):

```
pip install scipy
pip install statsmodels
pip install scikit-learn
```

2. Open Spyder or Jupyter Notebook to add the code ahead.

   Using `scipy.stats`, we can perform a simple Student's t-test. We will create two random dummy samples with four observations each and compare them using a t-test:

```
from scipy import stats
# sample data
data1 = [1, 2, 3, 4]
data2 = [5, 6, 7, 8]

# perform t-test
t_statistic, p_value = stats.ttest_ind(data1, data2)

# print results
print(f"t-statistic: {t_statistic}")
print(f"p-value: {p_value}")
```

   As can be seen in the code, we are using random data. To create random data, you can use brackets (`[...]`), create a list of numbers, and compare the data using a t-test. After that, we print the results, using the `print()` command.

3. Now, check the result in the console:

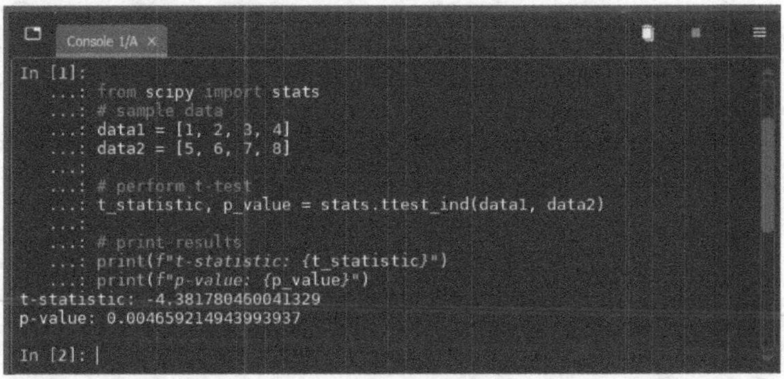

Figure 5.4 – A simple Student's t-test

> We can see two main outputs: `t-statistic` and `p-value`. The t-statistic is the main output of the t-test, but since it's difficult to interpret directly, it is used to calculate the p-value and use it for the main interpretation of the result. The p-value is `0.004`, which is less than the threshold mentioned previously of `p=0.05`. This means we can accept Na (the alternate hypothesis) and say there is a statistically significant difference between these two sets of numbers. More accurately, there is a statistically significant difference between the means of these two sets of numbers. If the situation was different and the p-value was >0.05, we could say that N0 (the null hypothesis) was accepted and a statistical significance was not found between these two numbers.

Now, what if we have counts and we cannot present them using means? Then, we can use a Chi-squared test. This time, we have a null hypothesis that there is no significant difference between observed and expected counts. Let's start:

```
from scipy.stats import chisquare

# sample data
observed = [10, 20]
expected = [15, 15]

# perform chi-squared test
chi2_statistic, p_value = chisquare(observed, f_exp=expected)

# print results
print(f"chi2-statistic: {chi2_statistic}")
print(f"p-value: {p_value}")
```

Here is the output:

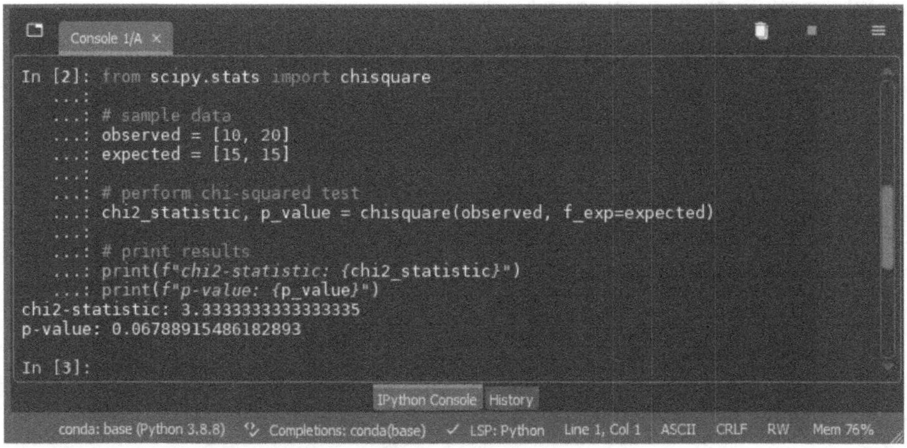

```
Console 1/A ×
In [2]: from scipy.stats import chisquare
   ...:
   ...: # sample data
   ...: observed = [10, 20]
   ...: expected = [15, 15]
   ...:
   ...: # perform chi-squared test
   ...: chi2_statistic, p_value = chisquare(observed, f_exp=expected)
   ...:
   ...: # print results
   ...: print(f"chi2-statistic: {chi2_statistic}")
   ...: print(f"p-value: {p_value}")
chi2-statistic: 3.3333333333333335
p-value: 0.06788915486182893

In [3]:
IPython Console  History
conda: base (Python 3.8.8)  Completions: conda(base)  ✓ LSP: Python  Line 1, Col 1  ASCII  CRLF  RW  Mem 76%
```

Figure 5.5 – Chi-squared test example

The p-value is `0.067`, which is larger than 0.05, and we accept the null hypothesis, meaning we didn't find a significant difference.

Artificial data (simulated example) was used for the example here, but in real datasets, the situation can be more complex and the data more diverse.

Now, let's try this code on the real diabetes dataset we downloaded in the previous chapter:

```
#Load libraries
import pandas as pd
from scipy import stats

#Load the diabetes dataset
data=pd.read_csv(r'C:\Users\MEDIN\Downloads\Dataset of Diabetes .csv')


#Select subjects with confirmed diabetes
data=data[data['CLASS']=='Y']
```

The second step is to define groups for comparison. In this case, we are comparing female and male diabetics based on their **high-density lipoprotein** (**HDL**) levels. Let's define those groups in the code:

```
#Define the separate datasets for Females-F and Males-M
data1 = data[data['Gender']=='F']
data2 = data[data['Gender']=='M']

#Select the high density lipoprotein (HDL)
HDLf=data1['HDL']
HDLm=data2['HDL']

import matplotlib.pyplot as plt
```

Next, we can define `HDLf` and `HDLm` labels for female and male values of HDL:

```
# Create a list of labels for the x-axis
labels = ['HDLf', 'HDLm']

# Create a list of heights for the bars
heights = [HDLf.mean(), HDLm.mean()]

# Create a bar plot
plt.bar(labels, heights)

# Add labels and title
plt.xlabel('Gender')
```

```
plt.ylabel('HDL')
plt.title('Mean HDL levels by Gender')

# Display the plot
plt.show()
```

Here is the code output:

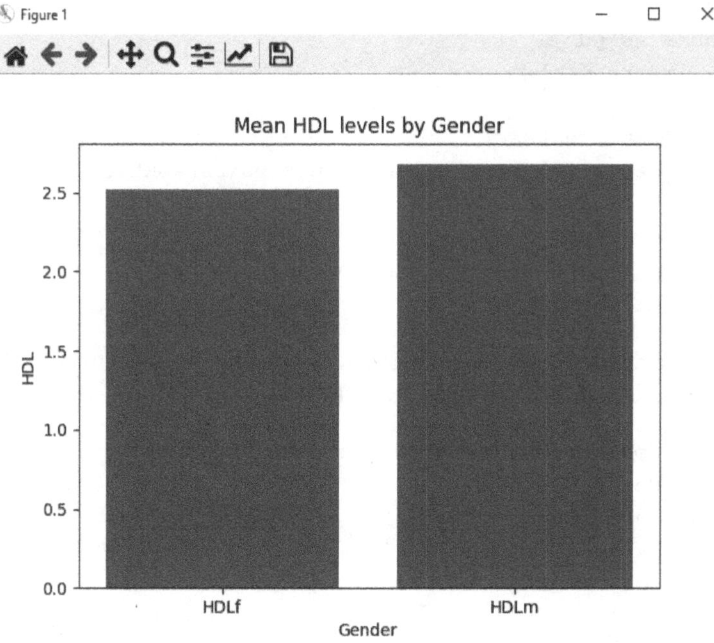

Figure 5.6 – Mean HDL comparison by gender

To perform the t-test, we can use the `ttest_ind()` function, which is a t-test for independent samples. Why do we use it in this case? Well, because males and females are independent of each other, they can be considered as different, independent samples:

```
# perform t-test
t_statistic, p_value = stats.ttest_ind(HDLf, HDLm)

# print results
print(f"t-statistic: {t_statistic}")
print(f"p-value: {p_value}")
```

Here is the code output:

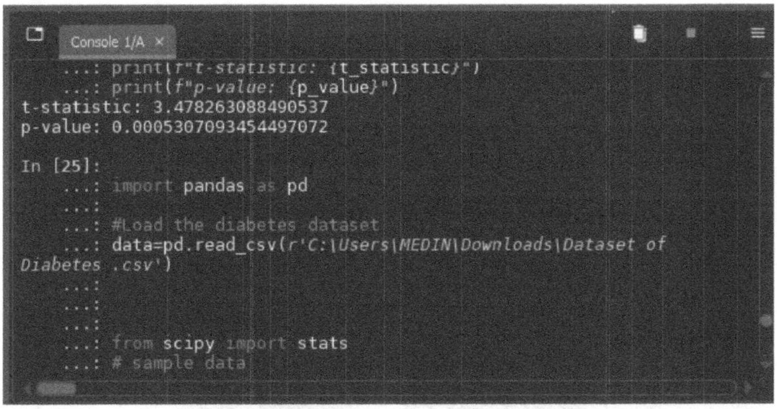

```
Console 1/A ×
    ...: print(f"t-statistic: {t_statistic}")
    ...: print(f"p-value: {p_value}")
t-statistic: 3.478263088490537
p-value: 0.0005307093454497072

In [25]:
    ...: import pandas as pd
    ...:
    ...: #Load the diabetes dataset
    ...: data=pd.read_csv(r'C:\Users\MEDIN\Downloads\Dataset of
Diabetes .csv')
    ...:
    ...:
    ...:
    ...: from scipy import stats
    ...: # sample data
```

Figure 5.7 – Console output of the Student's t-test results (HDL comparison)

The null hypothesis is that there was no significant difference in HDL between different genders in diabetic subjects, but we must reject the null hypothesis and accept Na since the p-value is `0.0005`, so we found a significant difference between genders in diabetic subjects.

Let's perform another t-test, this time for **low-density lipoprotein** (**LDL**) between genders in diabetics:

```
#Select the low-density lipoprotein (LDL)
LDLf=data1['LDL']
LDLm=data2['LDL']

# Create a list of labels for the x-axis
labels = ['LDLf', 'LDLm']

# Create a list of heights for the bars
heights = [LDLf.mean(), LDLm.mean()]

# Create a bar plot
plt.bar(labels, heights)

# Add labels and title
plt.xlabel('Gender')
plt.ylabel('LDL')
plt.title('Mean LDL levels by Gender')

# Display the plot
plt.show()
```

Here is the code output:

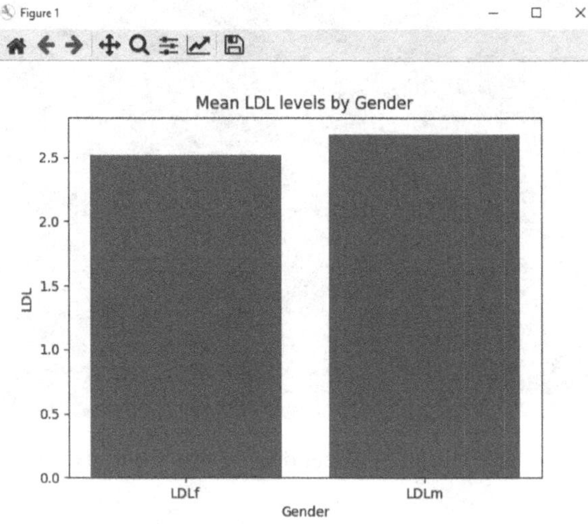

Figure 5.8 – Mean LDL comparison by gender

We will use the same function as before to perform an Independent Samples t-test and compare the mean LDL levels between females and males:

```
# perform t-test
t_statistic, p_value = stats.ttest_ind(LDLf, LDLm)

# print results
print(f"t-statistic: {t_statistic}")
print(f"p-value: {p_value}")
```

Here are the results:

```
Console 1/A

In [4]:
   ...:
   ...:
   ...:
   ...: LDLf=data1['LDL']
   ...: LDLm=data2['LDL']
   ...:
   ...: # perform t-test
   ...: t_statistic, p_value = stats.ttest_ind(LDLf, LDLm)
   ...:
   ...: # print results
   ...: print(f"t-statistic: {t_statistic}")
   ...: print(f"p-value: {p_value}")
t-statistic: -1.9292915793420666
p-value: 0.054032428239825535

In [5]: |

IPython Console  History
```

Figure 5.9 – Output for Student's t-test for LDL comparison between genders

The result is close to the threshold, 0.054. In this situation, we can formally accept the null hypothesis, which indicates there isn't a significant difference, but the situation is more complicated. Since the result is close to 0.05, we must be careful with the interpretation and say the results weren't conclusive as the p-value can change from study to study, and if it's close to 0.05, it could go either way if the study were repeated.

Comparisons such as this use the simple difference between average values and p-values, and hypothesis tests such as t-tests are also bound to them. This should be kept in mind when interpreting results; a t-test will mostly tell us about the average values of variables in terms of their difference. There are also statistical methods and hypothesis tests that tell us how much association can be identified among different variables, and these methods belong to a branch of biostatistics called predictive biostatistics.

Libraries for predictive biostatistics in Python

In predictive statistics, the hypothesis is a bit different. The main question is – Can we find associations between variables? To answer this, one of the best packages to start with is `statsmodels`. We can use it to create predictive models and perform predictive hypothesis tests. We will be using the diabetes dataset to try to find associations between cholesterol and triglycerides in diabetic subjects.

Let's start with coding:

```
import statsmodels.api as sm
import pandas as pd

data=pd.read_csv(r'C:\Users\MEDIN\Downloads\Dataset of Diabetes .csv')
```

The next step is to filter based on the diabetes presence status. After the filtering, only subjects with diabetes will remain in the data:

```
# Filter the data based on 'CLASS' column
filtered_data = data[data['CLASS'] == 'Y']

# Extract the Chol and TG variables
Chol = filtered_data['Chol']
TG = filtered_data['TG']

# Add a constant to the predictor variable
X = sm.add_constant(Chol)
Logistic regression model
```

The next step is to use **linear regression** based on **Ordinary Least Squares** (**OLS**). OLS regression is a statistical approach used to evaluate the linear relationship between independent variables and a target variable by finding the line that best fits the data based on the sum-of-squares criterion.

The sum-of-squares criterion is a measure of squared differences between observed and predicted values using the regression line. In this example, the target variable is `TG` (sometimes called dependent variable or `Y`), and `X` (independent variable) is defined as `Chol`:

```
# Fit the regression model
model = sm.OLS(TG, X)
results = model.fit()

# Print the regression results
print(results.summary())
```

Here is the code output:

```
                            OLS Regression Results
==============================================================================
Dep. Variable:                     TG   R-squared:                       0.112
Model:                            OLS   Adj. R-squared:                  0.111
Method:                 Least Squares   F-statistic:                     105.6
Date:                Mon, 12 Jun 2023   Prob (F-statistic):           2.06e-23
Time:                        14:41:12   Log-Likelihood:                -1444.9
No. Observations:                 840   AIC:                             2894.
Df Residuals:                     838   BIC:                             2903.
Df Model:                           1
Covariance Type:            nonrobust
==============================================================================
                 coef    std err          t      P>|t|      [0.025      0.975]
------------------------------------------------------------------------------
const          0.6261      0.184      3.402      0.001       0.265       0.987
Chol           0.3691      0.036     10.276      0.000       0.299       0.440
==============================================================================
Omnibus:                      397.423   Durbin-Watson:                   1.928
Prob(Omnibus):                  0.000   Jarque-Bera (JB):             2729.470
Skew:                           2.040   Prob(JB):                         0.00
Kurtosis:                      10.832   Cond. No.                         20.9
==============================================================================
```

Figure 5.10 – Linear regression output for association between TG and Chol variables

This is the output of the predictive model we created. It is called linear regression and provides many results to interpret. The most important ones are `R-squared` and `coef` (short for coefficients). `R-squared` is 0.112, and it tells us that approximately 11.2% of triglyceride variation can be explained by cholesterol variation, and this is called the coefficient of determination. The second important result is under `coef`. It tells us that the regression coefficient for cholesterol is 0.36, and if we observe `p>ItI`, this value corresponds to the statistical significance or the p-value, which is <0.05, so the association between cholesterol and triglycerides is found to be statistically significant in this case.

Now, let's visualize this using the `seaborn` data visualization package in Python:

```
import seaborn as sns
import matplotlib.pyplot as plt

# Create a DataFrame with HDL and LDL variables
df = pd.DataFrame({'Chol': Chol, 'TG': TG})
```

```
# Plot the scatter plot with regression line
sns.lmplot(x='Chol', y='TG', data=df)

# Set plot labels
plt.xlabel('Chol')
plt.ylabel('TG')
plt.title('Regression Analysis: Chol vs TG')

# Display the plot
plt.show()
```

Here is the output for this:

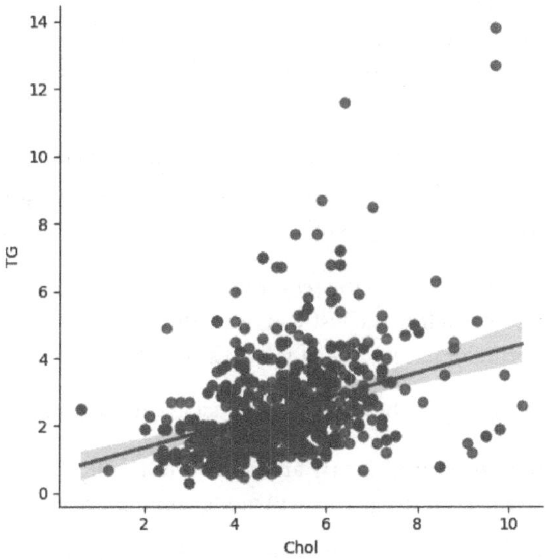

Figure 5.11 – Linear regression plot with trendline

Now, we can see the nature of the association, using a trendline. As cholesterol increases, the triglycerides also increase. The central blue line (for e-books) and black line (for hardcover books) show this as a trend.

The light blue area is called the confidence interval. In this case, it's a 95% confidence interval, and this is the area where there is a 95% probability of a true trendline being there.

In Python, there are several packages commonly used for predictive hypothesis tests. Here are some popular ones:

- `scipy`: The `scipy.stats` module provides a wide range of statistical functions, including various hypothesis tests. It works with different types of data, including continuous and categorical.
- `statsmodels`: `statsmodels` is a comprehensive statistical modeling package. It offers a variety of hypothesis tests, including t-tests, Chi-squared tests, **Analysis of Variance** (**ANOVA**), but also regression-based tests, and more. It is particularly useful for linear regression models and time series analysis. Time series analysis is a statistical technique used to analyze sequential data points collected over time. In biological and clinical studies, time aspects can be essential, and time series analysis packages play an important role in Python-based biostatistics and work with different types of data, including continuous and categorical.
- `scikit-learn`: `scikit-learn` is primarily a **machine learning** (**ML**) library but also includes some predictive hypothesis tests. This works with continuous data.
- `pingouin`: `pingouin` is a newer statistical package that provides a user-friendly interface for performing various statistical tests. It offers a wide range of predictive hypothesis tests, including t-tests, ANOVA, correlation tests, non-parametric tests, and more, and it works with both continuous and categorical data.
- `pymc3`: `pymc3` is a Bayesian modeling library that allows for hypothesis testing using Bayesian methods. It is useful when you want to incorporate prior knowledge or uncertainty into your analysis. It works with different types of data, including continuous and categorical.

The choice of package depends on the specific requirements of your research question and the type of data you have. Consider factors such as the test you need, the assumptions of the test, the complexity of the analysis, and any specific requirements of your research field.

Here's a workflow to help you choose the appropriate package for hypothesis testing in Python:

1. **Define your research question**: Clearly articulate the specific research question you want to answer through hypothesis testing. This will guide your choice of statistical test.
2. **Identify the type of data**: Determine whether your data is continuous or categorical. This will help you select the appropriate package to handle the data type.
3. **Understand the assumptions**: Different statistical tests have different assumptions. Identify the assumptions associated with the test you want to perform. For example, t-tests assume normality and equality of variances.
4. **Consider the complexity of analysis**: Assess the complexity of your analysis. Some packages offer more advanced statistical modeling capabilities, while others provide simpler and more straightforward tests. Decide whether you need a basic test or a more advanced analysis.
5. **Explore specific requirements of your research field**: Some research fields may have specific requirements or conventions for hypothesis testing. Familiarize yourself with any such requirements and consider them in your package selection.

6. **Research available packages**: Investigate the packages mentioned earlier (`scipy`, `statsmodels`, `scikit-learn`, `pingouin`, `pymc3`) and their documentation. Understand the tests they offer, their capabilities, and any limitations they may have.
7. **Match the test and data type**: Based on your research question, data type, assumptions, complexity, and specific field requirements, match the appropriate test to the suitable package. Consider the availability of the test in the package and whether it aligns with your needs.
8. **Implement and validate**: Implement the chosen package and perform the hypothesis test using your data. Validate the results by checking if the assumptions were met and whether they interpreted the outcomes correctly.
9. **Iterate if necessary**: If your initial choice of package does not meet your requirements or assumptions, re-evaluate and consider alternative packages or tests. Iteration may be necessary to find the best fit for your research question and data.

In this section, we learned what predictive analytics is and how a predictive hypothesis is different from a simple mean comparison such as a t-test. We learned about associations and how we can analyze them using linear regression. We also learned about the `statsmodels` library and how we can effectively use it in Python to perform linear regression and plot results using a scatterplot and a trendline. Finally, we learned about the principles of asking the right research questions, best practices in predictive analytics, and different Python packages for this area of biostatistics.

Choosing which method to use for answering different scientific or research questions

The method choice begins with the research question. This means the method choice should be made based on the fact that it can or cannot answer the scientific question.

Let's consider an example.

We have the average height of a group of people, and we want to compare that average against a hypothesized mean of people in that country. For this situation, we would use a one-sample Student's t-test. Why do we use the term *one-sample*? Well, because we are comparing just one sample against a certain mean we expected.

Let's consider another example.

Are there any differences in HDL values between different genders in diabetic subjects? To answer this question, we need to take into consideration how to represent HDL in diabetics. The best way is to use the average or mean value. Now, once we have decided the metric mean, we need to use the appropriate test. You can use a Student's t-test to do this. Since there are two groups (females and males), the appropriate test is a two-sample independent type of test. The most appropriate test for comparing means is a Student's t-test, so the final answer is an *Independent Samples Student's t-test*.

Let's consider another example. The hypothesized frequency of a disease in a study is 30%. But we got results that show it is 45%. How do we test this hypothesis? Since we are testing one sample against a hypothesized percentage, we can use a *one-sample Chi-squared test* because it's more appropriate for frequencies and counts. A one-sample Chi-squared test is used to test one sample against an expected frequency. There is another type of Chi-squared test that is used to compare two samples, two proportions, or two frequencies to each other, and this test is called the Chi-squared test of independence. Why independence? Because the results will tell us if those two samples' frequencies are related to each other, dependent or unrelated to each other, or independent.

Now, let's consider a more complex example. What if the goal of the study is to compare multiple means and medians? Then, we can use two specific hypothesis tests, again looking at *Figure 5.3*. For multiple means, we can use an ANOVA test, and for multiple medians, we can use a Kruskal-Wallis test.

Summary

In biostatistics, one of the most common ways of performing statistical tests is to use specific hypothesis test methods.

In this chapter, we learned what are the most important libraries for biostatistics hypothesis tests in Python and how to use them.

Additionally, we learned about libraries for predictive biostatistics in Python and how to write code for implementing linear regression methods.

Further, we learned how to interpret hypothesis tests and effect sizes. We learned how to differentiate statistical significance from the actual magnitude of the effect.

Finally, we learned how to choose which method to use to answer different scientific or research questions.

In the next chapter, we will learn more about predictive biostatistics, specific studies in which it is used, how to set up a research problem, and how to perform analysis practically using Python.

6

Biostatistical Inference Using Hypothesis Tests and Effect Sizes

Inferring statistical results using hypothesis tests and biological measurements is one of the most important aspects of life science research. This includes using both statistical and biological knowledge to make conclusions on the different biological hypotheses and research questions we make.

Research questions and their answers are vital parts of most life science research projects/studies. In this chapter, we will learn how to use hypothesis tests to test for statistical significance and effect sizes to evaluate magnitudes of effects. These are some of the main tools to answer various research questions statistically.

We're going to cover the following main topics in the chapter:

- Performing Student's t-test in Python
- Performing Wilcoxon signed-rank tests in Python
- Performing chi-squared tests in Python
- Analyzing associations among multiple variables – correlations in Python
- Analyzing multiple groups in Python – **analysis of variance** (**ANOVA**) and Kruskal–Wallis test

Technical requirements

Make sure you have installed the `scipy` and `researchpy` packages. You may run the following commands in your Anaconda command prompt or PowerShell:

```
pip install scipy
pip install researchpy
```

> **Note**
>
> We will be using either Spyder or JupyterLab to run the code and observe the output. If you use Spyder, you will have the option to use a variable explorer for ease of exploring different data objects. In JupyterLab, you may use the `print()` statements to show object values.

Performing Student's t-test in Python and interpreting the effect sizes

Student's t-test is a way of comparing the average values of two groups of data. It can help decide whether the difference between the groups is due to chance or a factor that is being studied.

Suppose we want to compare levels of metabolites, such as **high density lipoprotein** (**HDL**) or **low-density lipoprotein** (**LDL**), in two groups of subjects. Student's t-test can tell if the average difference in these metabolites is significantly different between the groups. Using this approach, one can tell that the level of confidence that the difference we measured reflects the difference in the larger sample and not just the data analyzed.

How does the t-test work?

Student's t-test works through the calculation of a number called the **t-statistic**, which measures how far apart the two group averages are, relative to the variability within each group. The larger the t-statistic, the more likely it is that the difference is not due to chance.

Student's t-test also uses a number of subjects included in the samples and infers credibility based on it. It is plain logic that an increased number of subjects means a larger credibility of the study result.

All these aspects are included in the calculation of p-values, which tell us how statistically significant the result is and how much we can trust it. But another very important aspect to consider is the effect size. A small reminder that p-values tell us about the sampling uncertainty regarding the results we see in the study and we call this measure **statistical significance**.

The effect size tells us about the magnitude of the effect. Magnitude is related to the actual strength of a difference or effect being studied, which is not the case for statistical significance, which is related to the trustworthiness of the effect. One way to express magnitude when performing Student's t-test on means is using **Cohen's d**. So, what is Cohen's d?

Let's first examine its formula:

$$Cohen's\ d = \frac{\mu 1 - \mu 2}{SD\ pooled}$$

As you can see, it includes calculating the raw difference between *mean 1* and *mean 2*. This accounts for the first part of the magnitude calculation – the raw mean difference. But every variable has a different interval of values and different possible variability, so the raw difference must be adjusted by this.

To perform the adjustment, we need to calculate the **standard deviation** (**SD**) pooled. SD pooled is calculated with the following formula:

$$SD\ pooled = \sqrt{(SD1^2 - SD2^2)/2}$$

This way, the formula accounts for the SDs of both groups and we get the SD pooled.

We can then consider Cohen's d for calculating the magnitude of difference between means of two groups as follows:

$$Cohen's\ d = \frac{\mu 1 - \mu 2}{\sqrt{(SD1^2 - SD2^2)/2}}$$

To summarize, Cohen's d is the metric for effect size that uses the raw difference between means and then adjusts it for the variability expressed as an SD to get the magnitude of the effect or difference that is being analyzed.

Cohen's d interpretation is done according to a Cohen's d scale, as depicted in the following table:

Cohen's d value	**Interpretation**
`d=0-0.19`	Trivial effect/difference
`d=0.2-0.49`	Small effect/difference
`d=0.50-0.79`	Medium effect/difference
`d=0.8-1`	Large effect/difference

Table 6.1 – Cohen's interpretation for Cohen's d scale

To proceed with the practice on Student's t-test and Cohen's d, we can use the diabetes dataset as in previous chapters, which may be found here, `https://data.mendeley.com/datasets/wj9rwkp9c2/1` with its code:

```
#Load libraries as in the previous chapter
import pandas as pd
import numpy as np
from scipy import stats

#Load the diabetes dataset from the downloads folder
data=pd.read_csv(r'C:\Users\MEDIN\Downloads\Dataset of Diabetes .csv')
```

The first couple of steps are the same as in the previous chapter; we can select different subjects based on their genders and test for differences in lipids, such as HDL or LDL:

```
#Select subjects with confirmed diabetes
data=data[data['CLASS']=='Y']
```

```
#Define the separate datasets for Females-F and Males-M
data1 = data[data['Gender']=='F']
data2 = data[data['Gender']=='M']

#Select the high density lipoprotein (HDL)
HDLf=data1['HDL']
HDLm=data2['HDL']

# Generate summary statistics for HDL among females
summary_HDLf = HDLf.describe()

# Generate summary statistics for HDL among males
summary_HDLm = HDLm.describe()

# Combine the summaries into a single DataFrame for comparison
summary_df = pd.DataFrame({
    'Female HDL Summary': summary_HDLf,
    'Male HDL Summary': summary_HDLm
})
```

Here is the result (please review `summary_df` in the variable explorer):

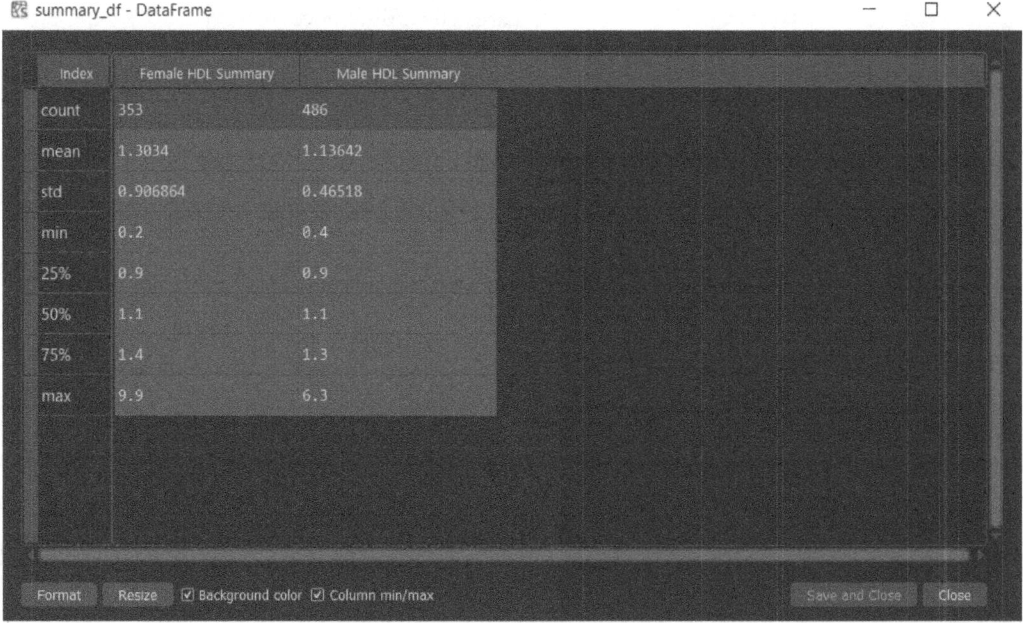

summary_df - DataFrame

Index	Female HDL Summary	Male HDL Summary
count	353	486
mean	1.3034	1.13642
std	0.906864	0.46518
min	0.2	0.4
25%	0.9	0.9
50%	1.1	1.1
75%	1.4	1.3
max	9.9	6.3

Figure 6.1 – Descriptive statistics DataFrame

The data shows us that females had a higher HDL mean of `1.30`, compared to males, who had a mean of `1.13`. So, that's a `0.17` difference. But how large is that difference in magnitude and is it significant statistically? HDL-c reference range is usually in the range of `1-2` mmol/L. The reference range is always important to consider when exploring the magnitude of differences. In this example, the difference between male and female diabetics was `0.17`, which is a small difference in terms of the magnitude when we consider the raw values. It should be noted that HDL-c is considered the "bad" cholesterol in biomedicine, due to being involved in cardiovascular diseases such as **coronary artery disease** (**CAD**), forming plaques that clog the arteries.

To answer that question, we will use Cohen's d calculation to tell us about the magnitude of difference and the Student's t-test p-value to tell us about the statistical significance of this difference.

Here is the code:

```
#Calculate the Cohen's d using the formula provided earlier
cohens_d = (np.mean(HDLf) - np.mean(HDLm)) /
           (np.sqrt((np.std(HDLf) ** 2 + np.std(HDLm) ** 2) / 2))

print(cohens_d)

# perform t-test
t_statistic, p_value = stats.ttest_ind(HDLf, HDLm)

# print results
print(f"t-statistic: {t_statistic}")
print(f"p-value: {p_value}")
```

Here is the output:

Name	Type	Size	Value
cohens_d	float64	1	0.23200343126795797
data	DataFrame	(840, 14)	Column names: ID, No_Pation, Gender, AGE, Urea, Cr, HbA1c, Chol, TG, H ...
data1	DataFrame	(353, 14)	Column names: ID, No_Pation, Gender, AGE, Urea, Cr, HbA1c, Chol, TG, H ...
data2	DataFrame	(486, 14)	Column names: ID, No_Pation, Gender, AGE, Urea, Cr, HbA1c, Chol, TG, H ...
HDLf	Series	(353,)	Series object of pandas.core.series module
HDLm	Series	(486,)	Series object of pandas.core.series module
p_value	float64	1	0.0005307093454497073
t_statistic	float64	1	3.478263088490537

Help | Variable Explorer | Plots | Files

Console 1/A

```
   ...:
   ...:
   ...: # perform t-test
   ...: t_statistic, p_value = stats.ttest_ind(HDLf, HDLm)
   ...:
   ...: # print results
   ...: print(f"t-statistic: {t_statistic}")
   ...: print(f"p-value: {p_value}")
0.23200343126795797
t-statistic: 3.478263088490537
p-value: 0.0005307093454497073
```

Figure 6.2 – Observing results of the Student's t-test and Cohen's d in variable explorer

As seen in the variable explorer, `cohens_d` is `0.232` and belongs to small effects (note the Cohen's d scale mentioned before). The `p_value` is `0.0005`, which is extremely significant, but statistically significant, not clinically or scientifically (while statistical significance may be a small part of scientific significance). Never confuse statistical significance with clinical significance or the magnitude of the effect. Statistical significance can be high but the clinical significance and effect size can still be small.

Next, we will see how to perform the Wilcoxon test to check statistical significance.

Performing Wilcoxon signed-rank test in Python

After interpreting the biostatistical results, it's always good to visualize them using histograms and check if the data follows a distribution, which is optimal for different tests and vice versa, such as, are specific tests optimal for the data we have? Why is this important? Because all statistical tests have certain assumptions and these need to be addressed. For example, Student's t-test has assumptions that are related to the normality of the data. Also, Student's t-test results can be affected by extreme values.

Let's check these by plotting the histograms of the variables:

```
import matplotlib.pyplot as plt

# Set up the figure and axis for plotting
plt.figure(figsize=(10, 6))

# Plot histogram for Female HDL
plt.hist(HDLf, bins=20, color='blue', alpha=0.7, label='Female HDL')

# Plot histogram for Male HDL
plt.hist(HDLm, bins=20, color='orange', alpha=0.7, label='Male HDL')

# Calculate means
mean_HDLf = np.mean(HDLf)
mean_HDLm = np.mean(HDLm)

# Calculate standard deviations
std_HDLf = np.std(HDLf)
std_HDLm = np.std(HDLm)

# Add vertical lines for means
plt.axvline(mean_HDLf, color='blue', linestyle='dashed',
    linewidth=2, label='Female Mean')
plt.axvline(mean_HDLm, color='orange', linestyle='dashed',
    linewidth=2, label='Male Mean')

# Add legend
plt.legend()

# Add labels and title
plt.xlabel('HDL Values')
plt.ylabel('Frequency')
plt.title('Histograms of HDL Values by Gender with Mean Lines')

# Display the plot
plt.show()
```

Here is the plot for it:

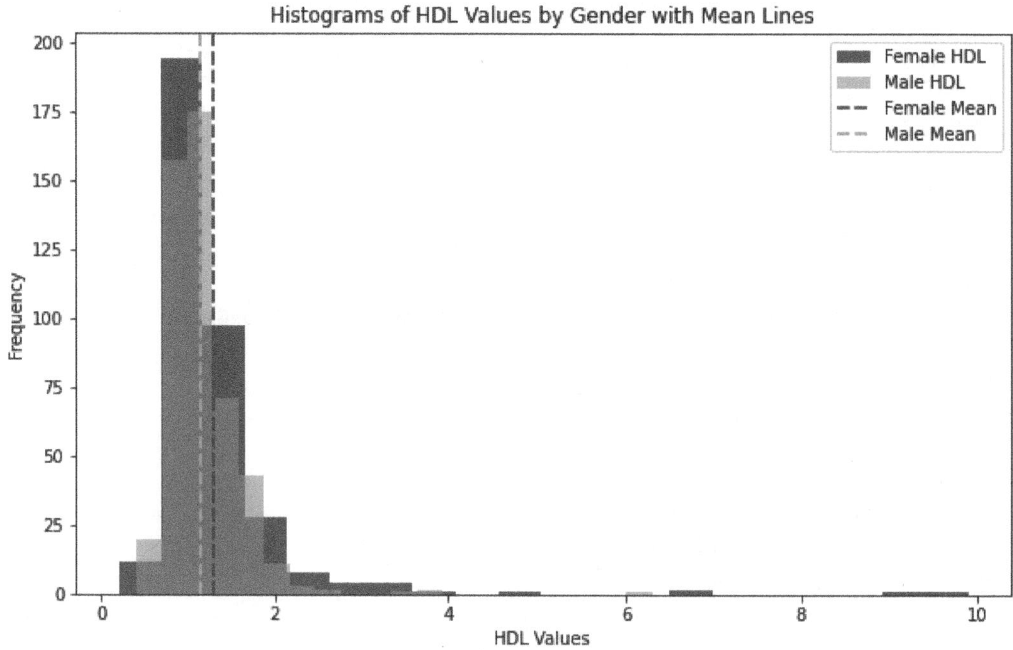

Figure 6.3 – Histograms of HDL values by gender

The histogram shows how small the difference is and it's in line with the small effect size identified using the Cohen's d metric. It does seem that both variables have a pattern of normal data but still, some outliers, extreme values, are shown on the plot, especially for *Female HDL*. Since the difference is small and the extreme values have been identified for *Female HDL*, this could indeed be the cause of the results we are seeing. In this situation, a good solution is to use a non-parametric test, one that does not have data normality assumption, and check the statistical significance.

We will use the Wilcoxon rank sum test. This test is also called the **Mann-Whitney** (**MW**) U test and both names are usually used in statistical analyses and research manuscripts.

To perform the MW U test, we will need to load the `ranksums` function from the `scipy.stats` module. Then, we just need to add the variable objects into the function and implement the MW U test, as shown here:

```
from scipy.stats import ranksums

# Perform the Mann-Whitney U test
statistic, p_value = ranksums(HDLf, HDLm)

# Print the results
```

```
print("Mann-Whitney U statistic:", statistic)
print("P-value:", p_value)

if p_value < 0.05:
    print("The difference in HDL cholesterol between females and males is statistically significant.")
else:
    print("There is no statistically significant difference in HDL cholesterol between females and males.")
```

Here is the output:

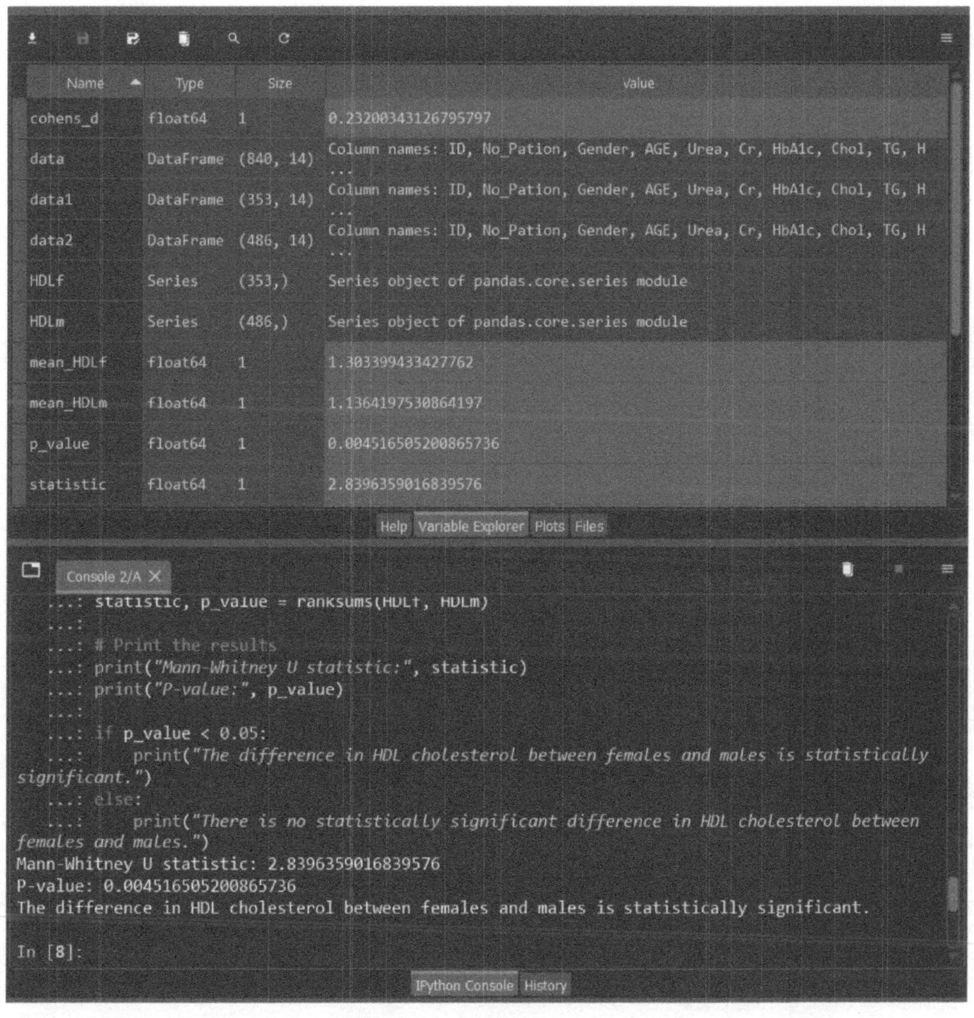

Figure 6.4 – MW test output

Now, let us make another comparison. This time, it's a bit different; instead of comparing female and male diabetics, let's compare diabetics versus non-diabetics based on their **triglycerides** (**TGs**) (instead of focusing on gender, focusing on the diabetes class).

Here is the code for it:

```
#Load the pandas library
import pandas as pd
from scipy import stats
import numpy as np
import matplotlib.pyplot as plt

#Using the pandas read_csv the file is loaded from its path
data=pd.read_csv(r'C:\Users\KORISNIK\Downloads\Dataset of Diabetes
.csv')

#Define the separate datasets for Females-F and Males-M
data1 = data[data['CLASS']=='Y']
data2 = data[data['CLASS']=='N']

#Select the high density lipoprotein (HDL)
TGd=data1['TG']
TGnd=data2['TG']

# Calculate the Cohen's d
cohens_d = (
    (np.mean(TGd) - np.mean(TGnd)) /
    (np.sqrt((np.std(TGd) ** 2 + np.std(TGnd) ** 2) / 2))
)

print(cohens_d)

# perform t-test
t_statistic, p_value = stats.ttest_ind(TGd, TGnd)
```

Also, let's describe the data:

```
# Generate summary statistics for TG among females
summary_TGd = TGd.describe()

# Generate summary statistics for TG among males
summary_TGnd = TGnd.describe()

# Combine the summaries into a single DataFrame for comparison
```

```
summary_df = pd.DataFrame({
    'Diabetes TG Summary': summary_TGd,
    'Control TG Summary': summary_TGnd
})
# Display the summary table
print(summary_df)
```

Here is the output:

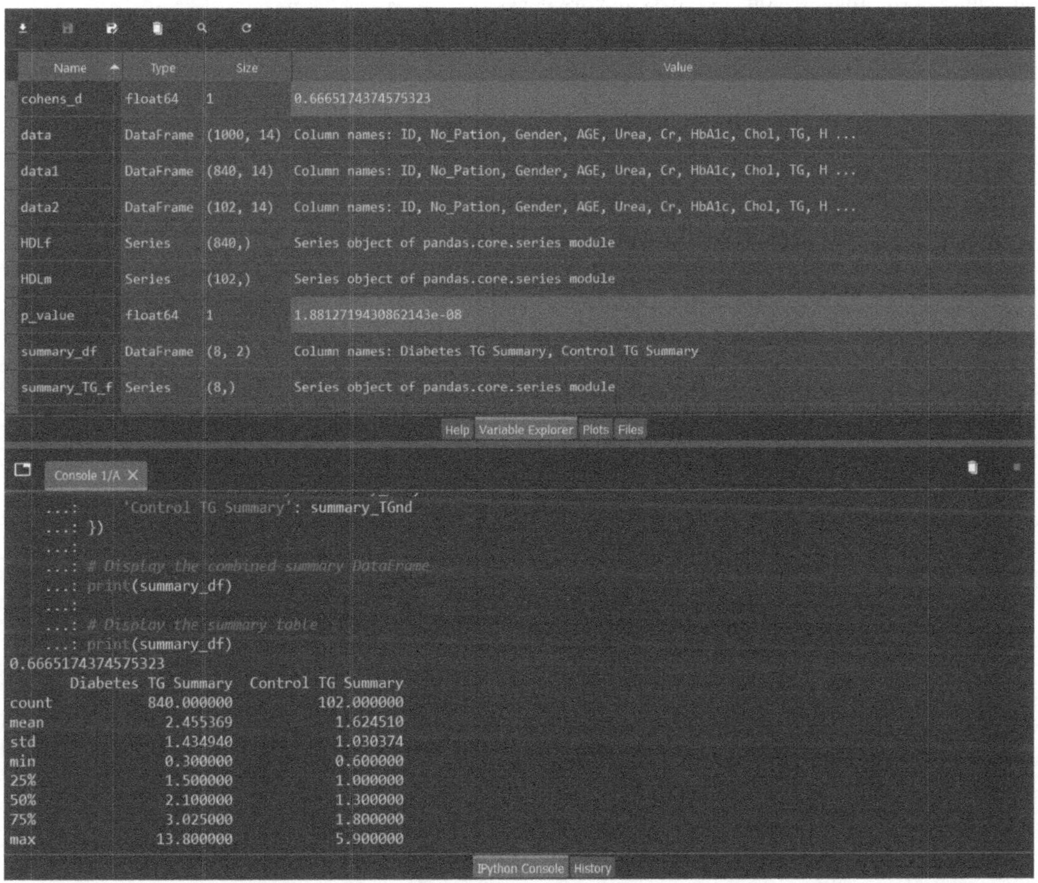

Figure 6.5 – Summary DataFrame (descriptives)

Next, we will perform a hypothesis test to compare different categories across the dataset.

Performing chi-squared tests in Python

In biostatistics, we often have categorical variables, and the data is frequently represented by the fact that it belongs to certain categories. In biostatistics, chi-square is often used for categorical variables (data). One such example is whether the lipids belong to a category within the normal or abnormal reference range.

According to the *American Board of Family Medicine*, the normal TG level is below 1.7 mmol/L, and all values above that are considered as increased TGs.

> **Note**
>
> You can refer to the aforementioned journal here: *Pejic RN, Lee DT (May–Jun 2006). "Hypertriglyceridemia". Journal of the American Board of Family Medicine*, `https://pubmed.ncbi.nlm.nih.gov/16672684/`.

To understand this, let's check the code where first, we will need to load the data and create a new column that will contain information about categories based on the reference threshold for the TG variable. Then, we can use that variable to perform the chi-squared test based on the newly created variable:

```
# Load the necessary libraries
import pandas as pd

# Load the data
data = pd.read_csv(r'C:\Users\KORISNIK\Downloads\Dataset of Diabetes
.csv')

# Create a new column 'increasedtg' based on conditions
data['increasedtg'] = data['TG'].apply(
    lambda x: 'yes' if x >= 1.7 else 'no')

# Display the updated DataFrame
print(data)
```

Here is the result:

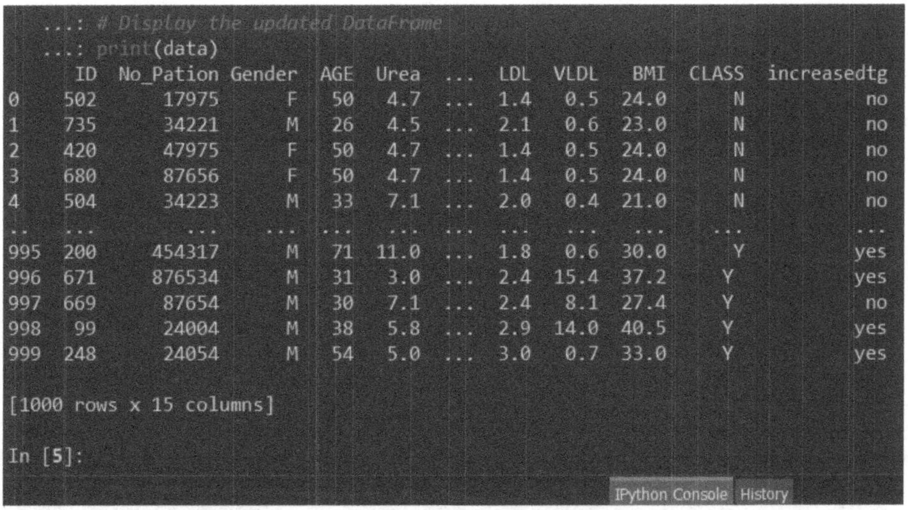

```
   ...: # Display the updated DataFrame
   ...: print(data)
      ID  No_Pation Gender  AGE  Urea  ...  LDL  VLDL   BMI  CLASS  increasedtg
0    502      17975      F   50   4.7  ...  1.4   0.5  24.0      N           no
1    735      34221      M   26   4.5  ...  2.1   0.6  23.0      N           no
2    420      47975      F   50   4.7  ...  1.4   0.5  24.0      N           no
3    680      87656      F   50   4.7  ...  1.4   0.5  24.0      N           no
4    504      34223      M   33   7.1  ...  2.0   0.4  21.0      N           no
..   ...        ...    ...  ...   ...  ...  ...   ...   ...    ...          ...
995  200     454317      M   71  11.0  ...  1.8   0.6  30.0      Y          yes
996  671     876534      M   31   3.0  ...  2.4  15.4  37.2     Y           yes
997  669      87654      M   30   7.1  ...  2.4   8.1  27.4     Y            no
998   99      24004      M   38   5.8  ...  2.9  14.0  40.5     Y           yes
999  248      24054      M   54   5.0  ...  3.0   0.7  33.0     Y           yes

[1000 rows x 15 columns]

In [5]:
```

IPython Console History

Figure 6.6 – DataFrame updated with the increasedtg variable

As can be seen from the preceding screenshot, the new column called `increasedtg` is now created in the DataFrame. This variable can now be used to compare the diabetes subject and controls, not just on the raw TG values, but also on the fact that they are or aren't within the clinically defined reference range. To make this comparison, we can use a specific test called the chi-squared test.

First, we will filter the data so that it contains only `Y` and `N` (`Y` for diabetes patients and `N` for non-diabetes subjects) classes for diabetes and `yes` and `no` for increased TGs. This way, we can test if there is an association between these variables and how strong they are. Let's start with the filtering (make sure to install `scipy` and `researchpie` using `pip install scipy` and `pip install researchpy` in your Anaconda prompt or command prompt):

```
import pandas as pd
from scipy.stats import chi2_contingency
import researchpy as rp

# Load the data
data = pd.read_csv(r'C:\Users\KORISNIK\Downloads\Dataset of Diabetes
.csv')

# Filter data for rows with 'Y' or 'N' values in 'CLASS' column
filtered_data = data[data['CLASS'].isin(['Y', 'N'])].copy()

# Create a new column 'increasedtg' based on conditions
filtered_data['increasedtg'] = filtered_data['TG'].apply(
    lambda x: 'yes' if x >= 1.7 else 'no')
```

Next, we need to create a 2 x 2 contingency table and convert it to percentages for easier interpretation. Here is the code:

```
# Create a contingency table
contingency_table = pd.crosstab(
    filtered_data['CLASS'], filtered_data['increasedtg'])

# Convert frequencies to ratios
contingency_table_ratios = contingency_table.apply(
    lambda r: r/r.sum(), axis=1)
```

Here is what we get when we run it:

Figure 6.7 – Output of the filtered data

Just by looking at the ratios in the table, we can see that subjects with class `Y` (`yes` for diabetes) have 71% increased TGs, which is much higher compared to the control group's (class `N`) only 34%. Now, let's perform the chi-squared test:

```
# Perform the chi-squared test
chi2, p, dof, expected = chi2_contingency(contingency_table)

# Calculate Cramer's Phi
n = contingency_table.values.sum()  # Total number of observations
```

```
min_dim = min(contingency_table.shape) - 1  # Minimum dimension minus
one
cramers_phi = (chi2 / (n * min_dim))**0.5

# Print the results
print(f"Chi-squared value: {chi2:.4f}")
print(f"P-value: {p:.4f}")
print(f"Degrees of freedom: {dof}")
print(f"Cramer's Phi: {cramers_phi:.4f}")
print(f"\nContingency table (ratios):\n{contingency_table_ratios}")
```

This is the result of the test:

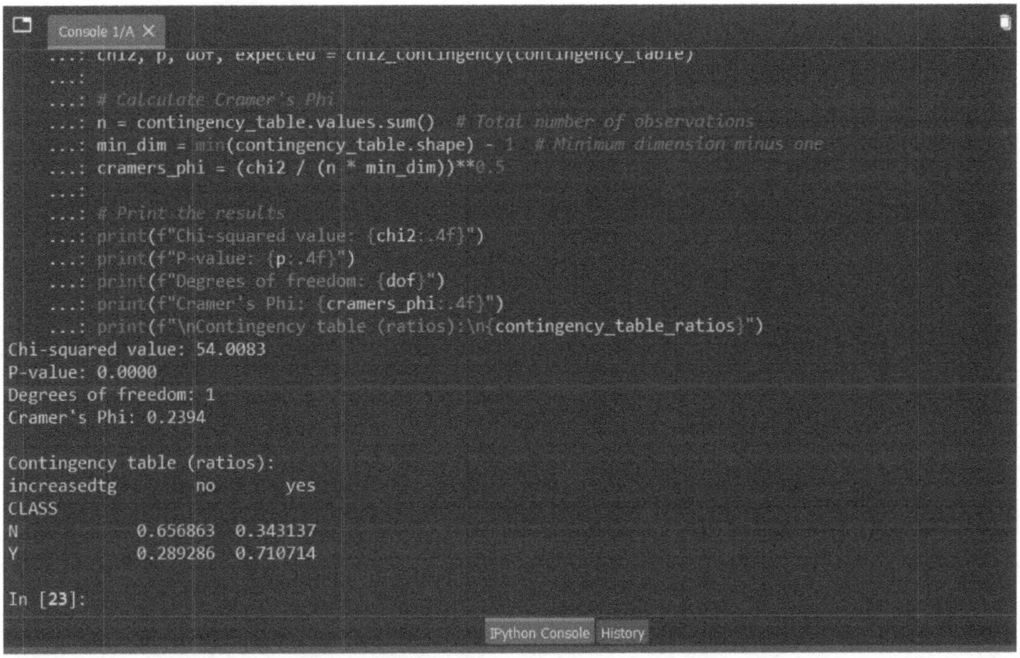

```
Console 1/A
   ...: chi2, p, dof, expected = chi2_contingency(contingency_table)
   ...:
   ...: # Calculate Cramer's Phi
   ...: n = contingency_table.values.sum()  # Total number of observations
   ...: min_dim = min(contingency_table.shape) - 1  # Minimum dimension minus one
   ...: cramers_phi = (chi2 / (n * min_dim))**0.5
   ...:
   ...: # Print the results
   ...: print(f"Chi-squared value: {chi2:.4f}")
   ...: print(f"P-value: {p:.4f}")
   ...: print(f"Degrees of freedom: {dof}")
   ...: print(f"Cramer's Phi: {cramers_phi:.4f}")
   ...: print(f"\nContingency table (ratios):\n{contingency_table_ratios}")
Chi-squared value: 54.0083
P-value: 0.0000
Degrees of freedom: 1
Cramer's Phi: 0.2394

Contingency table (ratios):
increasedtg        no       yes
CLASS
N            0.656863  0.343137
Y            0.289286  0.710714

In [23]:
IPython Console  History
```

Figure 6.8 – Code for accessing the chi-squared results in the console

In the variable explorer, you can see that a chi-squared value (`chi2`) is `54.008` and the p-value is outputted as `0.0000`. This means that this is an extremely low p-value and we can consider it as `p<0.001` (this is one of the academically accepted thresholds for low p-values). This p-value is much lower compared to the one that resulted from previous HDL comparisons in females and males.

This is indeed a statistically significant result but it still doesn't tell us much about the magnitude of the association between the fact that subjects have diabetes and the level of TGs. This means that we can use Cramer's Phi scale. **Cramer's Phi** is a measure of association between two categorical variables in a contingency table, calculated using the chi-squared statistic, but then adjusted for sample size and table dimensions. It yields a value between `0` and `1`, where `0` indicates no association and `1` indicates a perfect association between the variables. So, it is an alternative to correlations in terms of estimating the level of association between categorical variables.

Cramer's Phi value is `0.24`, which means the association between diabetes and increased TGs is moderate (check the following table).

Have a look at the following Cramer's Phi scale table for reference:

Cramer's Phi value	Interpretation
Φ=0-0.9	We see a trivial association
Φ=0.1-0.19	There is some small association
Φ=0.2-0.39	There is a medium association
Φ=0.4-59	There is a relatively strong association
Φ=0.6-79	This has a strong association
Φ=0.8-1	This means a very strong association

Table 6.2 – Cramer's Phi scale table

Cramer's Φ scale: `https://www.researchgate.net/publication/311335682_Alternatives_to_P_value_Confidence_interval_and_effect_size`.

Next, let's move on to the correlations in Python.

Analyzing associations among multiple variables – correlations in Python

We performed the chi-square test, which was ideal for categorical variables to test for independence between them. But what about testing the raw continuous variables and testing multiple associations between variables using a simple parametric approach (the one for continuous normal distribution)? One easy approach for this situation is the correlation analysis using the **Pearson correlation method**.

Let's perform the correlation analysis:

```
import pandas as pd

# Load the data
data = pd.read_csv(r'C:\Users\KORISNIK\Downloads\Dataset of Diabetes
.csv')

# Filter data for rows with 'Y' or 'N' values in 'CLASS' column
filtered_data = data[data['CLASS'].isin(['Y', 'N'])]

# Drop the first two columns
filtered_data = filtered_data.iloc[:, 2:]

# Compute the correlation matrix
corr_matrix = filtered_data.corr(
    method='pearson', numeric_only=True)
```

Here is the output:

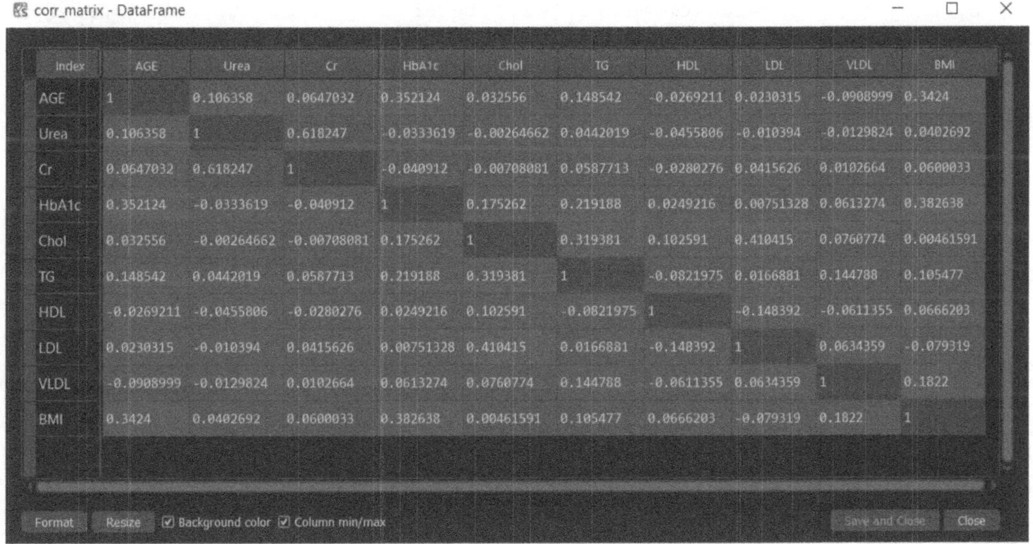

corr_matrix - DataFrame

Index	AGE	Urea	Cr	HbA1c	Chol	TG	HDL	LDL	VLDL	BMI
AGE	1	0.106358	0.0647032	0.352124	0.032556	0.148542	-0.0269211	0.0230315	-0.0908999	0.3424
Urea	0.106358	1	0.618247	-0.0333619	-0.00264662	0.0442019	-0.0455806	-0.010394	-0.0129824	0.0402692
Cr	0.0647032	0.618247	1	-0.040912	-0.00708081	0.0587713	-0.0280276	0.0415626	0.0102664	0.0600033
HbA1c	0.352124	-0.0333619	-0.040912	1	0.175262	0.219188	0.0249216	0.00751328	0.0613274	0.382638
Chol	0.032556	-0.00264662	-0.00708081	0.175262	1	0.319381	0.102591	0.410415	0.0760774	0.00461591
TG	0.148542	0.0442019	0.0587713	0.219188	0.319381	1	-0.0821975	0.0166881	0.144788	0.105477
HDL	-0.0269211	-0.0455806	-0.0280276	0.0249216	0.102591	-0.0821975	1	-0.148392	-0.0611355	0.0666203
LDL	0.0230315	-0.010394	0.0415626	0.00751328	0.410415	0.0166881	-0.148392	1	0.0634359	-0.079319
VLDL	-0.0908999	-0.0129824	0.0102664	0.0613274	0.0760774	0.144788	-0.0611355	0.0634359	1	0.1822
BMI	0.3424	0.0402692	0.0600033	0.382638	0.00461591	0.105477	0.0666203	-0.079319	0.1822	1

Format Resize Background color Column min/max Save and Close Close

Figure 6.9 – Correlation matrix

I can also output this data as a CSV file in Microsoft Excel, as follows:

```
filtered_data = filtered_data.select_dtypes(include='number')


# Compute the correlation matrix
corr_matrix = filtered_data.corr(method='pearson')

# Compute the p-values for the correlation matrix
p_values = pd.DataFrame(
    index=corr_matrix.index, columns=corr_matrix.columns)
#Go through the correlation matrix and find a p value
for i in range(len(corr_matrix)):
    for j in range(i, len(corr_matrix)):
        r, p = stats.pearsonr(filtered_data.iloc[:, i],
            filtered_data.iloc[:, j])
        p_values.iloc[i, j] = p
        p_values.iloc[j, i] = p

# Display the correlation matrix with results with p<0.05 marked with
an asterisk *
display_matrix = corr_matrix.applymap(lambda x: f"{x:.2f}")
for i in range(len(p_values)):
    for j in range(len(p_values)):
        if p_values.iloc[i, j] < 0.05:
            display_matrix.iloc[i, j] += "*"
print(display_matrix)


# Output the display_matrix DataFrame as a CSV file
# Use your own path to the correlation matrix csv file
display_matrix.to_csv(
    r'C:\Users\KORISNIK\Downloads\Correlation_Matrix.csv',
    index=True)
```

Here is the resulting CSV file:

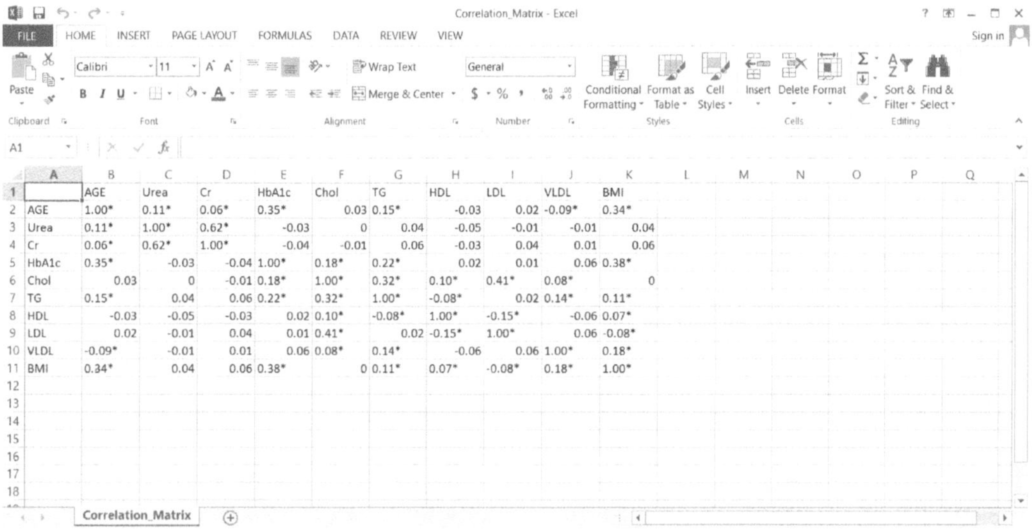

	AGE	Urea	Cr	HbA1c	Chol	TG	HDL	LDL	VLDL	BMI
AGE	1.00*	0.11*	0.06*	0.35*	0.03	0.15*	-0.03	0.02	-0.09*	0.34*
Urea	0.11*	1.00*	0.62*	-0.03	0	0.04	-0.05	-0.01	-0.01	0.04
Cr	0.06*	0.62*	1.00*	-0.04	-0.01	0.06	-0.03	0.04	0.01	0.06
HbA1c	0.35*	-0.03	-0.04	1.00*	0.18*	0.22*	0.02	0.01	0.06	0.38*
Chol	0.03	0	-0.01	0.18*	1.00*	0.32*	0.10*	0.41*	0.08*	0
TG	0.15*	0.04	0.06	0.22*	0.32*	1.00*	-0.08*	0.02	0.14*	0.11*
HDL	-0.03	-0.05	-0.03	0.02	0.10*	-0.08*	1.00*	-0.15*	-0.06	0.07*
LDL	0.02	-0.01	0.04	0.01	0.41*	0.02	-0.15*	1.00*	0.06	-0.08*
VLDL	-0.09*	-0.01	0.01	0.06	0.08*	0.14*	-0.06	0.06	1.00*	0.18*
BMI	0.34*	0.04	0.06	0.38*	0	0.11*	0.07*	-0.08*	0.18*	1.00*

Figure 6.10 – Excel file with the DataFrame output

As you can see, the CSV file can now be opened in Microsoft Excel and processed further, turned into a table, and used in publications or research reports.

We can also make a seaborn correlation plot for visual analysis for this. Here is the code for it:

```
import seaborn as sns
import numpy as np

# Create a mask for the upper triangle of the correlation matrix
mask = np.triu(np.ones_like(corr_matrix, dtype=bool))
```

Masking the triangle removes duplicate information in the symmetric correlation matrix, making the visualization aesthetically better and easier to interpret:

```
# Create a custom colormap to highlight significant p-values
cmap = sns.diverging_palette(230, 20, as_cmap=True)

# Create an array with the correlation coefficients and asterisks for
significant p-values
annot_array = np.vectorize(
    lambda x, y: f"{corr_matrix.iloc[x, y]:.2f}" +
    ("*" if p_values.iloc[x, y] < 0.05 else "")
)(
    np.arange(len(corr_matrix)),
```

```
    np.arange(len(corr_matrix))[:, None]
)
# Create a heatmap with the correlation coefficients and p-values
marked as significant
sns.heatmap(corr_matrix, mask=mask, cmap=cmap,
    annot=annot_array, fmt="s")

import matplotlib.pyplot as plt

# Set the figure size to be 4 times larger
fig, ax = plt.subplots(figsize=(12, 9))

# Create a heatmap with the correlation coefficients and p-values
marked as significant
sns.heatmap(corr_matrix, mask=mask, cmap=cmap,
    annot=annot_array, fmt="s", ax=ax)
```

Here is the seaborn report:

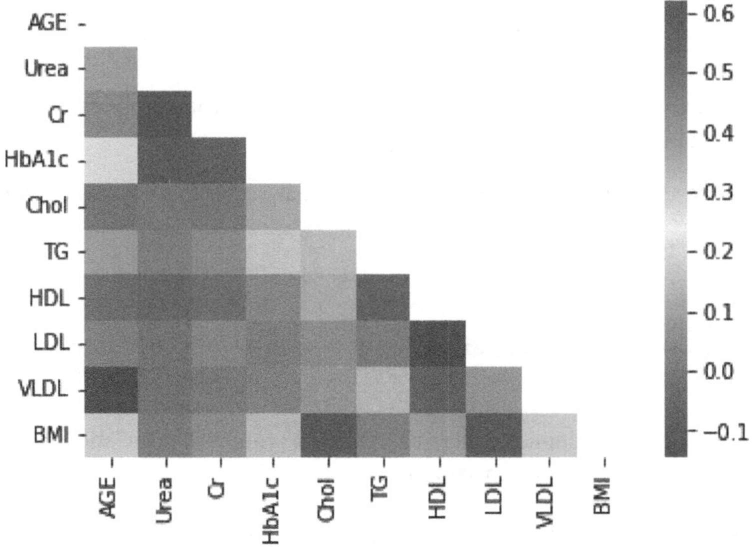

Figure 6.11 – The correlation output plot

The plot can be saved by right-clicking and pressing **Save as...** (choose the **PNG** file from the options).

Note that different versions of matplotlib and seaborn packages may cause the annotations to disappear; try upgrading or downgrading them according to your Python version to fix the problem.

Since we learned how to implement correlation analysis and create a correlation matrix, which shows indicators of associations between different variables, the next step is to add another angle: an angle of analyzing multiple groups within those variables.

Analyzing multiple groups in Python - ANOVA and Kruskal–Wallis test

So far, we have made comparisons between two groups based on variables such as TG. What if we have three or more categories to compare simultaneously? We can use ANOVA to perform this task.

This comparison can be used to compare multiple groups in the data. For example, for BMI, we can have underweight, normal, and overweight subjects and compare their lipid levels simultaneously.

Here is how we can try this:

```
import pandas as pd
import scipy.stats as stats

# Load the data
data = pd.read_csv(r'C:\Users\KORISNIK\Downloads\Dataset of Diabetes
.csv')

# Filter data for rows with 'Y'(diabetic) or 'N'(non-diabetic) values
in 'CLASS' column
filtered_data = data[data['CLASS'].isin(['Y', 'N'])]
```

Now, we need to create a new column called `weight_class`, which will contain the categories of weight class based on BMI. Those subjects with a BMI < `25` will be categorized as normal (in terms of no overweight or obese), a BMI between `25` and `30` will be categorized as overweight, and a BMI > `30` as obese:

```
# Create a new column 'weight_class' based on the 'BMI' column
def weight_class(bmi):
    if bmi < 25:
        return 'normal'
    elif bmi >= 25 and bmi < 30:
        return 'overweight'
    else:
        return 'obese'
```

```
filtered_data['weight_class'] = filtered_data['BMI'].apply(
    weight_class)

# Create a table object to display the descriptive statistics for each
class
normal_stats = filtered_data[
    filtered_data['weight_class'] == 'normal']['TG'].describe()
overweight_stats = filtered_data[
    filtered_data['weight_class'] == 'overweight']['TG'].describe()
obese_stats = filtered_data[
    filtered_data['weight_class'] == 'obese']['TG'].describe()

#Creating table for the weight classes
stats_table = pd.DataFrame({'Normal': normal_stats,
    'Overweight': overweight_stats,
    'Obese': obese_stats})
print('Descriptive Statistics for Each Class:')
print(stats_table)

# Perform ANOVA on the 'TG' column for the three classes
fvalue, pvalue = stats.f_oneway(
    filtered_data[filtered_data['weight_class'] == 'normal']['TG'],
    filtered_data[filtered_data['weight_class'] =='overweight']['TG'],
    filtered_data[filtered_data['weight_class'] == 'obese']['TG'])
```

The ANOVA takes into account the F values (based on means and variances of groups) and it is usually accompanied by results with p-values, which are calculated based on it:

```
# Create a table object to display the results of ANOVA
results = pd.DataFrame({'F-value': [fvalue],
                        'p-value': [pvalue]})
print('\nANOVA Results:')
print(results)
```

Here is the code output:

```
Descriptive Statistics for Each Class:
          Normal   Overweight       Obese
count  155.000000  267.000000  520.000000
mean     1.923226    2.370262    2.494712
std      1.081054    1.233916    1.566498
min      0.600000    0.600000    0.300000
25%      1.200000    1.550000    1.500000
50%      1.700000    2.100000    2.000000
75%      2.300000    2.850000    3.100000
max      5.900000    8.500000   13.800000

ANOVA Results:
    F-value   p-value
0  9.853715  0.000058
```

Figure 6.12 – ANOVA results

These results are actually two tables: descriptive statistics and ANOVA results. We can save both using the following code (in your code, use the path on your computer and the applicable folder instead of `C:\Users\KORISNIK\Downloads\`):

```
# Output the tables  as  CSV files
stats_table.to_csv(
    r'C:\Users\KORISNIK\Downloads\Descriptive_Statistics.csv',
    index=True)
results.to_csv(
    r'C:\Users\KORISNIK\Downloads\ANOVA_Results.csv',
    index=True)
```

But means are just one side of the story. What if we wanted to compare additional metrics, such as medians using a non-parametric approach (the approach that does not assume the normality of the data)?

This can be achieved using the method called the Kruskal–Wallis test. Here is the code for it:

```
import pandas as pd
import scipy.stats as stats
import matplotlib.pyplot as plt

# Load the data
data = pd.read_csv(r'C:\Users\KORISNIK\Downloads\Dataset of Diabetes
.csv')

# Filter data for rows with 'Y' or 'N' values in 'CLASS'
#column, 'Y' = diabetes, 'N' = control
```

```
filtered_data = data[data['CLASS'].isin(['Y', 'N'])]

# Create a new column 'weight_class' based on the 'BMI' column
def weight_class(bmi):
    if bmi < 25:
        return 'normal'
    elif bmi >= 25 and bmi < 30:
        return 'overweight'
    else:
        return 'obese'

filtered_data['weight_class'] = filtered_data['BMI'].apply(
    weight_class)

# Create a figure and axes with the specified size
fig, ax = plt.subplots(figsize=(12, 9))
```

Many hypothesis tests for categorical variables require the data to be ordered in a way that represents the biological characteristics. Also, the interpretation is easier if the values are ordered according to biological scales. Let's specify the order of categories based on the BMI obesity scale:

```
# Specify the order of the categories
order = ['normal', 'overweight', 'obese']

# Create a new DataFrame with the rows sorted in the desired order
sorted_data = filtered_data.copy()
sorted_data['weight_class'] = pd.Categorical(
    sorted_data['weight_class'], categories=order,
    ordered=True)
sorted_data.sort_values(by='weight_class', inplace=True)

# Create boxplots for each class based on the 'TG' column
sorted_data.boxplot(column='TG', by='weight_class', ax=ax)
plt.title('Boxplots of TG by Weight Class')
plt.suptitle('')

# Perform Kruskal-Wallis test on the 'TG' column for the three classes
hvalue, pvalue = stats.kruskal(
    sorted_data[sorted_data['weight_class'] == 'normal']['TG'],
    sorted_data[sorted_data['weight_class'] == 'overweight']['TG'],
    sorted_data[sorted_data['weight_class'] == 'obese']['TG'])

# Add the Kruskal-Wallis test results to the plot
```

```
plt.annotate(
    f'Kruskal-Wallis H-value: {hvalue:.2f}\np-value: {pvalue:.2e}',
    xy=(0.7, 0.9), xycoords='axes fraction')

# Show the plot
plt.show()
```

Note that we used a visualization with boxplots to present the Kruskal–Wallis test. This is because the boxplots will have medians presented as well as the quartiles, 25%, 50%, and 75%, of the data, so we can use the boxplots to see which values represent TG for these percentages in the data.

Here is the plot for it:

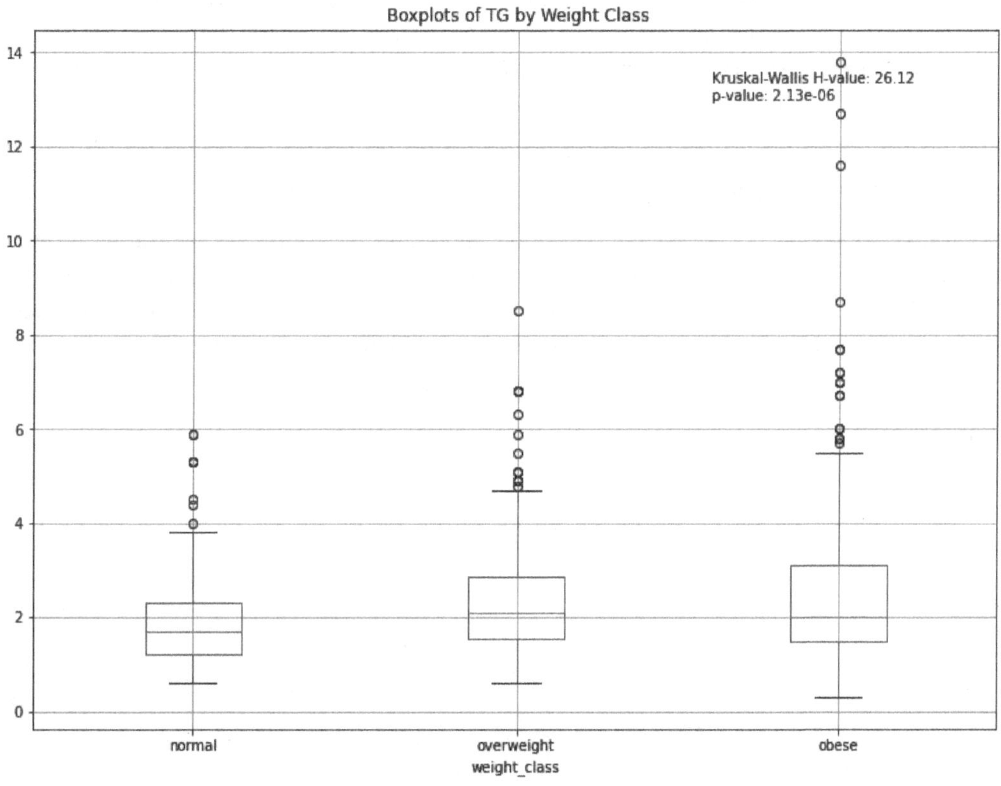

Figure 6.13 – Boxplot TG versus weight class

The green lines represent the medians and we can see that the medians for normal subjects' TGs are slightly lower compared to the obese and overweight. The Kruskal–Wallis H-value of `26.12` and p-value of `2.13e-6` means the result is 6 0s below `2.13` (`0.00000213`) and is statistically significant (`p<0.05`). The Kruskal-Wallis H-value measures the degree of difference between the groups in terms of category counts, with higher values indicating greater differences. So, the final interpretation of this result is that there is a small increase in TG across overweight and obese subjects but this increase is not large in magnitude. This is one of the statistical results that is interpreted as statistically significant but not large in the magnitude of increase or the effect size.

Summary

In this chapter, we learned how to perform Student's t-test in Python and interpret the magnitude of mean difference. This is a very important segment of implementing and interpreting central tendency in biostatistics. We also learned to perform the Wilcoxon rank test in Python and interpret the magnitude of mean difference as another approach to mean difference analysis.

Then, we learned how to perform a chi-squared test in analyses of associations among multiple categorical variables in Python.

Finally, we learned how to analyze multiple groups in Python, as group-level analysis is one of the most common approaches to statistical analysis in biostatistics. In the next chapter, you will learn about predictive biostatistics using Python and how to implement different functions from packages, such as `statsmodels` and `scikit-learn`, to create different predictive models.

7
Predictive Biostatistics Using Python

Most biostatistical studies rely on some form of predictive statistics. The ability to predict an outcome or predict a biological characteristic based on biomarkers is essential in life science. Predictive biostatistics is the key ingredient in many aspects of life science, with areas such as drug discovery and biomedicine being the most relevant examples. In addition to being able to predict outcomes, predictive biostatistics allow researchers to associate risk factors and other biological variables with specific diseases such as diabetes, as well as cardiological and oncology-related diseases.

In this chapter, we're going to cover the following main topics:

- Learning predictive biostatistics and their uses in different areas of life science
- Understanding dependent and independent variables and their relation to confounders and latent variables
- Implementing linear regression for biostatistics in Python
- Implementing logistic regression for biostatistics in Python
- Learning multiple linear and logistic regressions using Python

Learning predictive biostatistics and their uses in different areas of life science

Predictive biostatistics is a field that is used by both biologists and biostatisticians to make predictions or forecasts about biological phenomena, specifically in the context of healthcare and life science.

One of the focuses of predictive biostatistics is making predictions about future events or outcomes based on patterns and trends observed in historical data. However, predictive biostatistics is not just about making predictions on future events. It specializes in understanding the nature of associations between variables, such as *Can I predict biological variable B if I know biological variable A?*

Predictive analysis is one of the most important areas in biostatistics. It helps us find associations between variables and potentially predict new outcomes in areas such as biomedicine, the pharmaceutical industry, drug discovery, and public health.

An example use case of predictive biostatistics is predicting the diabetes outcome using risk factors such as obesity or nutrition. Have a look at the following figure:

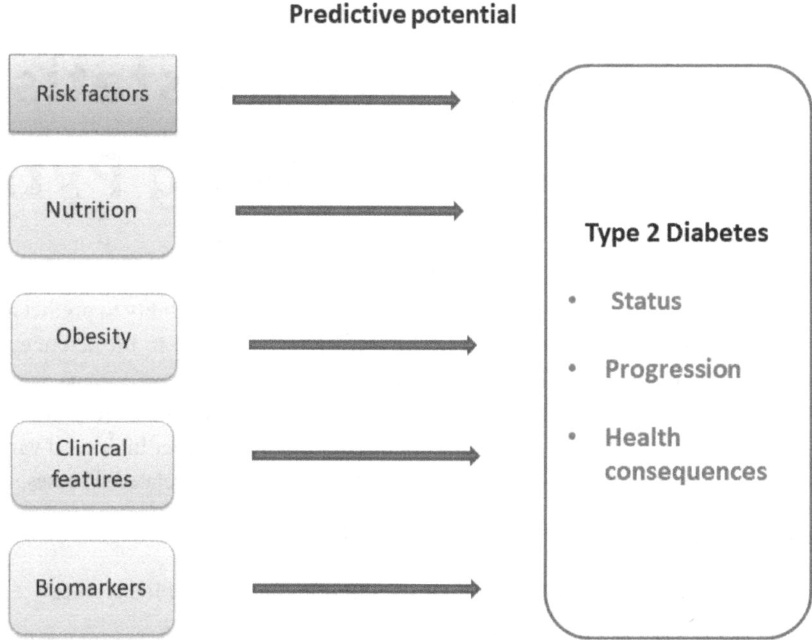

Figure 7.1 – The schemes for potential predictive models in type 2 diabetes

As you can see in the preceding figure, risk factors related to nutrition, obesity, clinical risk factors, or biomarkers can be used to predict target variables (in this case, **Type 2 Diabetes Mellitus** (**T2DM**) status), progression, or other health outcomes and consequences. In this case, the risk factors would be independent variables: the variables used to predict outcomes. Status, progression, and health consequences would be dependent or target variables.

There are many other areas in biomedicine, such as oncology and cardiology, that heavily rely on predictive biostatistics. Being able to predict outcomes is essential for prognosis and planning different treatment protocols. However, predictive biostatistics is not just about predicting future events. Another very important aspect of predictive biostatistics is creating models that can explain the nature of association between different biological variables using present or even past data. The nature of the association between these biological variables can be understood through mathematical equations. These equations are used to create what we call a model of association between variables. Term modeling is frequently used when we explain different aspects of predictive biostatistics.

So, how does a predictive model relate to biological variables? Models can explain how two or more biological variables change in relation to each other, in other words, how certain variables such as risk factors or biomarkers explain biological outcomes such as disease progression or the presence or absence of disease.

Predictive biostatistics is used in the healthcare and pharmaceutical industry to model and predict disease prevalence, patient outcomes, and treatment responses.

These models can be used to predict various biological metrics, such as the risk of developing a disease, the effectiveness of a new drug, or the progression of a particular condition.

The biostatistical predictive models are the foundation of modern biostatistics. In life science and especially in healthcare, trustworthiness is the key. This means that *accuracy* and *explainability* are very important. These two aspects are mainly achieved through creating predictive models that can accurately predict biological outcomes when using different biomedical treatments. Without predictive models, a significant portion of biomedical research would not be possible.

Predictive biostatistics is one of the most essential tools in public health research. Being able to create predictive models and forecasts is essential in making governance decisions and interventions in this area. *Survival analysis* is a set of statistical predictive approaches used to determine the time it takes for an event of interest to occur. In cardiology, this could be the time until a patient experiences a heart attack or stroke. This type of analysis is essential for biomedical prognosis research, providing medical practitioners and researchers with information on the disease's onset.

Some examples include trying to predict health outcomes such as diabetes or myocardial infarction based on risk factors such as obesity or nutrition.

To be able to predict the biological outcomes using statistical models, we must define the target variables (the ones we are trying to predict) and differentiate them from the predictors (variables used to predict target variables). In statistics, the terms *dependent* and *independent* are used to describe target variables and predictors. We will be discussing this in detail in the next section.

Dependent and independent variables

Every time we try to predict any outcome, that outcome can be considered as a target variable. The target variable that we are trying to predict is called the **dependent variable**. Also, other variables we use to predict the target variables are called **independent variables**. To be able to predict any variable based on any other variable, they must be associated with one another. This same principle applies to independent and dependent variables; they must have a level of association as a prerequisite for being able to predict the dependent variable.

Generally, in research, we have specific research questions. These research questions are related to what we call the endpoint or the target variable. Let's explore an example we already analyzed. In the previous three chapters, we analyzed the Diabetes dataset. For clinicians, HbA1C is one of the most important variables in diabetes. HbA1c tells us about the glycation of the hemoglobin over a period of three months, and as such, it tells us about the level of glucose that is directly responsible for the glycation of hemoglobin.

As such, HbA1c is frequently the dependent variable in diabetes research. Researchers try to predict how variables such as BMI, weight, LDL, HDL, and other variables affect HbA1c, and as a result, the status of diabetes. In other words, scientists are trying to see whether we can predict the change in HbA1c when other variables (BMI and others) change. This would be an example of how a predictive model with HbA1c and other variables such as BMI as independent variables are created.

In this section, we learned how independent variables can be used to predict biological outcomes, or dependent variables. We also learned that biological importance plays an important role in how we select predictors for predictive models. Independent variables should typically be biologically relevant to our target biological variables (dependent variables).

In the next section, we will learn how to actually create a scheme for predictive models and implement them in Python.

Linear regression for biostatistics in Python

We will continue evaluating the diabetes dataset from previous chapters, but this time, we will be setting the research questions related to predictive biostatistics and creating biostatical models. To proceed, make sure you previously downloaded the diabetes dataset as in previous chapters at `https://data.mendeley.com/datasets/wj9rwkp9c2/1`.

First, let's load the required libraries. Notice that we are loading the `stats` library from the `scipy` package. This library will be used to create the predictive models:

```
import pandas as pd
from scipy import stats
import statsmodels.formula.api as smf
import seaborn as sns
import matplotlib.pyplot as plt

# Load the data
data = pd.read_csv(r'C:\Users\MEDIN\Downloads\Dataset of Diabetes
.csv')
```

Sometimes, in the datasets, there may be inconsistencies with lower- and uppercase letters. In this dataset, this is the case with one `f` versus `F` letter in the `gender` column. Here is how to make the letters consistent by using `data.replace()` and converting all `f` letters to `F`:

```
data = data.replace('f','F', regex=True)
```

We will use the knowledge from the previous two chapters to create a simple model with HbA1c as the target (dependent) variable and BMI as its predictor (independent) variable.

Let's see how to implement this simple model creation in Python:

```
# Model 1: Univariate Linear Regression with  HbA1c and BMI
model1 = smf.ols(formula='HbA1c ~ BMI', data=data).fit()
print(model1.summary())
```

Here is the output:

```
                            OLS Regression Results
==============================================================================
Dep. Variable:                  HbA1c   R-squared:                       0.171
Model:                            OLS   Adj. R-squared:                  0.170
Method:                 Least Squares   F-statistic:                     205.7
Date:                Mon, 02 Oct 2023   Prob (F-statistic):           1.51e-42
Time:                        09:07:10   Log-Likelihood:                -2254.6
No. Observations:                1000   AIC:                             4513.
Df Residuals:                     998   BIC:                             4523.
Df Model:                           1
Covariance Type:            nonrobust
==============================================================================
                 coef    std err          t      P>|t|      [0.025      0.975]
------------------------------------------------------------------------------
Intercept      2.0380      0.441      4.617      0.000       1.172       2.904
BMI            0.2111      0.015     14.341      0.000       0.182       0.240
==============================================================================
Omnibus:                       29.395   Durbin-Watson:                   1.167
Prob(Omnibus):                  0.000   Jarque-Bera (JB):               31.214
Skew:                           0.421   Prob(JB):                     1.67e-07
Kurtosis:                       3.199   Cond. No.                         182.
==============================================================================
```

Figure 7.2 – Linear regression output

Now, let's plot the regression to see the association between HbA1c and BMI using a scatter plot and a regression trendline with the following code:

```
sns.regplot(x='BMI', y='HbA1c', data=data, ci=95)
plt.show()
```

Here is the plot for it:

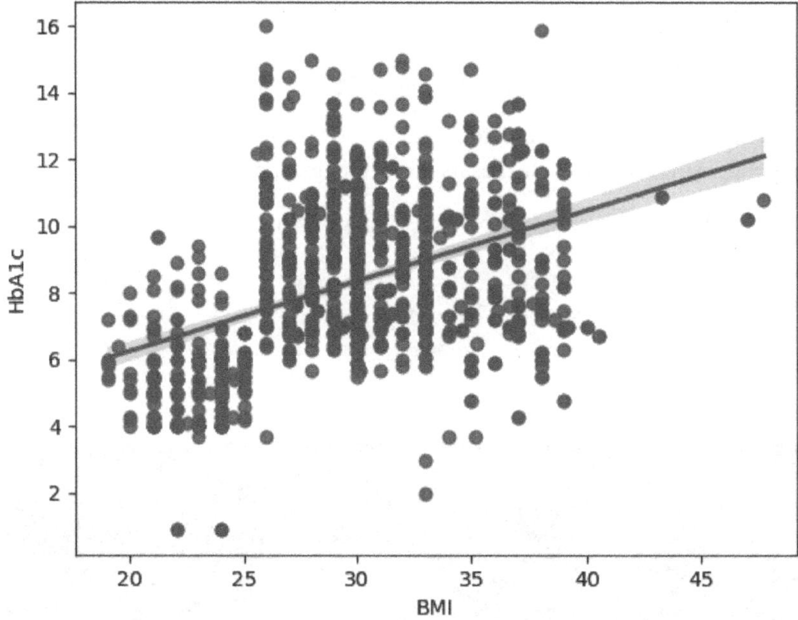

Figure 7.3 – Linear regression plot (BMI versus HbA1c)

You can see that the final `regplot` plot is created and you may see that the trendline shows an association between BMI and HbA1c (as BMI increases, HbA1c also increases, meaning that there is a positive association between them). In this section, we learned what linear regression represents in relation to biological variables in diabetes. Furthermore, we learned how to implement linear regression in Python and present outputs as scatter plots and trendlines. Finally, we learned how to complement numerical results with visual interpretation of results

Logistic regression in Python

In linear regression, we were predicting a continuous dependent variable such as HbA1c, but what if the variable is categorical and contains a certain diagnosis such as the presence or absence of diabetes in subjects? For this model, we can use the `CLASS` variable, which contains the `diagnosed diabetes` and `control` categories. However, we also need a different statistical method that is better adapted for this analysis. Linear model will not work well for categories because categorical data does not show linearity as normal continuous data. For this reason, it's better to use a specific statistical method called **logistic regression**. The logistic regression model will create the sigmoid probability function, which can be used to predict whether a subject has diabetes based on other parameters.

To see how the sigmoid function works, let's do a practical representation:

```
import pandas as pd
import statsmodels.api as sm
import seaborn as sns
import matplotlib.pyplot as plt
import numpy as np

# Load the data
data = pd.read_csv(r'C:\Users\MEDIN\Downloads\Dataset of Diabetes
.csv')
#Fix the lower-case value identified previously in CLASS variable
data = data.replace('f','F', regex=True)


# Filter data to include only 'Y' (diabetes present) and 'N' (diabetes
not present) in 'CLASS' variable
filtered_data = data[data['CLASS'].isin(['Y', 'N'])]

# Map 'Y' to 1 and 'N' to 0 in the 'CLASS' variable
filtered_data['CLASS'] = filtered_data['CLASS'].map({'Y': 1, 'N': 0})

# Define the formula for logistic regression
formula = 'CLASS ~ HbA1c'

# Create the logistic regression model
logit_model = sm.Logit.from_formula(formula, data=filtered_data)

# Fit the model
result = logit_model.fit()

# Print the model summary
print(result.summary())
```

Here is the output:

```
In [4]: print(result.summary())
                           Logit Regression Results
==============================================================================
Dep. Variable:                  CLASS   No. Observations:                  942
Model:                          Logit   Df Residuals:                      940
Method:                           MLE   Df Model:                            1
Date:                Sun, 01 Oct 2023   Pseudo R-squ.:                  0.6016
Time:                        18:06:06   Log-Likelihood:                -128.70
converged:                       True   LL-Null:                       -323.02
Covariance Type:            nonrobust   LLR p-value:                 1.638e-86
==============================================================================
                 coef    std err          z      P>|z|      [0.025      0.975]
------------------------------------------------------------------------------
Intercept     -8.4648      0.853     -9.921      0.000     -10.137      -6.793
HbA1c          1.7091      0.158     10.809      0.000       1.399       2.019
==============================================================================
```

Figure 7.4 – Logistic regression output

The model output gives you basic information about the model such as the dependent variable and the number of observations, as well as the fact that the model has converged (`converged : true`). This means that the model is successfully optimized by making the fit between the data and the model through iterations (repeated steps) until there is no need to optimize further.

What is most important to interpret here are coef values, p values, and confidence interval at the bottom of the output. The next step is to calculate the probabilities for the logistic function of HbA1c. This means that for each value of HbA1c, we will assign a probability of the presence of diabetes. In this case, the `CLASS` variable is `Y`. For the logistic regression model, when plotting on a graph, the line between the horizontal axis (HbA1c) and vertical axis (predicted probability of diabetes presence) should have an S-shape or similar. Such an S-shaped model is often called the **Sigmoid model**.

Let's check this using the code that follows:

```
# Calculate probabilities using the fitted model
fitted_probabilities = result.predict(filtered_data)

# Plot the probabilities against HbA1c
plt.figure(figsize=(10, 6))
sns.scatterplot(
    x=filtered_data['HbA1c'], y=fitted_probabilities)
plt.xlabel('HbA1c')
plt.ylabel('Predicted Probability')
plt.title('Predicted Probabilities vs HbA1c')
plt.show()
```

Here is the plot for it:

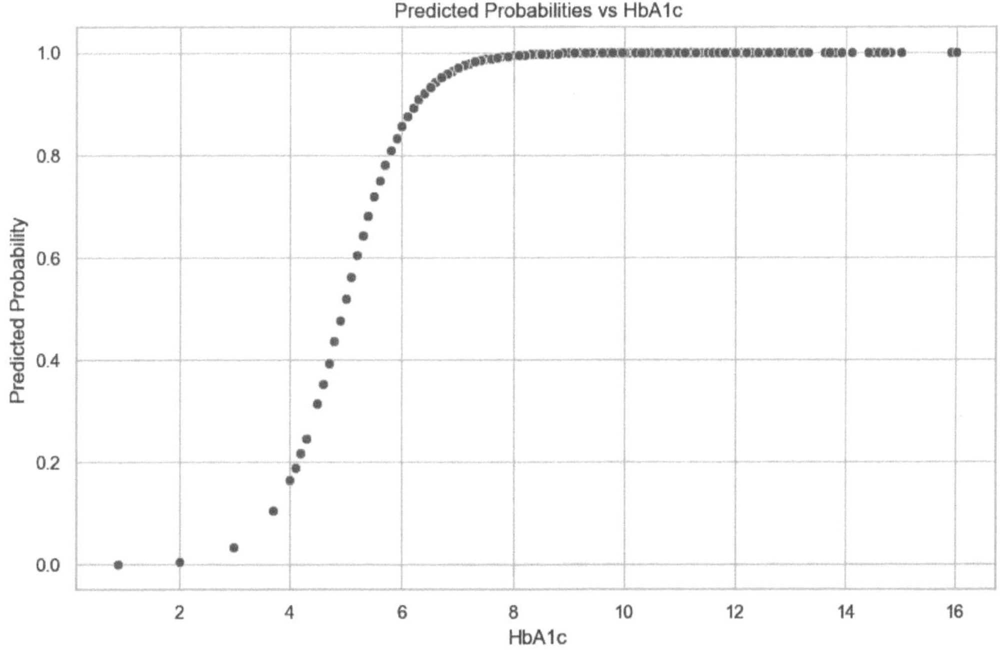

Figure 7.5 – Sigmoid model

In this section, we learned how to perform linear and logistic regression. We also learned how to interpret their outputs such as summary tables and Matplotlib plots.

In the next section, we will be learning how to expand the analysis to multivariate models with multiple independent variables and how to adjust for important biological variables.

Multiple linear and logistic regressions using Python

Having one dependent and one independent variable can provide a certain level of information in a predictive model. However, this information can be prone to the effects of other variables. In this case, the relationship between HbA1c and BMI could potentially be influenced by additional factors such as age and gender. These "other variables," which were not in the main framework composed of dependent and independent variables, are called **confounders**. To be able to account for the effects of confounders, we must include them in the model. Once they are included, the model will allow for their interaction with the main variables of interest.

In this case, our main dependent variable is HbA1c and our main independent variable is BMI, but we want to include the age and gender variables in the model too because they are known to be clinically important. Refer to the following figure:

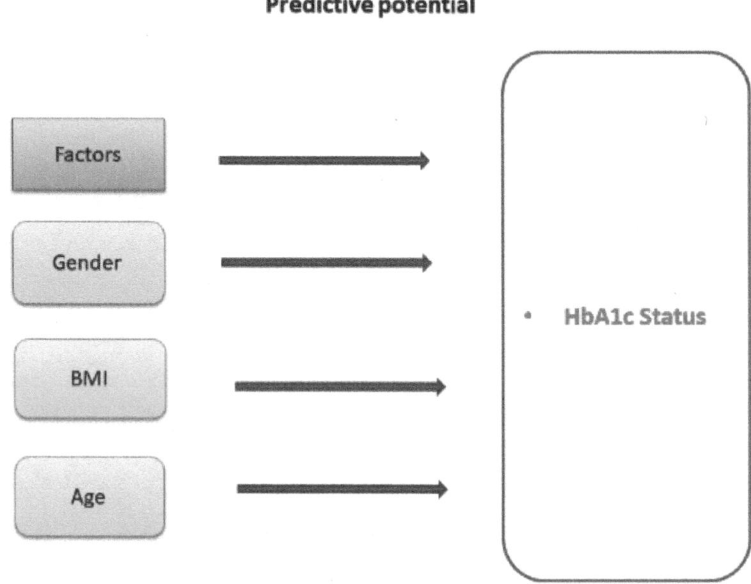

Figure 7.6 – HbA1c Status multivariate model scheme (four variables)

In the preceding figure, you can see a schematic of how to design this specific predictive model. On the left side, you can see the hypothesized risk factors clinically, but biologically, we will call them independent variables. On the right side, you can see the target variable we want to predict, which is the HbA1c Status (dependent variable).

We will be creating a model with three clinical hypothetically relevant independent variables, with HbA1c Status being the dependent variable. When I say *hypothetically*, that means we want to ask the specific research questions and answer them using the predictive linear regression model. The model will consist of a dependent variable, HbA1c, which is the target we are trying to predict, as well as three clinical factors as independent variables: `Gender`, `BMI`, and `Age`.

Let's see how to implement this model creation in Python:

```
# Model 2: Multivariate Linear Regression - HbA1c as #independend,
BMI, Age and Gender as dependent variables
model2 = smf.ols(formula='HbA1c ~ BMI + AGE + Gender',
    data=data).fit()
print(model2.summary())
```

Here is the output:

```
                            OLS Regression Results
==============================================================================
Dep. Variable:                  HbA1c   R-squared:                       0.231
Model:                            OLS   Adj. R-squared:                  0.228
Method:                 Least Squares   F-statistic:                     99.50
Date:                Mon, 02 Oct 2023   Prob (F-statistic):           2.47e-56
Time:                        09:15:57   Log-Likelihood:                -2217.2
No. Observations:                1000   AIC:                             4442.
Df Residuals:                     996   BIC:                             4462.
Df Model:                           3
Covariance Type:            nonrobust
==============================================================================
                 coef    std err          t      P>|t|      [0.025      0.975]
------------------------------------------------------------------------------
Intercept     -0.4285      0.519     -0.826      0.409      -1.447       0.590
Gender[T.M]   -0.1936      0.142     -1.360      0.174      -0.473       0.086
BMI            0.1625      0.015     10.584      0.000       0.132       0.193
AGE            0.0750      0.009      8.679      0.000       0.058       0.092
==============================================================================
Omnibus:                       23.651   Durbin-Watson:                   1.253
Prob(Omnibus):                  0.000   Jarque-Bera (JB):               25.210
Skew:                           0.350   Prob(JB):                     3.35e-06
Kurtosis:                       3.340   Cond. No.                         456.
==============================================================================
```

Figure 7.7 – Linear regression model output (four-variable model)

Now let us interpret the results. We can see that 1,000 observations are included in the model and that the **R-squared** value is **0.231**. R-squared is a metric that is called the coefficient of determination. This metric tells us a proportion of the variability of a target variable, which can be explained by predictors or independent variables. Multivariate models, which contain multiple independent variables, are best interpreted using the adjusted (**Adj. R-squared**) one as that is more relevant and in this case is `0.228`. The adjusted R-squared is an R-squared value that is adjusted for the number of variables in the model, which makes it better for multivariate models. That means that 22.8% of the variation in the HbA1c can be explained by the variation in **Gender**, **BMI**, and **Age**.

If we look at the p-values (**P> |t|**), we can see that **Gender** has a p-value of **0.174**, which means it is not a significant predictor of HbA1c. On the other hand, **BMI** and **Age** are outputted as **0.000**, which means that the p-values for them are <0.001. These are the significant predictors of this model. This is logical and it is biologically known that obesity and age are related to T2DM. We now have statistical results that are interpreted in the same way.

Now let's interpret the regression coefficients. **BMI** has a coefficient of **0.1625** and **AGE** has one of **0.075**, which means that there is a positive association. When these two increase, HbA1c also increases.

Let's explain the principles of the interpretation further. If a regression coefficient between the independent and dependent variables is above 0, it means that they increase and decrease together. In statistics, this is considered a positive association. If the regression coefficient is below 0, or it's negative, this means that a decrease in the independent variable is associated with an increase in the dependent (target) variable, and vice versa. Keep in mind that to make the claims mentioned earlier, it's important to have a statistically significant result too, $p < 0.05$ in most situations.

Gender and age are variables that are typically included in the multivariate biostatistical models. Statisticians often use the term *adjustment* when including them in the model. In that sense, this model is adjusted for gender and age. However, what about other parameters such as lipid-related parameters, TG, LDL, and HDL? Let's include them in the model draft. However, there is one statistical rule of thumb to keep in mind: the number of observations (n/10), which means at least one covariate per 10 observations. This means that we must have enough observations to create complex models. In this example, we have enough observations, a bit less than 1,000, but in any future analyses make sure to follow the one covariate per 10 observations rule of thumb.

Let's create the framework for adding new variables:

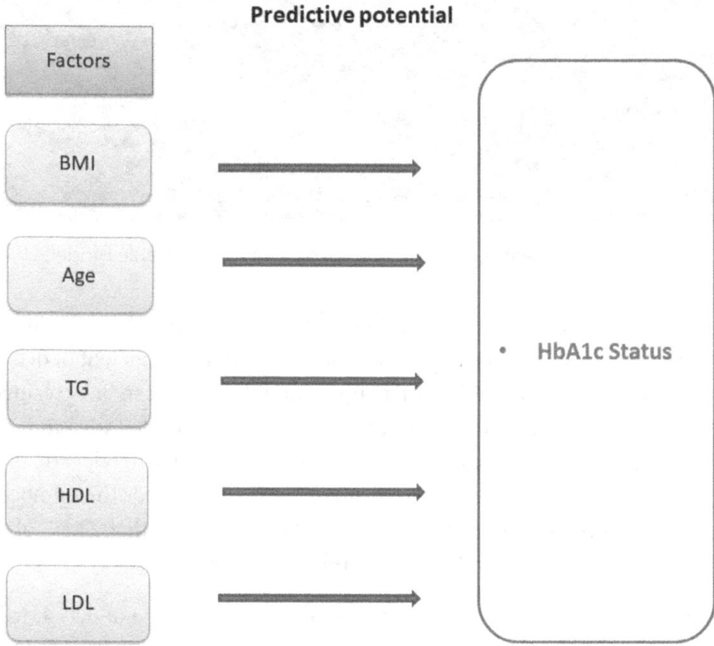

Figure 7.8 – Linear regression scheme (HbA1c status versus five independent variables)

Now, let's create the model in Python. We need to add TG, LDL, and HDL in addition to `Gender`, `BMI`, and `AGE` as independent variables:

```
# Model 3: Multivariate Linear Regression with all specified variables
model3 = smf.ols(
    formula='HbA1c ~ BMI + AGE + Gender + TG + LDL + HDL',
    data=data).fit()
print(model3.summary())
```

This code will create a more complex multivariate predictive model with multiple independent variables used to try to predict HbA1c in the subjects.

Refer to the following screenshot:

```
                            OLS Regression Results
==============================================================================
Dep. Variable:                  HbA1c   R-squared:                       0.254
Model:                            OLS   Adj. R-squared:                  0.249
Method:                 Least Squares   F-statistic:                     56.35
Date:                Tue, 03 Oct 2023   Prob (F-statistic):           5.26e-60
Time:                        14:26:49   Log-Likelihood:                -2201.7
No. Observations:                1000   AIC:                             4417.
Df Residuals:                     993   BIC:                             4452.
Df Model:                           6
Covariance Type:            nonrobust
================================================================================
                 coef    std err          t      P>|t|      [0.025      0.975]
--------------------------------------------------------------------------------
Intercept     -0.9214      0.561     -1.641      0.101      -2.023       0.180
Gender[T.M]   -0.2224      0.142     -1.567      0.117      -0.501       0.056
BMI            0.1580      0.015     10.340      0.000       0.128       0.188
AGE            0.0694      0.009      8.084      0.000       0.053       0.086
TG             0.2753      0.050      5.459      0.000       0.176       0.374
LDL            0.0711      0.063      1.126      0.261      -0.053       0.195
HDL            0.0874      0.108      0.810      0.418      -0.124       0.299
==============================================================================
Omnibus:                       34.735   Durbin-Watson:                   1.309
Prob(Omnibus):                  0.000   Jarque-Bera (JB):               40.604
Skew:                           0.401   Prob(JB):                     1.52e-09
Kurtosis:                       3.574   Cond. No.                         502.
==============================================================================
```

Figure 7.9 – Linear regression output (HbA1c status versus five independent variables)

We can see again that **Gender** is not a significant predictor, p>0.05 (P>|t|). We have a similar situation with LDL and HDL. Even though these variables are not significant as independent predictors, they can always affect other significant predictors. For that reason, it is a good idea not to exclude them from the model. We can see that coefficient is increased by 27% in diabetes subjects for each unit of increased TG (one of the new covariates we added; you can see in the figure that for TG, the coefficient is **0.27**), while for **BMI** and **AGE**, this coefficient is increased by 15.8 and 6 %, respectively, with all mentioned results being statistically significant, p<0.05.

OK, now let's create the multiple logistic regression models. We will be using a similar formula as before in this chapter, when we created the logistic regression model in the *Logistic regression in Python* section, but this time we will be adding multiple independent variables to try to predict the status of subjects in terms of presence or absence of T2DM.

Here is the framework:

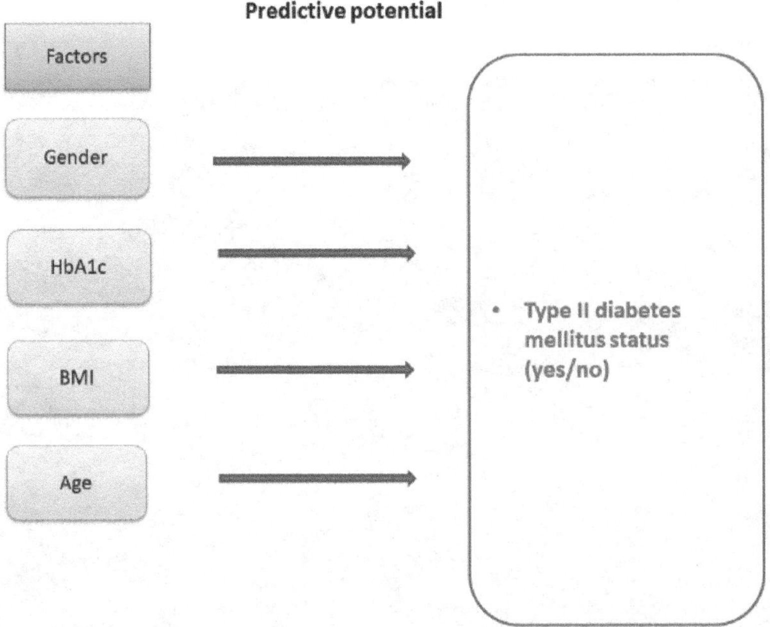

Figure 7.10 – Logistic regression scheme (T2DM status versus four independent variables)

In the next block of code, we need to define the formula with `CLASS` as the target dependent variable and the `HbA1c`, `AGE`, `Gender`, and `BMI` variables as independent variables. Then we can use the `sm.Logit.from_formula()` function to create the logistic regression model. Let's start:

```
# Define the formula for logistic regression
formula = 'CLASS ~ HbA1c + AGE + Gender + BMI'

# Create the logistic regression model
logit_model = sm.Logit.from_formula(
    formula, data=filtered_data)

# Fit the model
result = logit_model.fit()

# Print the model summary
print(result.summary())
```

Here is the output:

```
In [6]: print(result.summary())
                           Logit Regression Results
==============================================================================
Dep. Variable:                  CLASS   No. Observations:                  942
Model:                          Logit   Df Residuals:                      937
Method:                           MLE   Df Model:                            4
Date:                Sun, 01 Oct 2023   Pseudo R-squ.:                  0.7641
Time:                        18:07:22   Log-Likelihood:                -76.209
converged:                       True   LL-Null:                       -323.02
Covariance Type:            nonrobust   LLR p-value:                1.611e-105
==============================================================================
                 coef    std err          z      P>|z|      [0.025      0.975]
------------------------------------------------------------------------------
Intercept    -23.8281      3.456     -6.895      0.000     -30.602     -17.054
Gender[T.M]    0.8015      0.420      1.910      0.056      -0.021       1.624
HbA1c          1.4111      0.217      6.508      0.000       0.986       1.836
BMI            0.6883      0.126      5.444      0.000       0.440       0.936
AGE            0.0023      0.024      0.099      0.921      -0.044       0.048
==============================================================================
```

Figure 7.11 – Logistic regression output (T2DM status versus four independent variables)

To be able to control for more possible covariates that might affect the results, we should try to add as many covariates as possible. Keep in mind that the number of covariates that we should add should not be more than the number of observations (n/100), which means we must have at least one covariate per 10 observations. In this case, the number of observations is more than 900, so there is no problem in adding a few additional covariates to the model.

Now, let's create an even more complex multivariate logistic regression with diabetes class (`CLASS`) as the dependent variable and **Gender**, **HbA1c**, **HDL**, **LDL**, **BMI**, and **AGE** as independent variables:

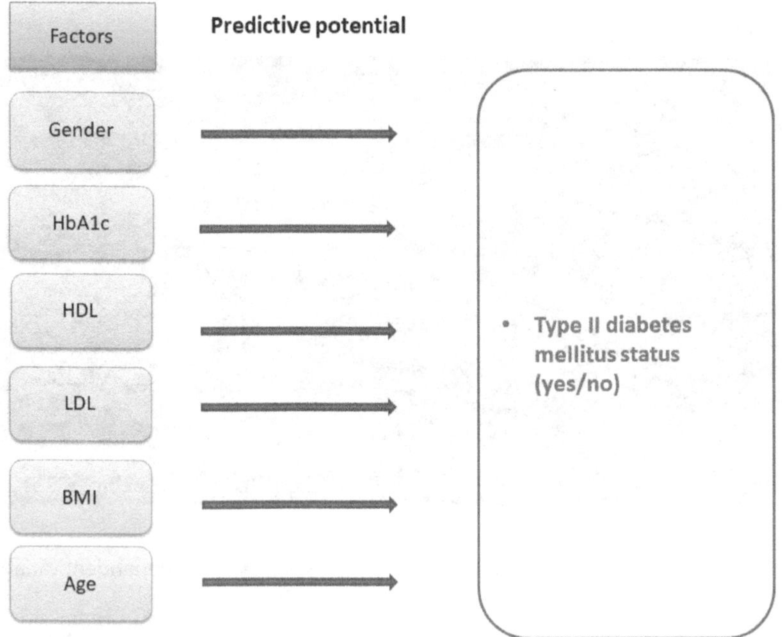

Figure 7.12 – Logistic regression scheme (T2DM status versus six independent variables)

As you can see, now there are six potential variables that can be used to predict T2DM. Some of these are known to have a relation to diabetes, such as HbA1c, because the higher levels of glucose in diabetics cause higher levels of HbA1c glycation. BMI as an obesity measure is also known to contribute.

However, other variables such as HDL, LDL, gender, and age might also affect the results, so let's test them all in the multivariate predictive model:

```
# Define the formula for logistic regression
formula = 'CLASS ~ Gender +  HbA1c + HDL + LDL + BMI + AGE '

# Create the logistic regression model
logit_model = sm.Logit.from_formula(formula, data=filtered_data)

# Fit the model
result = logit_model.fit()

# Print the model summary
print(result.summary())
```

Here are the results:

```
                           Logit Regression Results
==============================================================================
Dep. Variable:                  CLASS   No. Observations:                  942
Model:                          Logit   Df Residuals:                      935
Method:                           MLE   Df Model:                            6
Date:                Sun, 01 Oct 2023   Pseudo R-squ.:                  0.7800
Time:                        18:08:02   Log-Likelihood:                -71.054
converged:                       True   LL-Null:                       -323.02
Covariance Type:            nonrobust   LLR p-value:                1.201e-105
==============================================================================
                 coef    std err          z      P>|z|      [0.025      0.975]
------------------------------------------------------------------------------
Intercept    -27.3471      3.809     -7.179      0.000     -34.813     -19.881
Gender[T.M]    0.7530      0.444      1.696      0.090      -0.117       1.623
HbA1c          1.5000      0.234      6.405      0.000       1.041       1.959
HDL            0.3850      0.402      0.958      0.338      -0.403       1.173
LDL            0.6862      0.227      3.020      0.003       0.241       1.132
BMI            0.7356      0.130      5.664      0.000       0.481       0.990
AGE           -0.0071      0.027     -0.264      0.792      -0.060       0.045
==============================================================================
```

Figure 7.13 – Logistic regression output (T2DM status versus six independent variables)

First, let's look at the p-values (the **P>|z|** column). The p-values that are significant, $p<0.05$, are related to the following predictors: HbA1c, LDL, and BMI. What is interesting is that in this multivariate model, **AGE** is not a significant predictor, which is different compared to the previous model. While it is clinically known that age is relevant to T2DM, this situation tells us that even though it is not significant in this model, it interacts with other variables in the model and should be included.

To have some idea about the level of significance, we can also observe the z-value, which is the largest for HbA1c. The z-value in logistic regression is calculated when the coefficient is divided by the standard deviation. In other words, it tells us about the ratio of magnitude/variability of the coefficient estimate. However, the z-values are not typically intuitive to interpret, so we can use the confidence intervals instead for a better interpretation.

Observe the 95% confidence interval on the right of the p-value.

In this section, we learned how to implement multivariate linear regression and multivariate logistic regression in Python. Furthermore, we learned how to adjust for important biological variables using the diabetes dataset and variables HbA1c and T2DM status as target variables for predictive models.

Summary

In this chapter, we learned what predictive biostatistics is and why we need it in different areas of life science. We also learned about the dependent and independent variables and their relation to confounders and latent variables. We performed linear regression for biostatistics in Python and used it to create the predictive model for predicting the HbA1c in different subjects based on their anthropometric and biochemical parameters.

We learned how to create logistic regression models used to predict whether subjects have T2DM based on their anthropometric and biochemical parameters related to obesity and biomarkers such as HbA1c.

We learned how to include multiple anthropometric and biochemical parameters in a single model and use them to predict HbA1c and T2DM presence using multivariate models. Using these models, we learned how predictive modeling works in biostatistics and how to apply it to a real-world diabetes dataset.

In the next chapter, we will learn about the practical aspects of what we learned in *Chapters 6* and *7*. We will be both applying hypothesis tests and implementing predictive models using different practical examples. The next chapter will be a specific exercise to consolidate what we learned about hypothesis tests and predictive biostatistics.

8

Part 2 Exercise – T-Test, ANOVA, and Linear and Logistic Regression

In this chapter, we will gain more practical experience using different exercises and variations of Student's t-test, **Analysis of Variance** (**ANOVA**), and predictive analysis. Previously, we learned about typical examples of performing these tests and predictive analysis. However, evaluating different specific statistical situations is very important and it is often needed to modify the versions of the test to adapt it to the data that is being analyzed.

We will be using the same data and tests and data visualizations, but this time with more detailed analysis and different versions and angles of analysis applied.

In this chapter, we're going to cover the following main topics:

- Implementing different versions of Student's t-test
- Applying post-hoc tests using ANOVA
- Performing and visualizing linear regression in Python
- Performing and visualizing logistic regression in Python

By the end of the chapter, you will learn how to use different hypothesis tests in Python and make predictive models such as linear and logistic regression. Finally, you will learn how to visualize and interpret those results from a biostatistical angle.

Implementing different versions of Student's t-test

The diabetes dataset contains data from subjects that are with diagnosed diabetes (CLASS: Y), predicted diabetes (CLASS: P), and controls (CLASS: N). Using the available data, we will perform an analysis comparing average values of HbA1c between diagnosed diabetes and controls. The biological parameter we will be analyzing in this exercise is known as HbA1c, or glycosylated hemoglobin, which is higher in diabetes subjects due to the glycation of hemoglobin due to increased blood glucose. Let's explore how grouping can be made based on the CLASS variable, which contains information on the presence or absence of diagnosed **Type 2 Diabetes Mellitus** (**T2DM**).

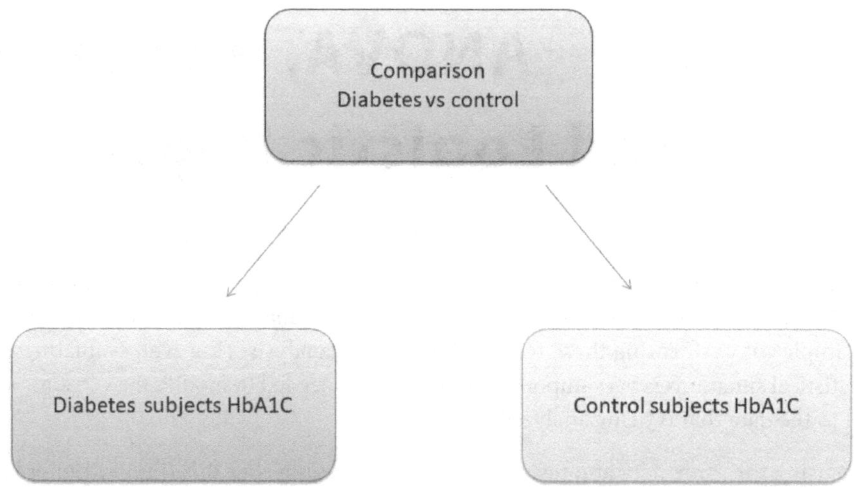

Figure 8.1 – Comparing HbA1C in diabetes subjects and control group

For this exercise, the main setting is that we are comparing two groups, diabetes subjects (CLASS: Y) and control subjects (CLASS: N). There is a third CLASS, P, or predicted diabetes. We are not including this because it is not within the scope of the exercise. We will be using the **Scientific Development Environment for Python** (**Spyder IDE**), but you may also use Jupyter Notebook if preferred. Finally, we will be comparing the HbA1C average values for the two groups and evaluating the statistical significance using the Student's t-test.

Here is how we do it.

First, let's load the libraries and the dataset:

```
import pandas as pd
import numpy as np
import matplotlib.pyplot as plt
```

```
# Load the diabetes dataset from your folder
data = pd.read_csv(r'C:\Users\MEDIN\Downloads\Dataset of Diabetes
.csv')
```

The second step is to separate the `CLASS` variables for the presence or absence of diabetes. We will be looking at `CLASS=='Y'` versus `CLASS=='N'`:

```
# Separate the datasets according to CLASS variable
data1 = data[data['CLASS'] == 'Y']
data2 = data[data['CLASS'] == 'N']
```

Next, select the glycoside hemoglobin from the variables. We are selecting the `HbA1c` variable:

```
# Select the HbA1c values
HbA1c_d = data1['HbA1c']
HbA1c_c = data2['HbA1c']
```

Let's also calculate the means and standard deviations of HbA1c. We will need these both for the descriptive statistics and the plotting part:

```
# Calculate the average and standard deviation for HbA1c values
avg_HbA1c_d = HbA1c_d.mean()
std_HbA1c_d = HbA1c_d.std()
avg_HbA1c_c = HbA1c_c.mean()
std_HbA1c_c = HbA1c_c.std()
```

OK, we are now ready to create some data visualization using the averages, standard deviations, and bar plot in Python:

```
# Create a barplot with error bars
categories = ['T2DM', 'Controls']
values = [avg_HbA1c_d, avg_HbA1c_c]
std_devs = [std_HbA1c_d, std_HbA1c_c]

plt.bar(categories, values, yerr=std_devs, capsize=5)
plt.ylabel('HbA1c')
plt.title('Average HbA1c Values')
plt.show()
```

Here is the plot for it:

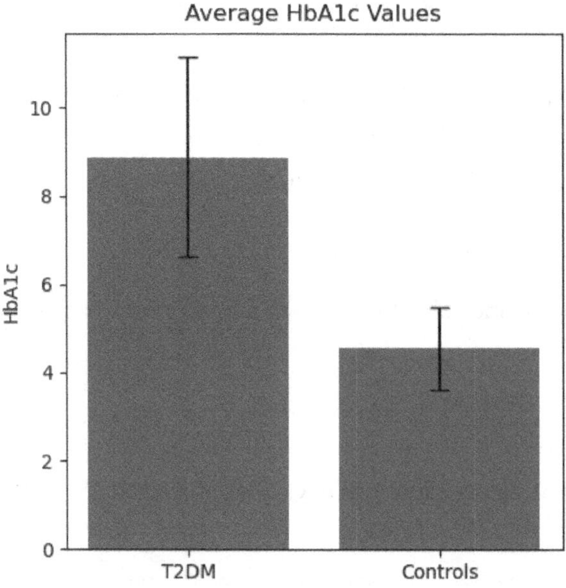

Figure 8.2 – Comparison of average HbA1c average values between diabetes subjects and controls

We can visually see that diabetes subjects not only have higher HbA1c values on average but also that the standard deviation error is above the controls. This is a good indication that we could have an interesting difference in the analysis, in terms of it being a large difference.

Now, let's perform the t-test:

```
# Generate summary statistics for HbA1c diabetics
summary_HbA1c_d = HbA1c_d.describe()

# Generate summary statistics for HbA1c controls
summary_HbA1c_c = HbA1c_c.describe()

# Combine the summaries into a single DataFrame
summary_df = pd.DataFrame({
    'Diabetes HbA1c Summary': summary_HbA1c_d,
    'Control HbA1c Summary': summary_HbA1c_c
})
#Calculate the Choen's d
cohens_d = (
    np.mean(HbA1c_d) - np.mean(HbA1c_c)
) / (np.sqrt((np.std(HbA1c_d) ** 2 + np.std(HbA1c_c) ** 2) / 2))
```

```
print(f"cohens d: {cohens_d}")
# perform t-test
t_statistic, p_value = stats.ttest_ind(HbA1c_d, HbA1c_c)
# print results

print(f"t-statistic: {t_statistic}")
print(f"p-value: {p_value}")
```

Here is the output of the test:

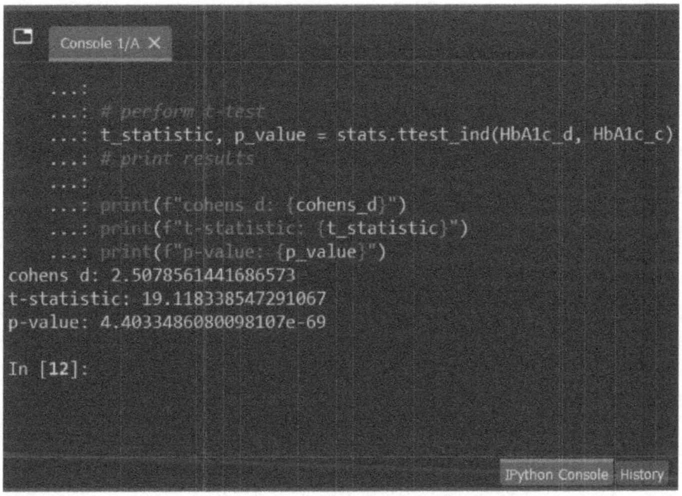

```
Console 1/A

   ...:
   ...: # perform t-test
   ...: t_statistic, p_value = stats.ttest_ind(HbA1c_d, HbA1c_c)
   ...: # print results
   ...:
   ...: print(f"cohens d: {cohens_d}")
   ...: print(f"t-statistic: {t_statistic}")
   ...: print(f"p-value: {p_value}")
cohens d: 2.5078561441686573
t-statistic: 19.118338547291067
p-value: 4.4033486080098107e-69

In [12]:

IPython Console  History
```

Figure 8.3 – HbA1C t-test output

We printed three results in this order, `cohens_d (2.507)`, which means that this is a large effect size (refer to the previous chapter for Cohen's d interpretation). This result is perfectly in line with the biological knowledge that HbA1c is a common biomarker used to diagnose T2DM, which adds more credibility to the fact that we have a statistically significant result too for the value being higher in diabetes subjects (p-value <0.001). We printed the t-value, but in terms of interpretation, the p-value is more relevant. That's because the t-value was used to calculate the p-value as the final interpretation output.

We will start by loading the required libraries and loading the diabetes dataset:

```
import pandas as pd
import numpy as np
import matplotlib.pyplot as plt
from scipy.stats import ttest_1samp
```

```
# Load the diabetes dataset from your folder
data = pd.read_csv(r'C:\Users\MEDIN\Downloads\Dataset of Diabetes
.csv')
```

Then we need to select the Y class or rows in the data only for subjects with diabetes subjects:

```
# Select the HbA1c values for diabetes subjects (T2DM)
HbA1c_d = data[data['CLASS'] == 'Y']['HbA1c']

#The variables are now prepared for a One sample t-test.
```

Now let's consider another example in which we want to compare the HbA1c values of subjects relative to a specific threshold. The usual reference threshold for diagnosing diabetes (type 2) is 7%, which means that 7 % of the hemoglobin in these subjects is glycated. To compare the HbA1c values of subjects against this threshold, we can use the one-sample t-test:

```
# Perform a one-sample t-test comparing HbA1c values to the value 7
t_statistic, p_value = ttest_1samp(HbA1c_d, 7)

# Display the results
print("T-Statistic:", t_statistic)
print("P-Value:", p_value)

# Defining the average
# values and standard deviation objects
avg_HbA1c_d = HbA1c_d.mean()
std_HbA1c_d = HbA1c_d.std()

categories = ['T2DM']
values = [avg_HbA1c_d]
std_devs = [std_HbA1c_d]
```

In this section, we learned how to conduct the Student t-test (one sample and two sample forms) and how to implement it in Python. We used a real-world example using the clinical data (diabetes dataset).

In the next section, we will learn how to implement ANOVA and the associated post-hoc tests.

Applying post-hoc tests using ANOVA

Select the data from the diabetes dataset subject with diagnosed diabetes (`CLASS: Y`). Using the available data, perform an analysis comparing average values of glycated hemoglobin (HbA1C) between underweight, normal, and overweight subjects.

> **Note**
> Perform the ANOVA analysis for comparing multiple groups and present the results visually.

In this analysis, we will be comparing multiple groups' HbA1C based on their BMI level categories.

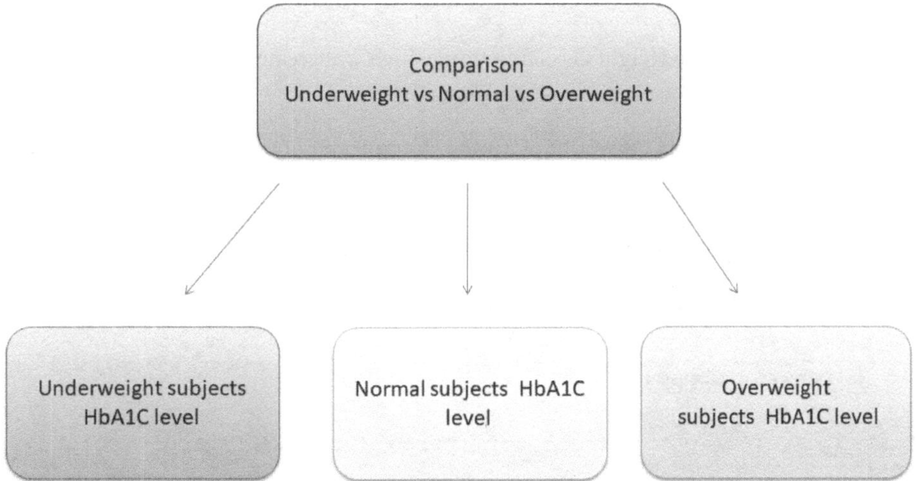

Figure 8.4 – Comparison of HbA1c between underweight, normal weight, and overweight subjects

In the initial steps, we will load the data and separate subjects into groups based on their BMI levels:

- Those with a BMI of < 25 belong to Normal
- Those with a BMI of >= 25 and < 30 belong to Overweight
- BMI > 30 is for Obese

Let's start with the coding part:

```
import pandas as pd
import scipy.stats as stats

# Load the data
data = pd.read_csv(r'C:\Users\MEDIN\Downloads\Dataset of Diabetes
.csv')
```

First, the data needs to be filtered in a way that removes the predicted diabetes class. Why? We are not sure whether these are indeed diabetes subjects. We want to use only the credible diabetes and control classes, `Y` and `N`:

```
# Filter data for rows with 'Y'  in 'CLASS' column
filtered_data = data[data['CLASS'].isin(['Y'])]
```

The next step is to define the weight class function, which will create categories for normal weight, overweight, and obese subjects:

```
# Create a new column 'weight_class' based on the 'BMI' column
def weight_class(bmi):
    if bmi < 25:
        return 'normal'
    elif bmi >= 25 and bmi < 30:
        return 'overweight'
    else:
        return 'obese'

filtered_data['weight_class'] = filtered_data['BMI'].apply(weight_
class)

# Create a table object to display the descriptive statistics for each
class
normal_stats = filtered_data[
    filtered_data['weight_class'] == 'normal']['HbA1c'].describe()
overweight_stats = filtered_data[
    filtered_data['weight_class'] == 'overweight']['HbA1c'].describe()
obese_stats = filtered_data[
    filtered_data['weight_class'] == 'obese']['HbA1c'].describe()
```

`stats_table` can now be used to create a data frame that will be easier to store and read when interpreting:

```
stats_table = pd.DataFrame({'Normal': normal_stats,
                            'Overweight': overweight_stats,
                            'Obese': obese_stats})
print('Descriptive Statistics for Each Class:')
print(stats_table)
```

Finally, the ANOVA analysis can now take place using the variables and the data filtered:

```
# Perform ANOVA on the 'HbA1c' column for the three classes
fvalue, pvalue = stats.f_oneway(
    filtered_data[
```

```
        filtered_data['weight_class'] == 'normal'
    ]['HbA1c'],
    filtered_data[
        filtered_data['weight_class'] == 'overweight'
    ]['HbA1c'],
    filtered_data[
        filtered_data['weight_class'] == 'obese'
    ]['HbA1c']
)
```

Output is best stored by creating another result data frame, which will contain both the `F` values and the `p` values that will later be used to interpret the statistical significance of the results:

```
# HbA1ceate a table object to display the results of ANOVA
results = pd.DataFrame({'F-value': [fvalue],
                        'p-value': [pvalue]})
print('ANOVA Results:')
print(results)
```

Here is the output:

```
Console 1/A ×
filtered_data[filtered_data['weight_class'] == 'obese']['HbA1c'])
   ...:
   ...: # HbA1ceate a table object to display the results of ANOVA
   ...: results = pd.DataFrame({'F-value': [fvalue],
   ...:                         'p-value': [pvalue]})
   ...: print('ANOVA Results:')
   ...: print(results)
Descriptive Statistics for Each Class:
          Normal  Overweight       Obese
count  53.000000  267.000000  520.000000
mean    6.409434    9.132809    9.004808
std     1.806853    2.250516    2.165360
min     4.000000    3.700000    2.000000
25%     4.900000    7.500000    7.300000
50%     6.800000    9.000000    8.950000
75%     7.900000   10.650000   10.500000
max     9.700000   16.000000   15.900000
ANOVA Results:
     F-value       p-value
0  36.935556  4.245019e-16
```

Figure 8.5 – ANOVA, HbA1c between underweight, normal weight, and overweight subjects

The results clearly show that both overweight and obese subjects have increased BMI compared to the subjects with normal BMI. We can also see that quartile 1 (marked as **25%**), quartile 2 (marked as **50%**), and quartile 3 (marked as **75%**) have also been increased for the overweight and obese subjects. What is most important in terms of statistical significance for this comparison is the p-value, which is 4.24 x 10-16. This is a very low value and we have a statistically significant result here. However, this p-value is for a simultaneous comparison of all three groups.

To get more details into different combinations of the group comparisons, we can use the Tukey-Kramer post-hoc test. This test is post-hoc, which means that we can use it after the initial ANOVA analysis to identify the differences between the group pairs in multi-group data. The analysis can be performed as shown here:

```
# Perform Tukey-Kramer post hoc test
from statsmodels.sandbox.stats.multicomp import MultiComparison
mc = MultiComparison(
    filtered_data['HbA1c'], filtered_data['weight_class'])
result = mc.tukeyhsd()
print('Tukey-Kramer Post Hoc Test:')
print(result)

# Display p-values in a more readable format
print('P-Values in Scientific Notation:')
for i, val in enumerate(result.pvalues):
    print(f'Comparison {i + 1}: {val:.2e}')

print('P-Values in Decimal Format:')
for i, val in enumerate(result.pvalues):
    print(f'Comparison {i + 1}: {val:.15f}')
```

Here is the result of the test:

```
Tukey-Kramer Post Hoc Test:
 Multiple Comparison of Means - Tukey HSD, FWER=0.05
=====================================================
group1   group2   meandiff p-adj  lower  upper  reject
-----------------------------------------------------
normal      obese   2.5954    0.0 1.8599 3.3309   True
normal overweight   2.7234    0.0 1.9563 3.4904   True
 obese overweight    0.128 0.7139 -0.256  0.512  False
-----------------------------------------------------
P-Values in Scientific Notation:
Comparison 1: 2.95e-13
Comparison 2: 2.94e-13
Comparison 3: 7.14e-01
P-Values in Decimal Format:
Comparison 1: 0.000000000000295
Comparison 2: 0.000000000000294
Comparison 3: 0.713858688672214
```

Figure 8.6 – Post-hoc tests for ANOVA

Here, we can see the results of the Tukey-Kramer post-hoc comparison, so we are comparing HbA1c values directly for all combinations of groups according to their BMI. Observe the **group1** versus **group2** comparisons. There are all combinations of the groups: normal versus obese, normal versus overweight, and obese versus overweight. The first metric to compare is the **meandiff**, or the mean difference. You can see it is **2.59 kg/m2** for normal versus obese, which is a large average difference. The situation is similar for normal versus overweight, **2.72 kg/m2** (refer to *Figure 8.6*). For the obese versus overweight comparison, the difference is much smaller at 0.128.

Now let us observe p-Values (see the **p-adj** column). Since we are making multiple comparisons, the p-values are adjusted for those multiple comparisons and are more robust, more credible p-values compared to the basic p-value calculation.

The adjustments of p values mean that they are made more conservative by actually increasing them based on the number of tests made.

After learning how to implement the Student's t-test and ANOVA, the next step is to learn how to perform predictive analysis in Python. We will be using the same diabetes data.

Performing and visualizing linear regression in Python

Using the diabetes dataset, we will now perform analysis with the main research question of checking if there is an association between urea and HbA1C in diabetes subjects (`CLASS: Y`). Then we will present the results visually using the regression plot and also check whether gender, age, or **High-Density Lipoprotein** (**HDL**) influence the results. Before starting with the linear regression coding, let's define the model using a simple diagram:

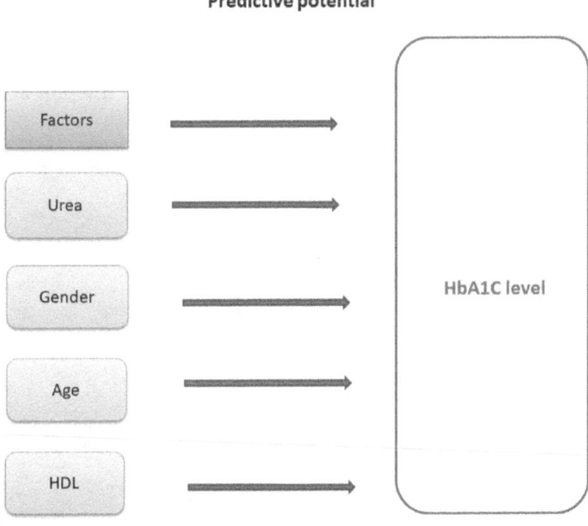

Figure 8.7 – Predictive analysis, linear regression of HbA1C as dependent variable

You can see that gender and age are added to the model scheme. This will adjust the results for age and gender, which is routinely done in most biostatistical analysis where such data is available and where these factors could influence the results.

The variables that are interesting from the perspective of predictor potential are HDL and urea in this example. They are also seen on the preceding schematic as independent variables.

In the coding part, we will be adding the variables defined in *Figure 8.7*, urea, gender, age, and HDl, as independent variables and HbA1c level as a dependent variable for the linear regression model.

As usual, we first need to load the libraries and the data. Here is the code for it:

```
import pandas as pd
import statsmodels.formula.api as smf
import seaborn as sns
import matplotlib.pyplot as plt

# Load the data
data = pd.read_csv(r'C:\Users\MEDIN\Downloads\Dataset of Diabetes
.csv')
data = data.replace('f','F', regex=True)
```

Here is how to create the linear regression model, as specified in the preceding scheme:

```
# Model 1: Multiriate Linear Regression with Age and Gender
model1 = smf.ols(
    formula='HbA1c ~ Urea + Gender+ AGE + HDL', data=data).fit()
sns.regplot(x='Urea', y='HbA1c', data=data, ci=95)
plt.show()
```

The `ci=95` argument means that we set the confidence interval in the plot at the level of 95%.

Here is the resulting plot:

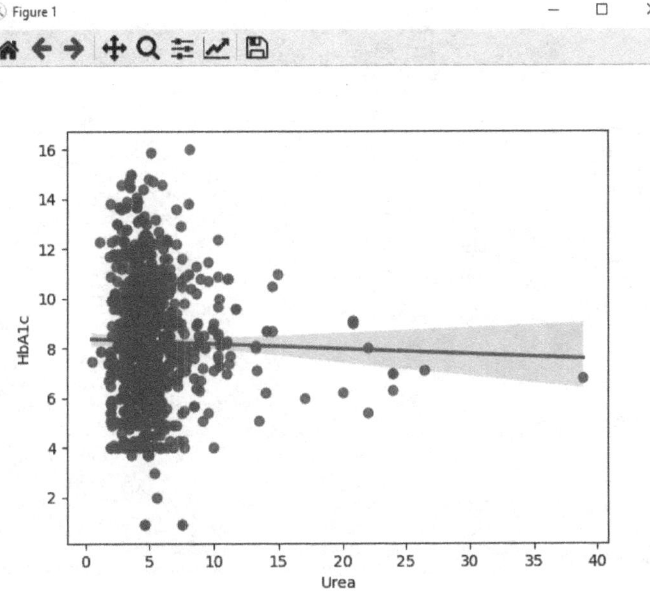

Figure 8.8 – Linear regression (urea versus HbA1c)

By observing the regression plot, we can see that the regression line is almost horizontal. That is the first indicator of the absence of association between urea and HbA1c. The second indicator is the fact that no association is seen as the individual points do not follow the trendline. This means that using the regression, the appropriate trendline could not be drawn. This is a frequent situation for variables that have no association.

Now let's print the result:

```
print(model1.summary())
```

Here is the output for it:

```
==============================================================================
Dep. Variable:                  HbA1c   R-squared:                       0.149
Model:                            OLS   Adj. R-squared:                  0.146
Method:                 Least Squares   F-statistic:                     43.56
Date:                Sun, 15 Oct 2023   Prob (F-statistic):           1.02e-33
Time:                        16:40:13   Log-Likelihood:                -2267.6
No. Observations:                1000   AIC:                             4545.
Df Residuals:                     995   BIC:                             4570.
Df Model:                           4
Covariance Type:            nonrobust
==============================================================================
                 coef    std err          t      P>|t|      [0.025      0.975]
------------------------------------------------------------------------------
Intercept      2.4612      0.494      4.978      0.000       1.491       3.431
Gender[T.M]   -0.0309      0.152     -0.204      0.838      -0.329       0.267
Urea          -0.0538      0.026     -2.103      0.036      -0.104      -0.004
AGE            0.1113      0.008     13.140      0.000       0.095       0.128
HDL            0.1289      0.113      1.138      0.255      -0.093       0.351
==============================================================================
Omnibus:                       12.466   Durbin-Watson:                   1.166
Prob(Omnibus):                  0.002   Jarque-Bera (JB):               12.579
Skew:                           0.260   Prob(JB):                      0.00186
Kurtosis:                       3.179   Cond. No.                         365.
==============================================================================
```

Figure 8.9 – Multivariate linear regression output

By exploring the values in the output, it can be seen that two results with $p<0.05$ are related to age and urea as predictors. Surprisingly, HDL was not significant. HDL is known to be decreased in diabetes subjects and this result should be taken with caution. Just because we did not find significant results, which doesn't mean that some other study may not find them. This means that the results we get in the study are a source of evidence, but do not lead to final conclusions about the specific field. This is especially the case when specific biological knowledge exists that doesn't conform to the results of the study. Studies frequently vary. So, we interpret this result as follows: we didn't find a significant difference in this analysis but it remains to be further analyzed in other studies. This doesn't mean that, in reality, there is no difference.

No significant association was found (in this analysis) between HDL and gender as independent variables and HbA1C dependent variable, while a significant difference was found for age and urea. What about the nature of associations? Urea has a negative association, which means its reversed association in terms of decreasing urea trend is associated with increased HbA1c. For age, it is the other way around. As age increases in this model, HbA1c also increases. Explore the adjusted R squared value in the top-right angle and you can see it is 0.146, which means approximately 14.6% of the variation in the HbA1c can be explained by this model. We implemented the linear regression. We can now move to the next exercise, which is implementing a different approach and method: logistic regression to make a predictive model related to T2DM status in different subjects.

Performing and visualizing logistic regression in Python

When the association between variables is linear as in the previous examples in this chapter, linear regression was used. However, in this section, we will be analyzing a potentially non-linear association, so we need a different regression method. What is implemented in this specific scenario is a method called **binary logistic regression**. This type of variable can take different inputs and have a binary outcome (e.g., a **Yes** and **No** variable) and the association between them can (but does not need to) have a normal distribution. For this reason, it is more appropriate for binary outcomes.

Using the diabetes dataset, you will learn how to perform analysis with the main research question of checking whether there is an association between BMI and diabetes (`CLASS : Y`). You will also use controls for binary comparison (`CLASS: N`).

Further in this section, we will be focusing on the implementation of the data visualization methods using the regression plot. We will also check whether gender, age, urea, or **Creatinine** (**Cr**) influence the results.

Refer to the following figure:

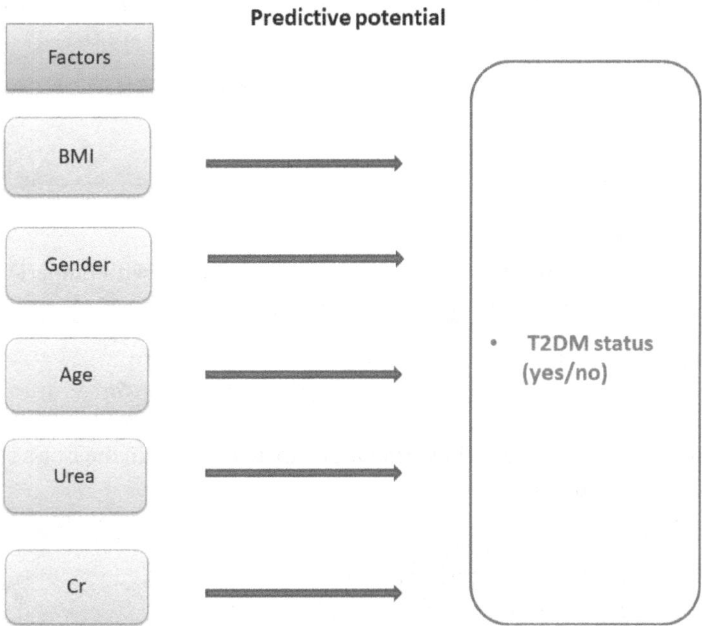

Figure 8.10 – Predictive model scheme, multiple independent variables, and T2DM status as a dependent target variable

We will be creating the predictive model for predicting the diabetes status of T2DM subjects based on the BMI, but also adding the gender, age, urea, and Cr covariates. So, the multivariate model will be created allowing the BMI to interact with other covariates in the predictive model where the target for prediction is the **yes/no** status for the presence of T2DM.

In the following code, we will be loading the data and using the `statsmodels` package to implement the logistic regression (binary).

First, let's load the required libraries and the dataset. We will be using `statmodels` for creating the model and we can import it as `sm`:

```
import pandas as pd
import statsmodels.api as sm
import seaborn as sns
import matplotlib.pyplot as plt
import numpy as np

# Load the data
data = pd.read_csv(r'C:\Users\MEDIN\Downloads\Dataset of Diabetes
.csv')
```

Now, we can perform the filtering again to make sure that only `Y` and `N` or `yes` and `no` classes for diabetes are present:

```
# Filter data to include only 'Y' and 'N' in 'CLASS' variable
filtered_data = data[data['CLASS'].isin(['Y', 'N'])]
```

Converting the `Y` and `N` to `1` and `0` will make it easier for us to work with numeric variables. The `1` will correspond to presence of disease and 0 to absence of disease:

```
# Map 'Y' to 1 and 'N' to 0 in the 'CLASS' variable
filtered_data['CLASS'] = filtered_data['CLASS'].map({'Y': 1, 'N': 0})
```

Here is how to create the logistic regression formula and then include it in the `statsmodels` logistic regression model, as defined in the scheme:

```
# Define the formula for logistic regression
formula = 'CLASS ~ BMI + Gender + AGE + Urea + Cr'

# Create the logistic regression model
logit_model = sm.Logit.from_formula(formula, data=filtered_data)

# Fit the model
result = logit_model.fit()
```

Then, we can calculate the probabilities for the fitted model using `result.predict()`. The probability of diabetes will be presented on a scale from 0-1; this is the standard way of presenting probability. A value of 0 would mean the lowest probability and 1 would be the highest possible probability:

```
# Calculate probabilities using the fitted model
fitted_probabilities = result.predict(filtered_data)
```

Now we can plot the logistic regression model and explore the visual aspects of the results:

```
# Plot the probabilities against HbA1c
plt.figure(figsize=(10, 6))
sns.scatterplot(x=filtered_data['BMI'], y=fitted_probabilities)
```

In the preceding code, `x` represents the BMI values and `y` represents the probability of diabetes presence (the `CLASS` variable being `Y` or `1` when recorded). Now, the plot can be made to show this association:

```
plt.xlabel('BMI')
plt.ylabel('Predicted Probability')
plt.title('Predicted Probabilities vs BMI')
plt.show()
```

Here is the output for it:

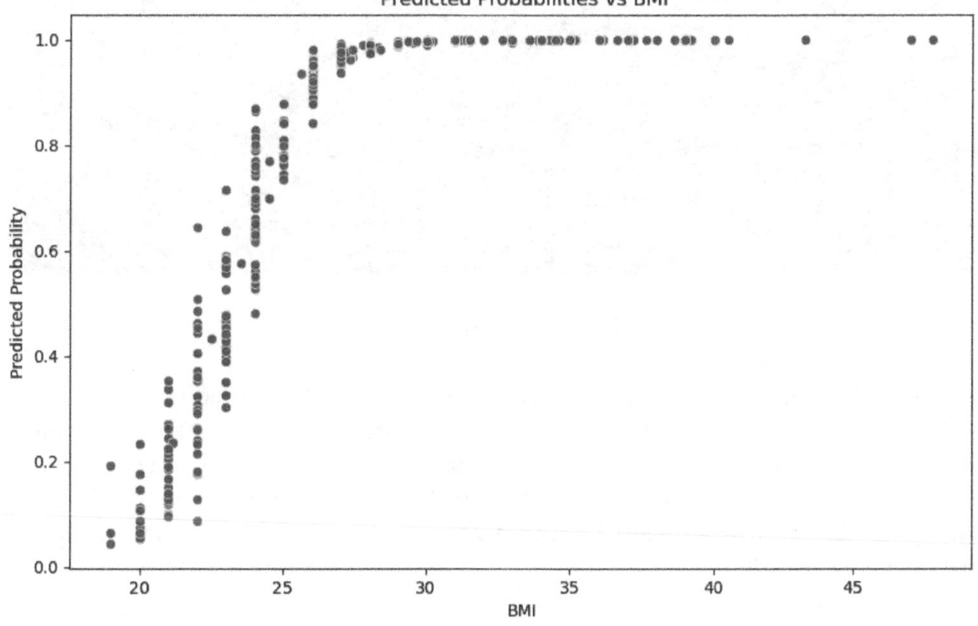

Figure 8.11 – Logistic regression plot

We can see that the BMI and predicted probability resemble a partially sigmoid shape that is a bit widespread. This is due to including the age, gender, urea, and Cr covariates. Observing the plot, we can see that as the BMI increases from 20 to 25, the probability of diabetes (y axis) increases drastically, but the largest probability is when the 25 threshold is reached, 80% or more (0.8 > on the graph as the probability of diabetes is presented on a scale of 0-1). This shows the nature of how obesity might influence the probability of subjects being classified as diabetes and tells us that obesity is indeed related to diabetes even when we include covariates. Models with covariates are generally considered to be more credible compared to univariate models (models with only one1 independent variable).

Now, let's print the model summary:

```
print(result.summary())
```

Here is the resulting output:

```
                           Logit Regression Results
==============================================================================
Dep. Variable:                  CLASS   No. Observations:                  942
Model:                          Logit   Df Residuals:                      935
Method:                           MLE   Df Model:                            6
Date:                Sun, 15 Oct 2023   Pseudo R-squ.:                  0.6090
Time:                        16:50:25   Log-Likelihood:                -126.30
converged:                      False   LL-Null:                       -323.02
Covariance Type:            nonrobust   LLR p-value:                 7.251e-82
==============================================================================
                 coef    std err          z      P>|z|      [0.025      0.975]
------------------------------------------------------------------------------
Intercept    -19.9510      1.993    -10.009      0.000     -23.858     -16.044
Gender[T.M]    0.6643      0.348      1.910      0.056      -0.017       1.346
Gender[T.f]    4.0630    364.846      0.011      0.991    -711.022     719.148
BMI            0.7895      0.083      9.458      0.000       0.626       0.953
AGE            0.0303      0.017      1.760      0.078      -0.003       0.064
Urea          -0.0812      0.088     -0.922      0.357      -0.254       0.091
Cr             0.0088      0.008      1.121      0.262      -0.007       0.024
==============================================================================

Possibly complete quasi-separation: A fraction 0.20 of observations can be
perfectly predicted. This might indicate that there is complete
quasi-separation. In this case some parameters will not be identified.
```

Figure 8.12 – Multivariate logistic regression output

We can see that the only two p-values less than 0.05 are intercept and BMI (note that p=0.000 means p<0.0001). We usually don't consider intercept when interpreting p-values, so the only significant predictor in this analysis is the **BMI**. A coefficient of **0.789** for **BMI** means that on average, subjects with diabetes have increased **0.789** units of BMI (kg/m3). Note that results are adjusted for age and gender by including them in the model. **Urea** and **Cr** have not been found to be significant predictors of diabetes in this analysis.

Summary

In this chapter, we performed exercises using the diabetes dataset and practiced how to implement different versions of Student's t-test. We also practiced using ANOVA to test multiple groups simultaneously using the diabetes data. Finally, we performed a practical exercise for both linear and logistic regression, visualized the results using linear and sigmoid regression plots, and practiced how to adjust for different covariates such as age and gender.

In the next chapter, we will be applying what we have learned so far in predictive analysis on an example project from the cardiology biostatistics subject. We will be discussing the cardiology domain, electrocardiograms, and coronary artery disease, as well as how to use predictive models in this area to answer different research questions.

9

Biostatistical Inference and Predictive Analytics Using Cardiovascular Study Data

In this chapter, we will implement an exemplar project from cardiology biostatistics. We learned about biostatistical inference and predictive analysis in previous chapters and now, we will implement specific statistical solutions and answer different research questions on real-life data from cardiology. We will be using new data this time, but we will use the same principle we learned about in the previous four chapters.

In this chapter, we're going to cover the following main topics:

- The Cleveland dataset
- Loading and examining cardiovascular data in Python
- Hypothesis tests applied to evaluate mean differences
- Linear regression for cardiovascular predictive analysis
- Using logistic regression to derive an **odds ratio** (**OR**) for categorical variables

Technical requirements

You need to have Spyder or JupyterLab installed (preferably through Anaconda Navigator).

You need to have the following packages installed for Python:

- `pandas`
- `numpy`
- `statsmodels`

- `seaborn`
- `Matplotlib`

The Cleveland dataset

The Cleveland Heart Disease dataset (`http://archive.ics.uci.edu/dataset/45/heart+disease`) is a dataset used in exemplar data analysis and machine learning for predicting the presence or absence of heart disease in patients. In this case, we will use it for biostatistical example purposes.

To proceed with this chapter, please download the dataset using the link provided (the downloaded file should be in `.data` format). Here is the name of the dataset found in the `.zip` file you downloaded: `processed.cleveland.data`

Here is the citation for the dataset: *Janosi, Andras, Steinbrunn, William, Pfisterer, Matthias, and Detrano, Robert. (1988). Heart Disease. UCI Machine Learning Repository. https://doi.org/10.24432/C52P4X.* Before proceeding with analyzing the variables, let's first explore the main topic in this project, which is **coronary artery disease** (**CAD**). This specific form of heart disease is characterized by 50% or more congestion of the heart's blood vessels (arteries). This may lead to many complications in cardiovascular patients, such as myocardial infarction or even death. CAD is one of the most frequent and most dangerous death causes around the world and most of the research in cardiology is related to either preventing CAD or treating its consequences. This is what a CAD artery looks like:

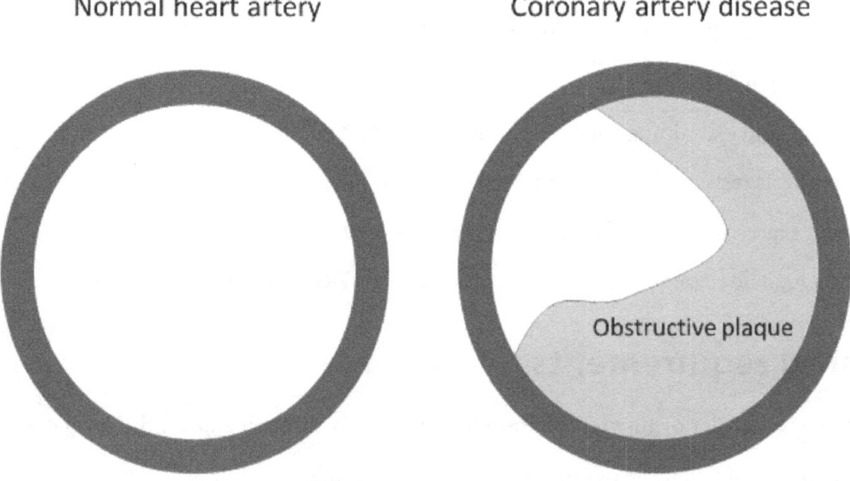

Figure 9.1 – Arterial plaque in CAD

In this illustrative example, we can see that a normal heart artery does not have any obstructions. On the right side, you may see another artery cross-section, which is obstructed more than 50% by **obstructive plaque** (*Figure 9.1*). This plaque can be made of different biological compounds and cellular components. Sometimes, it's made of cholesterol, sometimes of triglycerides, or sometimes of calcium. Most plaques have all these mixed up and even the cellular debris and components may be present. Some plaques even contain fibrous tissue, as plaque formation is frequently associated with inflammatory processes on the arterial walls. Regardless of the biological processes leading to the formation of arterial plaques, they all have one aspect in common: They reduce the flow of blood through the arteries and this reduction may lead to many cardiovascular complications. The most dangerous complications associated with CAD are myocardial infarction, fibrosis of the heart tissue, and at times, even atrial fibrillation of the heart tissue, which may lead to death.

First, let's explore the variables (features) that describe the various subject characteristics in the data. The main variable, or the target variable, in this dataset is typically the presence or absence of heart disease. In the Cleveland dataset, this target variable is usually represented as a binary outcome with the name `num`:

Target variable: This variable (typically named `target` or `num`) is binary and indicates whether a patient has heart disease or not. It is often coded as follows:

`0`: No heart disease (negative class)

1-4: Presence of heart disease (positive class)

In addition to the target variable, the dataset includes several other variables that serve as features for predicting the presence of heart disease. Some of these features include the following:

Here are the independent variables (potential predictors):

- `age`: The age of the subject
- `sex`: The sex of the patient (`0` for female, `1` for male)
- `cp`: A categorical variable describing the type of chest pain experienced by the patient (for example, typical angina, atypical angina, non-anginal pain)
- `trestbps`: The patient's resting blood pressure in mmHg
- `chol`: The patient's serum cholesterol level in mg/dL
- **Fasting blood sugar**: A categorical variable indicating whether the patient's fasting blood sugar level is above 120 mg/dL (`1` for yes, `0` for no)
- **Resting electrocardiographic results**: A categorical variable describing the results of the resting electrocardiogram
- `thalach`: The maximum heart rate achieved during a stress test
- `exang`: A binary variable indicating exercise-induced angina (`1` for yes, `0` for no)

- `oldpeak`: The ST segment depression induced by exercise relative to rest
- `ca`: An integer indicating the number of major vessels colored by fluoroscopy
- `thal`: A categorical variable describing the results of the thallium stress test

The Cleveland dataset is often used to build machine learning models to predict whether a patient is likely to have heart disease based on these features. Researchers and data scientists use this dataset for tasks such as logistic regression, decision trees, random forests, and support vector machines to develop predictive models for heart disease diagnosis.

In the Cleveland Heart Disease dataset, the `num` variable represents the presence and severity of heart disease, coded from `0` to `4`, where stages `1` to `4` are different levels of severity. We will consider `0` as there being no CAD and all other classes as CAD being present.

The main predictor of CAD in this project will be ST depression. **ST depression** is a pattern in the electrocardiogram used to diagnose heart disease, such as myocardium infarction. Before learning about ST depression, we must discuss the electrocardiograms first.

What is electrocardiogram? This is the set of electrical signals detected by electrodes and translated into a 2D representation in electrocardiogram paper. It is used by doctors to detect and monitor different changes in the heart muscle and the heart arteries and can be a very useful tool in detecting serious conditions, such as myocardial infarction.

The following is the typical heartbeat wave as seen on an electrocardiogram:

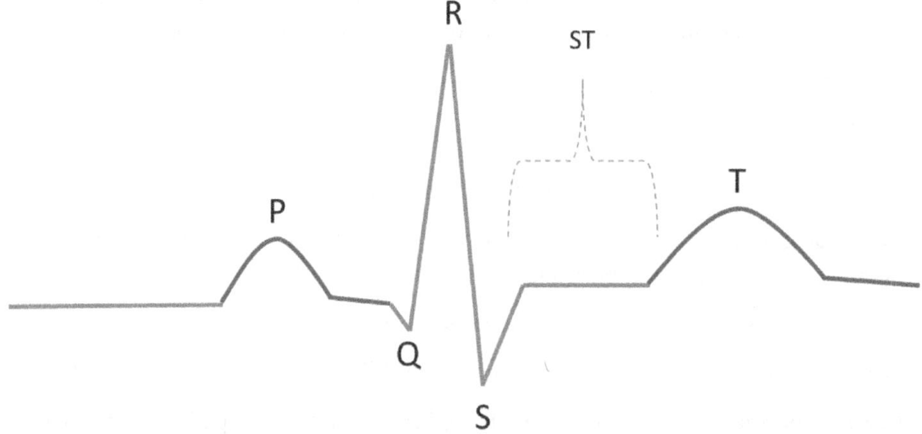

Figure 9.2 – PQRST segments of the heartbeat on an electrocardiogram

As you can see, it has several distinct parts which are marked as *P*, *Q*, *R*, *S*, and *T*. Also, note how the *P* and *T* parts resemble small hills and the *R* signifies the peak of the heartbeat wave.

As can be seen, there are many patterns that could be drawn here, such as *PQ*, *RS*, or *ST*. In this project, we are interested in the ST pattern. It is an important diagnostic tool in diagnosing not just CAD but also serious cardiovascular conditions, such as myocardial infarction (in combination with chest pain and biomarkers, such as serum troponin).

If the ST is lower than normal and has a specific depression pattern, it is called **electrocardiogram (ECG) ST depression** on a heartbeat. Now, let's see what an ST depressed ECG looks like:

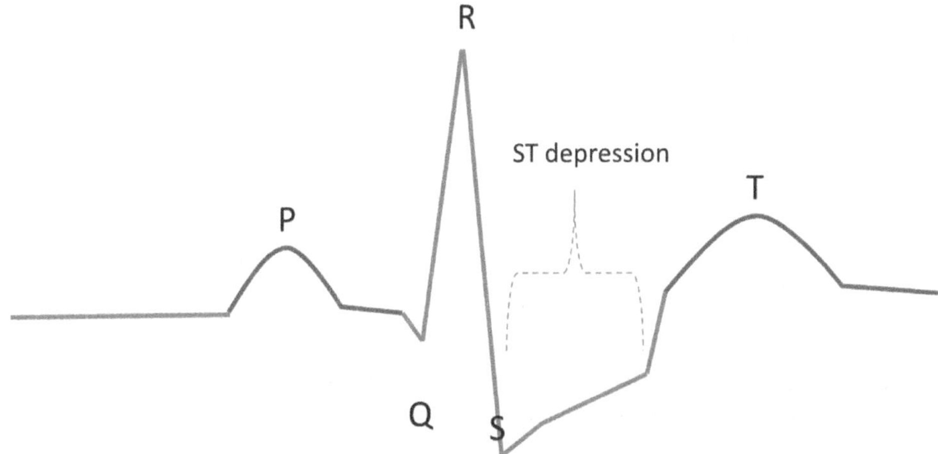

Figure 9.3 – ST depression on an electrocardiogram

Now, you can see what ST depression looks like and we will be using this predictor to try to find an association between ST depression as a predictor and the CAD status of subjects.

Loading and examining the cardiovascular data in Python

Now, let's see how all this looks from the coding perspective. Open your Spyder or JupyterLab application and let's load the `pandas` library and the data.

Even before loading the data, we will use the information available on the dataset documentation website (`http://archive.ics.uci.edu/dataset/45/heart+disease`) to set 14 specific variable names.

Once you have opened Python in your IDE (Spyder or JupyterLab; this book uses Spyder IDE), first, let's set the names of the variables according to the dataset documentation. This is needed because the original dataset column names are not added to the dataset, so we need to add them manually:

```
#Set the column names according to information from the dataset
documentation
column_names = [
    "age", "sex", "cp", "trestbps", "chol", "fbs", "restecg",
```

```
    "thalach", "exang", "oldpeak", 'slope', 'ca', 'thal', 'cad'
]
# Load the dataset and set the variable names
```

Then, we can load the dataset we are downloading using the link I provided before for the Cleveland dataset for the UCI machine learning repository (`http://archive.ics.uci.edu/dataset/45/heart+disease`). Make sure to set the `names` argument to correspond to the column names list we created before. Notice how we use lists within these brackets to specify the variable names. This is a very frequent procedure in biostatistics:

```
#Load numpy and pandas libraries
import pandas as pd
import numpy as np
#Load the dataset
dataset = pd.read_csv(
    r"C:\Users\Korisnik\Downloads\processed.cleveland.data",
    names=column_names, header=None, na_values=["?"]
)
```

`na_values=["#"]` tells the `read_csv()` `pandas` function to interpret any `?` in the data as a missing value (NaN) when reading the file:

```
# Display the first few rows of the dataset
print(dataset.head())
```

The `head()` function will be the first five rows if a number is not specified in the function. If the number is specified as an argument between the brackets, the function will return that number of rows. This is very useful to quickly explore the structure of our data and see if the values are as expected.

Since our research question is related to the comparison of subjects with CAD versus subjects without CAD, we can re-code those with `1` or more to represent subjects with CAD, and `0` remains and represents subjects without CAD.

Now, let's group the subjects based on the new `cad` variable and perform descriptive statistics. First, the `np.where()` function is used to dichotomize the `cad` variable (set values to `1` for disease presence and `0` for disease absence). Here is how to do it:

```
#Dichotomize the 'cad' variable
dataset['cad'] = np.where(dataset['cad'] > 0, 1, 0)
```

Next, the data is grouped by the dichotomized `cad` variable. This means that the statistics we make in the next segments of the code will be grouped specifically according to `cad` (subjects with CAD versus subjects without it):

```
# Group the dataset by the "cad" column
grouped_data = dataset.groupby("cad")
```

```
# Calculate descriptive statistics for each group
statistics = grouped_data.describe()

# Transpose the statistics dataframe
transposed_statistics = statistics.T
```

Finally, we can transpose the DataFrame with statistics for easier exploration and visualization in the Variable Explorer:

```
# Display the transposed statistics
print(transposed_statistics)
```

Notice how I transposed the descriptive statistics to be visually more intuitive.

Let's see the results:

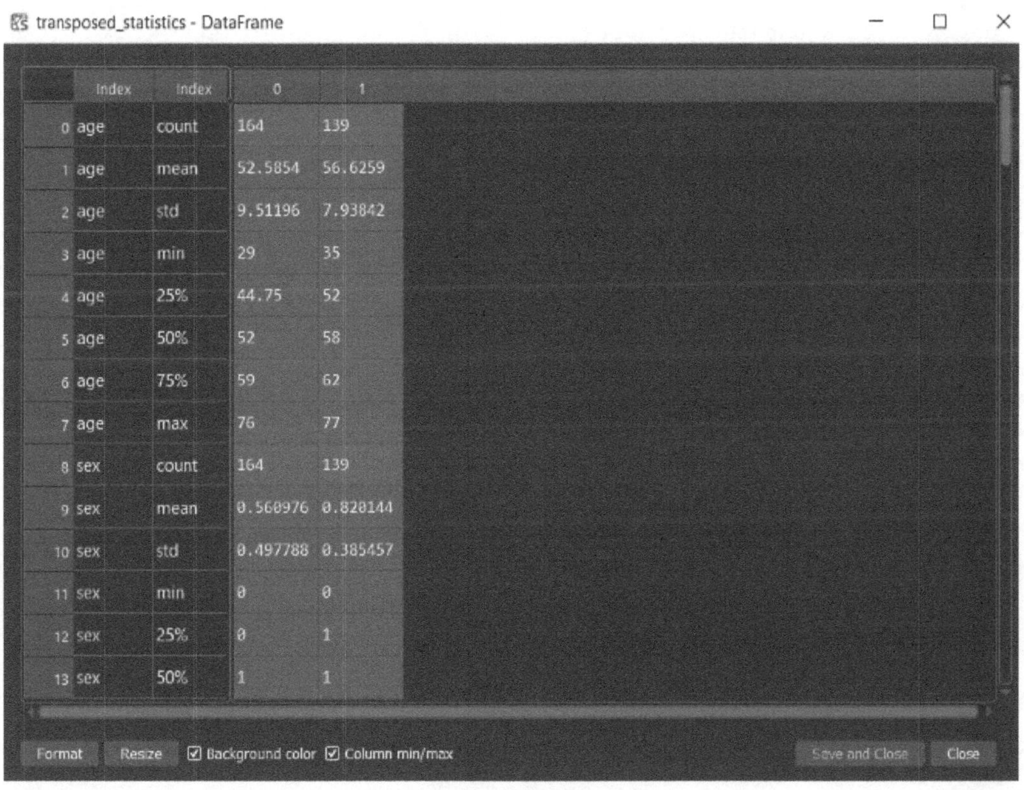

transposed_statistics - DataFrame

	Index	Index	0	1
0	age	count	164	139
1	age	mean	52.5854	56.6259
2	age	std	9.51196	7.93842
3	age	min	29	35
4	age	25%	44.75	52
5	age	50%	52	58
6	age	75%	59	62
7	age	max	76	77
8	sex	count	164	139
9	sex	mean	0.560976	0.826144
10	sex	std	0.497788	0.385457
11	sex	min	0	0
12	sex	25%	0	1
13	sex	50%	1	1

Format Resize Background color Column min/max Save and Close Close

Figure 9.4 – Exploring the data descriptives

As you can see, the descriptives, such as count, mean, standard deviation (`std`), minimum (`min`), maximum (`max`), and different quartiles (25%, 50%, 75%), are shown for both `0` and `1` groups separately. As noted before, `1` corresponds to subjects who have CAD and `0` to those who don't.

Hypothesis tests applied to evaluate mean differences

The first aspect of descriptive statistics in any clinical research data analysis is the age of the subjects. Age is known to affect many diseases, especially cardiovascular diseases. We can see that the mean age of subjects without CAD is `52.58` years and for subjects with CAD, it is `56.62` years (observe *Figure 9.4* again). Standard deviations are `9.51` and `7.93` for these groups, respectively. Taking this into consideration, age could play a role in future analyses, as the CAD subjects are older.

Let us use a hypothesis test to check if this difference is statistically significant:

```
#Separate the dataset according to 'cad' groups of interest
datacad = dataset[dataset['cad']==1]
datacontrol = dataset[dataset['cad']==0]

#Separate the dataset according to 'age' groups of interest
cadage=datacad['age']
contage=datacontrol['age']

from scipy import stats

# perform t-test
t_statistic, p_value = stats.ttest_ind(cadage, contage)
print(f"t-statistic: {t_statistic}")
print(f"p-value: {p_value}")
```

Here is the output:

```
t-statistic: 3.971100226293655
p-value: 8.955636917529706e-05
```

We will need to adjust any analysis we perform for `age` to avoid any potential problems with the confounding bias of `age` as a variable.

You may scroll down and find the `oldpeak` variable, which is the ST depression data.

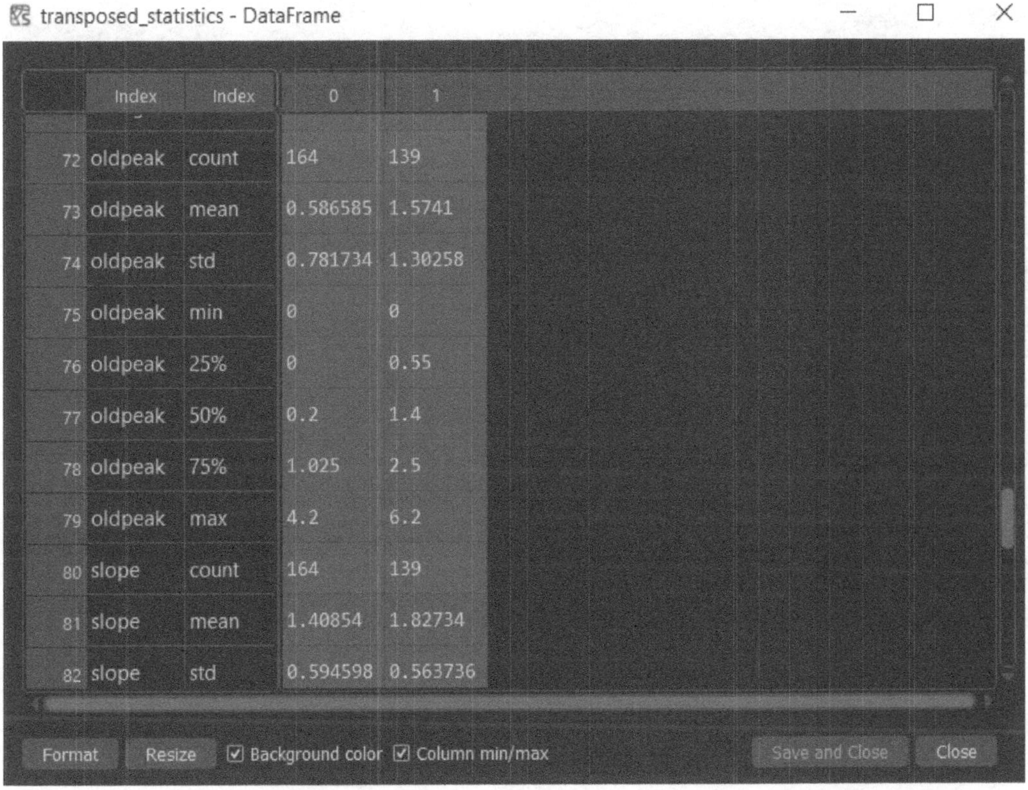

transposed_statistics - DataFrame

	Index	Index	0	1
72	oldpeak	count	164	139
73	oldpeak	mean	0.586585	1.5741
74	oldpeak	std	0.781734	1.30258
75	oldpeak	min	0	0
76	oldpeak	25%	0	0.55
77	oldpeak	50%	0.2	1.4
78	oldpeak	75%	1.025	2.5
79	oldpeak	max	4.2	6.2
80	slope	count	164	139
81	slope	mean	1.40854	1.82734
82	slope	std	0.594598	0.563736

Format Resize ☑ Background color ☑ Column min/max Save and Close Close

Figure 9.5 – Exploring the ST depression data

As you can see, `oldpeak` (ST depression magnitude in mm on an electrocardiogram) is quite different from the other subjects with and without CAD. On average, the ST depression magnitude is interestingly higher in the CAD group. This is another indicator of the importance of including this variable in the analysis.

You can also see the mean slope (slope of the ST peak on an electrocardiogram), which is the angle of the ST depression, which is also increased in CAD subjects (those with CAD). This immediately tells us that this parameter is an interesting variable to evaluate further. You may scroll through the DataFrame to observe the descriptive statistics of other variables too.

transposed_statistics - DataFrame

	Index	Index	0	1
55	restecg	max	2	2
56	thalach	count	164	139
57	thalach	mean	158.378	139.259
58	thalach	std	19.1991	22.5932
59	thalach	min	96	71
60	thalach	25%	148.75	125
61	thalach	50%	161	142
62	thalach	75%	172	156.5
63	thalach	max	202	195
64	exang	count	164	139
65	exang	mean	0.140244	0.546763
66	exang	std	0.348303	0.499609
67	exang	min	0	0
68	exang	25%	0	0

Format Resize Background color Column min/max Save and Close Close

Figure 9.6 – Exploring the stress test data

Now, let's perform a Student t-test to check for the statistical significance of the ST depression magnitude mean difference:

```
cadst=datacad['oldpeak']
contst=datacontrol['oldpeak']
# perform t-test
t_statistic, p_value = stats.ttest_ind(cadst, contst)

# print results
print(f"t-statistic: {t_statistic}")
print(f"p-value: {p_value}")
```

Here is the output:

```
t-statistic: 8.134292027638805
p-value: 1.0976053396984802e-14
```

You may see that the maximum heart rate is increased for subjects without CAD (*Figure 9.6*) and the result is statistically significant; the p-value in the output is `<0.05`. This is not expected, as subjects with CAD are known to have an increased stress test heartbeat rate. However, there is an explanation for this. Subjects with CAD in this dataset are on average older and that is the reason for their decreased stress test heart rate. This is yet another indicator of the need to adjust any further analysis for age.

Now that we describe the data, let's form the main research questions for predictive analytics.

The main research questions

In this section, we will first define the research questions biologically. This means we have a specific biological variable, such as ST depression and angina, and a question in relation to different comparisons for each variable:

1. Is ST depression magnitude associated with the maximum heart rate achieved during a test set?
2. Is ST depression magnitude associated with the presence of CAD?
3. Is exercise-induced angina related to CAD?

Linear regression for cardiovascular predictive analysis

OK, now, let's formulate the model based on research question number 1: Is ST depression magnitude associated with the maximum heart rate achieved during a test set? Why is it not easy to formulate this model?

Well, this is because the heart rate is known to be affected by age. Namely, the younger subjects typically have higher heart rates when compared to the older subjects. So, age might affect the value of maximum heart rate and have a confounding effect on the final results. This is especially important when considering the fact that we would like the CAD subjects to be slightly older in our sample.

Now, let's create the diagram for the potential predictive model:

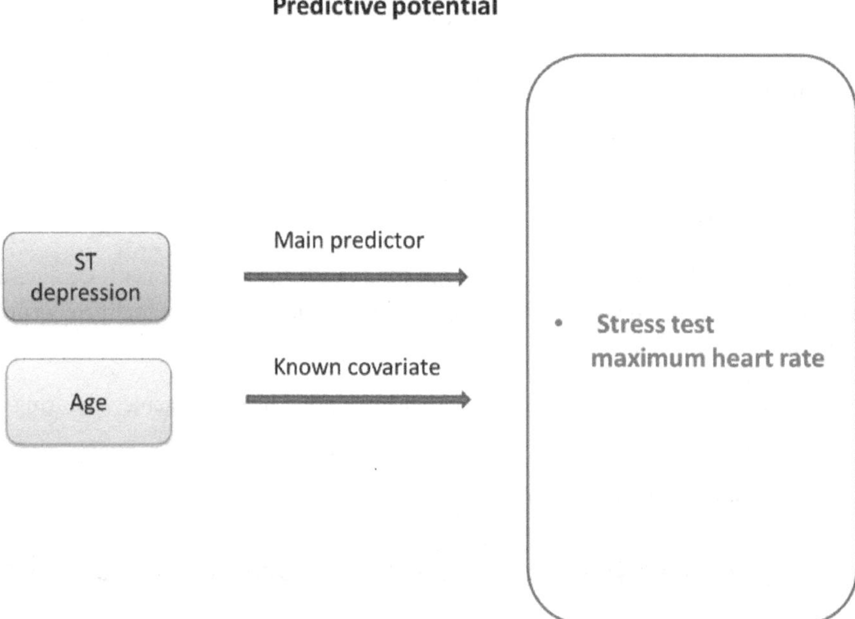

Figure 9.7 – The basic stress test model

As you can see, the ST depression is the independent variable and the target variable is the stress test maximum heart rate. However, what is also added is a known **covariate** and age as an independent variable. This will adjust the analysis for age and limit its effect on the overall results. In biostatistics, we would say that the results are adjusted for age.

Let's start with the coding part.

First, we need to load the `pandas`, `statsmodels`, `seaborn`, and `matplotlib` libraries. We will need these during the analysis.

Then, we will use `statsmodels.formula.api` to create a linear regression model and visualize the results:

```
import pandas as pd
import statsmodels.formula.api as smf
import seaborn as sns
import matplotlib.pyplot as plt

# Load the data (replace 'dataset' with the shorter name 'data').
data = dataset
```

```
# Filter rows where 'oldpeak' is equal to 1
data_filtered = data[data['cad'] == 1]

# Model 2: Multivariate Linear Regression with olpeak and
# age as independent variables
```

The `thalach` column represents the maximum heart rate of the subjects. We will use this variable as a target variable, the one we are trying to predict using the `age` and `olpeak` (ST depression on ECG) columns:

```
linmod = smf.ols(
    formula='thalach ~ oldpeak + age ', data=data_filtered
).fit()
print(linmod.summary())
```

Here is the output of the code:

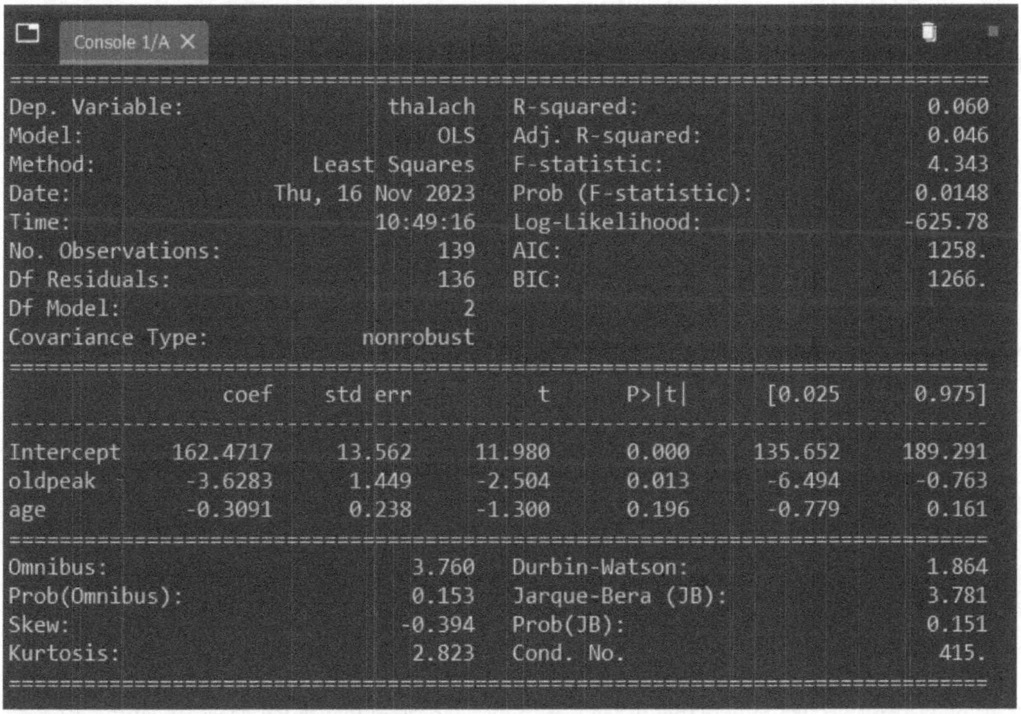

```
Console 1/A
==============================================================================
Dep. Variable:                thalach   R-squared:                       0.060
Model:                            OLS   Adj. R-squared:                  0.046
Method:                 Least Squares   F-statistic:                     4.343
Date:                Thu, 16 Nov 2023   Prob (F-statistic):             0.0148
Time:                        10:49:16   Log-Likelihood:                -625.78
No. Observations:                 139   AIC:                             1258.
Df Residuals:                     136   BIC:                             1266.
Df Model:                           2
Covariance Type:            nonrobust
==============================================================================
                 coef    std err          t      P>|t|      [0.025      0.975]
------------------------------------------------------------------------------
Intercept    162.4717     13.562     11.980      0.000     135.652     189.291
oldpeak       -3.6283      1.449     -2.504      0.013      -6.494      -0.763
age           -0.3091      0.238     -1.300      0.196      -0.779       0.161
==============================================================================
Omnibus:                        3.760   Durbin-Watson:                   1.864
Prob(Omnibus):                  0.153   Jarque-Bera (JB):                3.781
Skew:                          -0.394   Prob(JB):                        0.151
Kurtosis:                       2.823   Cond. No.                         415.
==============================================================================
```

Figure 9.8 – Linear regression stress test model

Now, let's interpret the results. The adjusted R-squared is just `0.046`, which means that only 4.6% of the variance in the predicted variable is explained via predictors. This is a very small value, so very little variance is explained via this model. However, the `oldpeak` variable has a p-value (`(P>|t|)=0.013`). This means that `ST_depression` is probably a significant predictor of the stress test results, and is associated with it, but does not determine it. The conclusion is that the ST depression magnitude is associated with the stress test of the subjects but not as a dominant predictor. This means there could be other predictors too and further analysis could be recommended in a study or a research paper.

Let's interpret the results visually now with the following code:

```
# Plot the regression line with confidence interval
sns.regplot(x='oldpeak', y='thalach', data=data_filtered, ci=95)
plt.title('Linear Regression: Stress test maximum bpm vs. ST
depression')
plt.xlabel('ST depression level (mm)')
plt.ylabel('Stress test max bp')
plt.show()
```

Here is the output:

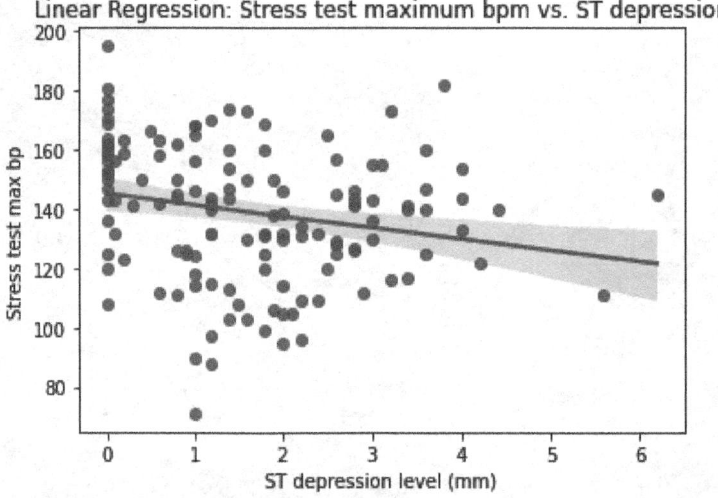

Figure 9.9 – Linear regression plot, max bp versus ST depression level

If we plot the results, we can see that indeed, as the ST depression level decreases (meaning more ST depression), the stress test max bp also increases. We can see this by observing the trendline. Still, a lot of individual points are away from the trendline and the shaded 95% confidence interval, which is another confirmation that the model is not optimal and that other predictors/confounders are probably affecting the results in a meaningful and dominant way.

Using logistic regression to derive odds ratios for categorical variables

OK, now, let's make a logistic regression model for research question number 2: Are ST depression level, age, and cholesterol associated with the presence of CAD?

For this model, we will use the CAD status coded as `1` and `0` or `yes/no` variables. We will set ST depression, age, and serum cholesterol as the independent variables:

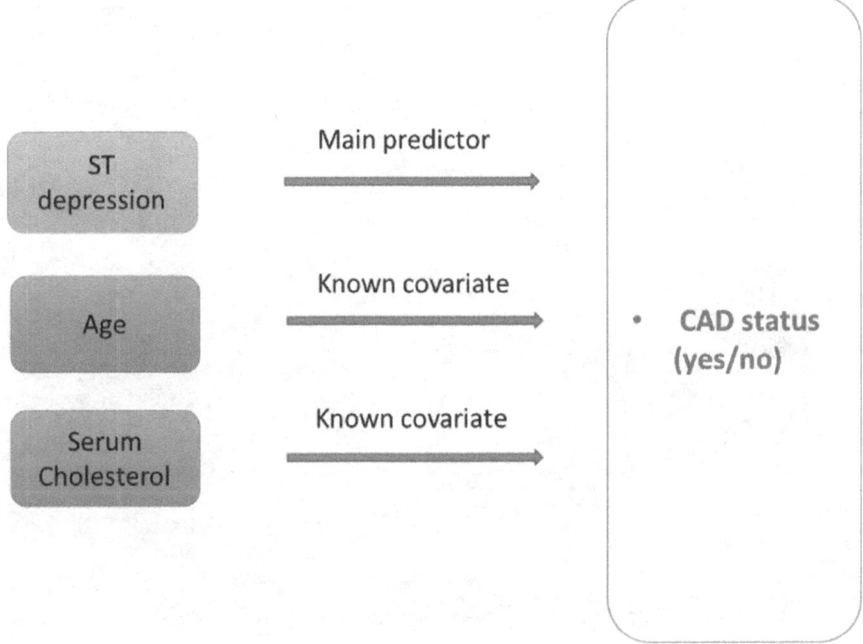

Figure 9.10 – CAD status prediction model with added serum cholesterol

This time, we will be using the binary logistic regression, because the dependent variable is a categorical binary variable coded as `1` and `0`. Let's see the code for this:

```
import pandas as pd
import statsmodels.api as sm
import seaborn as sns
import matplotlib.pyplot as plt
import numpy as np

# Define the formula for logistic regression
formula = 'cad ~ oldpeak + age + chol'
```

```
# Create the logistic regression model
logit_model = sm.Logit.from_formula(formula, data=dataset)

# Fit the model
result = logit_model.fit()
print(result.summary())
```

Here is the output:

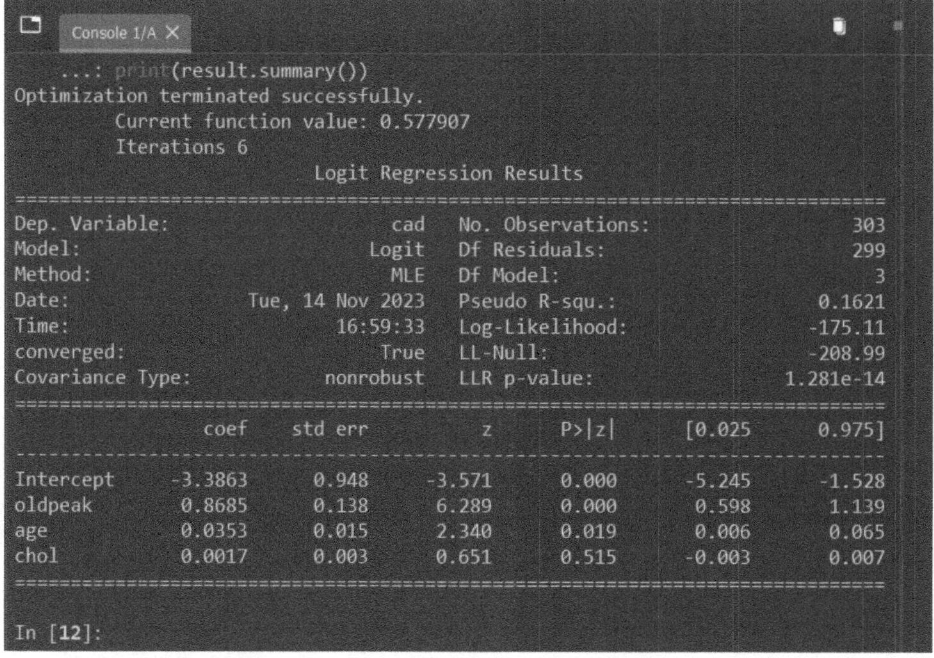

```
Console 1/A
   ...: print(result.summary())
Optimization terminated successfully.
         Current function value: 0.577907
         Iterations 6
                           Logit Regression Results
==============================================================================
Dep. Variable:                    cad   No. Observations:                  303
Model:                          Logit   Df Residuals:                      299
Method:                           MLE   Df Model:                            3
Date:                Tue, 14 Nov 2023   Pseudo R-squ.:                  0.1621
Time:                        16:59:33   Log-Likelihood:                -175.11
converged:                       True   LL-Null:                       -208.99
Covariance Type:            nonrobust   LLR p-value:                 1.281e-14
==============================================================================
                 coef    std err          z      P>|z|      [0.025      0.975]
------------------------------------------------------------------------------
Intercept     -3.3863      0.948     -3.571      0.000      -5.245      -1.528
oldpeak        0.8685      0.138      6.289      0.000       0.598       1.139
age            0.0353      0.015      2.340      0.019       0.006       0.065
chol           0.0017      0.003      0.651      0.515      -0.003       0.007
==============================================================================

In [12]:
```

Figure 9.11 – CAD status prediction linear regression model

We will start again with interpreting the R-squared. Since this is a logistic regression, we have a specific metric called **pseudo-R-squared**. It explains the determination coefficient in a similar way.

Even though the pseudo-R-squared is not the same as linear regression R-squared, it provides a similar evaluation of the variability explained, which is around 16%. This is much more than the stress test linear regression, which was around 4.6%. We can immediately see that this model is better. Now, let's interpret the individual variables' p-values and coefficients. Interestingly, we can see that both ST depression and age are significant predictors. However, if you look at the field on the right of `P>|z|  [0.025 0.095]`, which corresponds to the 95% confidence interval lower and upper bounds, the story is a bit different. The lower bound for age is `0.006`, which is very close to `0`, and this predictor is quite uncertain in terms of this factor. The lower interval for ST depression is `0.598`, which is far above `0`, and this is a good indicator of a significant predictor. Both ST depression and age have positive coefficients, which means as they increase, the probability of CAD increases as well. To summarize, if a lower bound of a confidence interval for a regression coefficient is far above `0`, we can be more confident that the magnitude of the effect is stronger, and we are also more confident that the effect is above `0` in reality for that specific population (the one studied).

Now, let's add the `exang` variable (exercise-induced angina) into the model as it's another known biological covariate for CAD. Subjects with CAD frequently exhibit exercise-induced angina and this variable could also be a significant predictor:

```
# Define the formula for logistic regression
formula = 'cad ~ oldpeak + age + exang + chol'

# Create the logistic regression model
logit_model = sm.Logit.from_formula(formula, data=dataset)

# Fit the model
result = logit_model.fit()

print(result.summary())
```

Here is the output for it:

```
Console 1/A
Optimization terminated successfully.
         Current function value: 0.519379
         Iterations 6
                           Logit Regression Results
==============================================================================
Dep. Variable:                    cad   No. Observations:                  303
Model:                          Logit   Df Residuals:                      298
Method:                           MLE   Df Model:                            4
Date:                Tue, 14 Nov 2023   Pseudo R-squ.:                  0.2470
Time:                        16:58:16   Log-Likelihood:                -157.37
converged:                       True   LL-Null:                       -208.99
Covariance Type:            nonrobust   LLR p-value:                 2.010e-21
==============================================================================
                 coef    std err          z      P>|z|      [0.025      0.975]
------------------------------------------------------------------------------
Intercept     -3.7771      1.022     -3.697      0.000      -5.780      -1.774
oldpeak        0.7355      0.144      5.113      0.000       0.454       1.017
age            0.0368      0.016      2.282      0.023       0.005       0.068
exang          1.7333      0.304      5.710      0.000       1.138       2.328
chol           0.0012      0.003      0.447      0.655      -0.004       0.007
==============================================================================

In [11]:
```

Figure 9.12 – CAD status prediction model with the exercise-induced angina variable added

As it can be seen, again, the ST depression and age are significant, with similar values as in the previous model. However, the pseudo-R-squared increased from 0.16 to 0.247, which is a sharp increase. This is an indicator that the model with `exang` included is much better and we should keep it in our final model.

Finally, the `exang` variable itself is highly significant; `p=0.000` is actually `p<0.001` and the lower bound of the confidence interval is `1.138` for a positive coefficient (much above `0`). All this contributes to the credibility of exercise-induced angina also being a strong predictor. In summary, ST depression and exercise-induced angina are shown to be credible predictors of CAD in this analysis, while age also contributed but as a weaker predictor.

As you have noticed, we are interpreting the probability of CAD as the target variable in the logistic regression, but the coefficients themselves are not that interpretable if we don't adjust the probability interpretation to a more intuitive scale. To achieve this, we can convert the scale to OR. Before going through the code, let's discuss the odds and ORs. What is odds? The odds is basically a comparison of events versus non-events. The event/non-event division will provide the odds of the event occurring. Now, if we do the same in the experimental and control groups in a study and divide two odds, we will get the OR. ORs will then have the following formula:

event experimental / non-event experimental) / (event control / non-event control. Or, simply, *odds experimental event / odds control event.* This ratio can be used to express how likely the event in the experimental group is versus the event in the control group. This event can be disease, mortality, side effects, or any other biomedical event analyzed.

The regression coefficients we observed in *Figure 9.12* were `log(OR)`. In other words, logistic regression coefficients are `log` values of OR. To convert the logistic regression coefficient table to the OR scale, we simply need to do the opposite of `log`, which is to exponentiate the coefficients and the associated results.

We can now proceed with the code. By adding the OR information, we get the following:

```
# Calculate odds ratio scale by exponentiating
odds_ratios = np.exp(result.params)
conf_intervals = np.exp(result.conf_int())
p_values = result.pvalues

# Combine odds ratios, confidence intervals, and p-values into a
DataFrame
summary_df = pd.DataFrame({
    'Odds Ratio': odds_ratios,
    'CI Lower': conf_intervals[0],
    'CI Upper': conf_intervals[1],
    'P-value': p_values})

# Display the summary DataFrame
```

```
print("\nSummary with Odds Ratios, Confidence Intervals, and
P-values:")
print(summary_df)

#Save results to a csv file
summary_df.to_csv('logistic_regression_results.csv', index=True)
```

Here is the output for it:

```
   ...: print("\nSummary with Odds Ratios, Confidence Intervals, and P-values:")
   ...: print(summary_df)

Summary with Odds Ratios, Confidence Intervals, and P-values:
           Odds Ratio  CI Lower   CI Upper       P-value
Intercept    0.022889  0.003089   0.169571  2.184752e-04
oldpeak      2.086511  1.573865   2.766139  3.177817e-07
age          1.037464  1.005200   1.070765  2.250705e-02
exang        5.659501  3.121817  10.260035  1.126643e-08
chol         1.001214  0.995902   1.006554  6.549900e-01

In [9]: summary_df.to_csv('logistic_regression_results.csv', index=True)

In [10]:
```

Figure 9.13 – The final predictive CAD status model converted to OR scale

Now, the ORs are easier to interpret in terms of the magnitude of the predictor and target variable (CAD) changes. How can we interpret these? Well, the interpretation is based on the units. An increase in 1 mm of ST depression (`oldpeak`) is associated with an increase of OR from `1` to `2.08` (note that the baseline for OR is `1`). This increase means `2.08` increased odds of CAD for each millimeter increase in ST depression. Also, the presence of exercise-induced angina (`exang`) is associated with `5.65` times increased odds of CAD.

In this section, we learned how to create logistic regression models and how to convert between regression coefficients and ORs. We also learned how to present and interpret the results of multivariate logistic regression and translate the statistical results into conclusions and biological explanations.

Summary

In this chapter, we performed a real-life biostatistical analysis in cardiology using the Cleveland heart disease dataset. First, we learned about the biological foundation of cardiological aspects, such as coronary arterial disease, maximum heart rate in stress tests, electrocardiogram, and ST depression. Further, we discussed the relation between these cardiovascular events and complications, such as myocardial infarction.

We developed linear-regression-based predictive models with the goal of evaluating the association between ST depression and maximum heart rate during a cardiovascular stress test. Finally, we developed a predictive model where the target variable was CAD and the predictor was ST depression on an electrocardiogram.

In the next chapter, we will discuss clinical study design and how to pay attention to both biological and statistical aspects when designing studies in clinical research.

Part 3: Clinical Study Design, Analysis, and Synthesizing Evidence

In *Part 3*, the main topic is learning the principles of clinical study design. Furthermore, you will learn how to practically perform survival analysis and then how to synthesize evidence using meta-analysis. The exemplar hands-on project for this chapter is about oncology survival data in clinical research.

This part contains the following chapters:

- *Chapter 10, Clinical Study Design*
- *Chapter 11, Survival Analysis in Biomedical Research*
- *Chapter 12, Meta-Analysis – Synthesizing Evidence from Multiple Studies*
- *Chapter 13, Survival Predictive Analysis and Meta-Analysis Practice*
- *Chapter 14, Part 3 Exemplar Project – Meta-Analysis of Survival Data in Clinical Research*

10
Clinical Study Design

Clinical studies are some of the most important studies in modern biostatistics. From academic institutions to biomedical institutions and the pharmaceutical industry, clinical studies are the most important tool for drug discovery, patient safety, medical affairs, epidemiology, and public health. For this reason, biostatistics professionals and anyone performing biostatistics should understand how clinical trial designs such as observational studies and clinical trials work and what research questions we can answer by implementing those designs.

This chapter is focused on a basic understanding of clinical trial design, how to differentiate different types of clinical studies, and how to relate all of this to statistics. Another very important aspect will be discussed, which is defining the essential documents for clinical studies: clinical trial protocols.

In this chapter, we're going to cover the following main topics:

- Understanding clinical studies and their relationship with biostatistics
- Understanding observational studies
- Learning about the principles of clinical trials
- Calculating sample size for clinical studies
- Defining the protocols for clinical studies

Let's get started!

Understanding clinical studies and their relationship with biostatistics

Clinical studies are primarily used in two specific areas today: the pharmaceutical industry and medical research. Most drug development companies use clinical studies not just to evaluate present drugs but also to discover and develop new ones. The drug discovery process typically involves a series of clinical studies such as clinical trials to make sure that the drugs are safe and effective before they proceed to the medical industry areas.

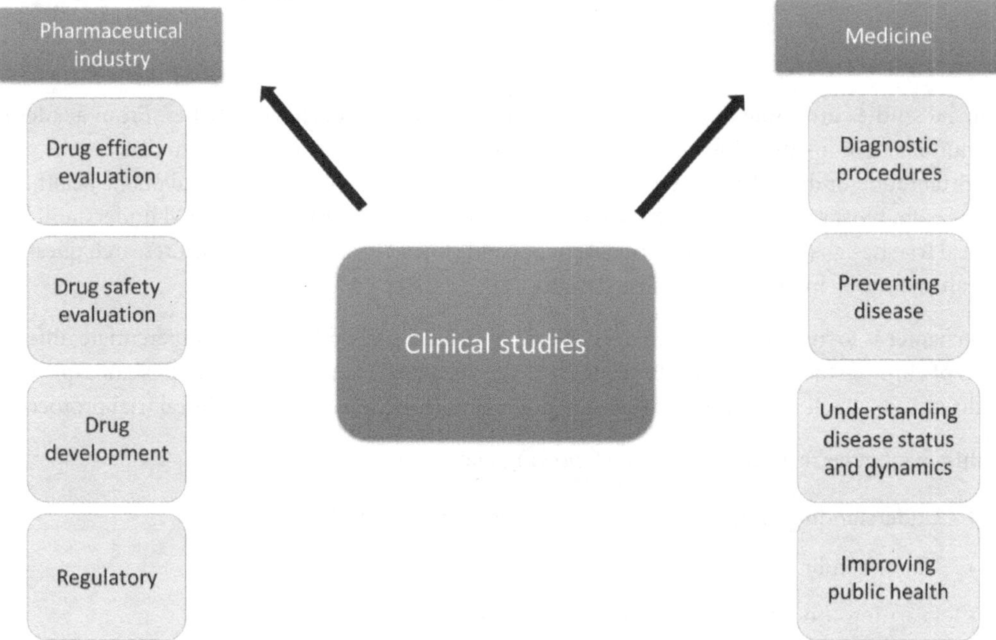

Figure 10.1 – Clinical studies in the pharmaceutical industry and medicine

Regulations in many countries are a key aspect to consider. Very often, pharmaceutical companies conduct their own clinical studies such as Phase I to Phase IV clinical trials in compliance with regulatory frameworks of agencies such as the United States of America-based **FDA** (short for **Food and Drug Administration**; you may find more information on FDA at `fda.gov`).

These clinical studies generally have the main goals of drug efficacy evaluation, drug safety evaluation, and drug development as a means of gathering important evidence for regulatory approvals for medical use.

Typically, regulatory frameworks are very strict to make sure that the drugs approved for patients are as safe and as effective as possible. Pharmaceutical companies spend billions of dollars during these processes on conducting various clinical research and only a fraction of the drugs tested are approved.

Clinical studies are also the basis for medical diagnostic procedures. Many diagnostic tests require clinical studies and biostatistical procedures to determine metrics such as sensitivity and specificity and show how much we can trust different diagnostic procedures and tests.

Clinical studies serve as the basis for doctors and epidemiologists, as well as governments and other decision-makers, to develop strategies for preventing different diseases. Overall, clinical studies are the foundation for improving public health, starting from individual subjects and then being transposed to entire populations.

Clinical studies are not used just to understand biological and medical aspects of human biology but also for drug discovery, understanding the safety and efficacy of different drugs, and improving health outcomes for billions of patients yearly. Clinical studies are the foundation of today's medicine and the way doctors diagnose, treat, and, very importantly, help patients prevent different diseases.

Biostatistics is one of the foundation blocks for all clinical studies. It helps researchers design clinical studies, analyze the data, and interpret the results while keeping in mind both statistical and biological aspects. There is a specific branch of biostatistics called **clinical biostatistics** that specializes in clinical study problems and methods used as a solution for them.

Clinical studies can have different designs, and for most of these designs, we use a combination of statistical and biological knowledge to define the problems and answer different research questions. One very important aspect of clinical studies is that they are often used to make decisions in both the medical and pharmaceutical industries. Biostatistical evaluation is the key to making these aforementioned decisions. Billions of dollars worth of project decisions are made every month based on the results analyzed and interpreted by biostatisticians in clinical studies.

Let's explore the areas in which both clinical biostatistics and clinical research integrate systematically.

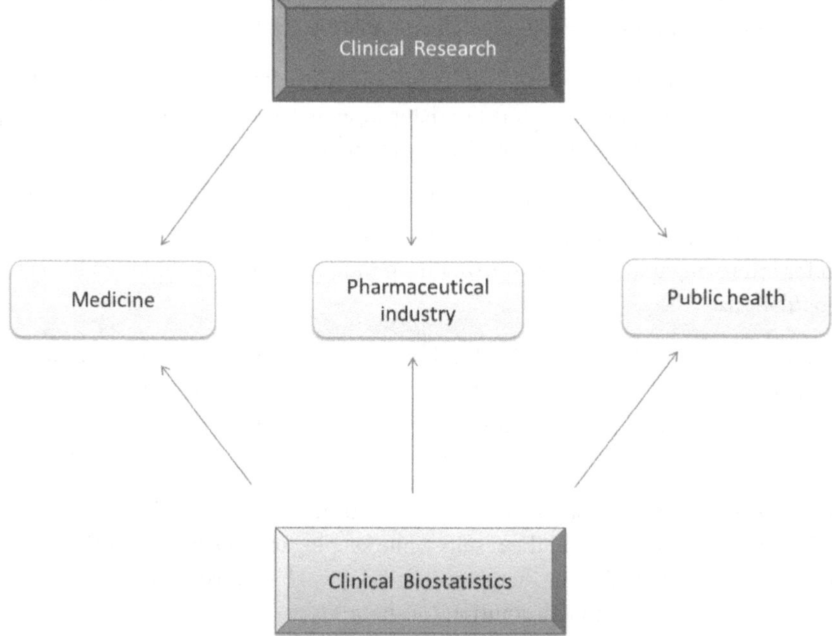

Figure 10.2 – Clinical biostatistics and research in medicine, pharmaceutical industry, and public health

Biostatistics helps us understand what the data tells us. If we apply a certain statistical method, we must know exactly what the output tells us and what it doesn't. The workflow in clinical studies is very specific and it is up to biostatisticians to understand each step and the type of study.

One of the most important aspects of clinical studies is that they are typically bound to specific clinical research questions that we must understand in detail and define before clinical studies.

Clinical study design and research questions

The research questions we can answer in a specific clinical research domain predominantly depend on the type of clinical study design we create. These research questions also depend on the domain knowledge we have in that clinical research area. This is one of the reasons why it is a big advantage for any biostatistician to have some experience and understanding of clinical research.

In this chapter, we are also going to discuss how to reason the research questions in different clinical research areas. What is it that we are typically looking for in clinical research? Let's start with some examples. Most research questions in clinical studies have to do with the phenomena of causality.

In drug discovery, scientists try to find drugs that will have a beneficial effect on various diseases. They also try to find drugs that will not cause serious side effects and are safe to be used in a wide population.

In medicine, doctors also use causal reasoning. Trying to find the causes of disease, as well as relating those causes to different risk factors, is among the main biomedical features of not just the clinical studies but also the medicine itself. As you can see, one of the main questions to ask in most clinical research has to do with the causal effects. Causal research questions can also help us understand human biology and advance medicine overall.

On the other hand, many clinical studies aim to describe different diseases, exposures, and risk factors, without explicitly finding the causal relation.

Most clinical research questions are formed using two scientific phenomena: causality and association. These two are frequently mistakenly considered as the same phenomena, but they are actually quite different.

Causality versus association

Before understanding causality in clinical research, let's explore some fundamental principles of causality.

Causality is a deterministic phenomenon. This means that the cause determines the effect. Another very important aspect of causality is direction. The cause defines the effect and not the other way around. This means that the direction is *Cause → Effect*. The third important aspect of causality is the time aspect; the cause precedes the effect. In many studies, there are hundreds of different variables of interest. Some of them are measured, some not, but generally, a lot of them influence each other. Not all these associations are causal. Sometimes, the variables are just correlated to some third kind of variables or are under the influence of some external variables. This means it's very difficult to differentiate between cause and effect association or just correlation.

Only experimental designs that can isolate the cause-and-effect variable from external influence can be a good empirical way of concluding the causal relation.

Now, let's translate this into the world of real clinical studies.

The type of study where we cannot control the variables that could influence the analysis of interest and all the confounders or variables that could influence the results would not be appropriate for causal analysis. These are the types of studies where we can just observe or measure the data without setting up the experimental design and are called **observational studies**. However, observational studies can provide some great inputs about the associations of variables in real-world settings and provide large amounts of useful descriptive, observational, and associative data on different diseases. Observational studies can provide some insights into the potential causal angles, but they should not be considered as the absolute source of causal evidence.

Retrospective clinical study design

In retrospective clinical study design, the data collected is from the time points preceding the actual implementation of the study. This type of study often involves collecting data that is already stored in medical systems, such as electronic health records or other medical databases. Because the data is already stored from the past, conducting retrospective studies is much easier compared to other types of studies. The limitation is that we cannot collect any present or future data and we cannot affect the methodology of the data collection.

Snapshot (cross-sectional-based) study design

Cross-sectional design, often called a **snapshot study**, is very important because it can capture the situation with health conditions and diseases at a particular point in type. This type of study is typically used in epidemiology. Its limitations are in the fact that it's more difficult to follow the dynamics of health conditions over time, but multiple snapshot studies can be a way to overcome this problem.

Prospective study design

Prospective studies are most credible from the standpoint of the results. This is due to the reason that experimenters or clinicians included in the study can define how the data is collected, in what follow-up periods, and what specific variables are of interest. The ability to control various aspects of clinical studies is essential from the standpoint of confounder effects and bias reduction. For this reason, prospective studies such as randomized clinical trials are considered the gold standard for clinical research overall.

Now we have learned the basics of different clinical study designs, let's explore the most basic type, observational studies, in more detail.

Understanding observational studies

The most fundamental type of study in clinical research is the type with data collected from subjects in the present or even in the past where we just *observe* or measure the parameters without having control over those parameters. It is much easier to just observe and measure the biomedical parameter than it is to set up experiments, so most of the data in clinical research comes in this form. This type of study is called observational study.

Observational studies have the main goal of helping us understand patterns in the biological data and answer research questions in the context of real-world settings, without influencing the subjects of the environment being studied.

As seen from the previous explanation, observational studies have a big advantage over experimental ones when it comes to availability and ease of collecting the data. In addition, fewer ethical concerns and financial aspects such as being cheaper in terms of budget needed to complete the study make the observational study much easier to conduct and create observational data. For this reason, the majority of data in clinical research is in the form of observational data.

Observational studies have their disadvantages, too. Without experimental control, these studies are not well optimized for controlling the confounders, or variables that affect our main data. This means that the results we get from observational studies are prone to being affected by variables not accounted for.

Observational studies are very important tools to provide clinical researchers and medical doctors with a valuable and wide spectrum of knowledge. However, because of the confounder effects and inability to control all the variables of interest, observational studies should not be considered as definite proof or evidence of causal results. Having said that, it should also be noted that in a path toward finding the causal effects in clinical research, observational studies can serve as a good indicator of where to look for those and conduct experimental research, including clinical trials, later.

The main topics studied in observational studies can be different. These can include simply describing different characteristics of subjects and their diseases, comparing them to control subjects, and finding potential associations between different exposures and risk factors and their outcomes. Control subjects are usually subjects whose data can be used as a reference group for the study.

Let's explore different types of observational studies using the following diagram:

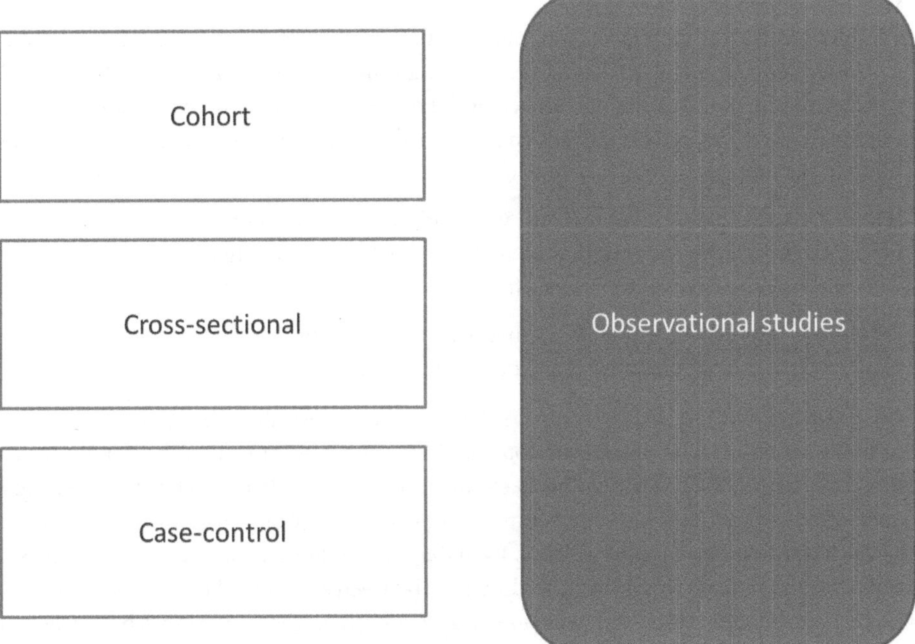

Figure 10.3 – Observational studies

As we can see from the figure, there are three main types of observational study:

- **Cohort studies**: One of the most useful types of study in clinical research is clinical cohort studies. These studies involve collecting data on a specific group or groups of subjects (there can be multiple cohorts) and following up on the specific characteristics, exposures, risk factors, and clinical aspects. Typically, they are used to try to establish a connection between different exposures and risk factors and their outcomes.

- **Cross-sectional studies**: Cross-sectional studies are observational studies that are performed at one point in time, meaning they are a snapshot of variables for the population that is being studied. Being a snapshot in time means we don't have longitudinal data, and that is a limitation, especially when evaluating the causal factors for diseases. However, cross-sectional studies are very easy to conduct, timely, have better data collection feasibility parameters, and are also cheaper to conduct, which makes them feasible in many scenarios where other study types could not be implemented.

  Some examples where cross-sectional studies are very effective include studies involving the prevalence and incidence of diseases.

  Prevalence is the proportion of a population having a specific disease in a snapshot in time. Consider an example of 10,000 subjects having type II diabetes in a population of 100,000, which would mean that the prevalence of the disease is 10,000/100,000 and it would be 0.1. But for ease of interpretation, this is converted to a percentage, so 0.1 prevalence would be 0.1 x 100 in percentage or 10%.

  If multiple such snapshots were made, we could evaluate the incidence or the rate at which the newly diagnosed disease was identified. An example would be that for a specific population, 1,000 new cases per week were identified.

  Prevalence and incidence are the key metrics that enable epidemiologists to evaluate the spread of diseases in the population and the rate of new spread.

- **Case-control studies**: In case-control studies, as the name states, a group of cases, most often subjects having a certain disease, are compared against controls that don't have it. By performing this comparison, the differences between those cases and controls can be evaluated. However, unlike cross-sectional studies, case-control studies focus on the outcome, which is most frequently a health condition or a specific disease. Even the controls are evaluated so that outcomes can be studied with less bias. This is the main difference between cohort studies and case-control studies (note: cohort studies focus on exposure and follow-up on the relationship between exposure and outcomes). Case-control studies allow for both prospective and retrospective designs.

The following table explains these three study types in tabular form:

Type of study	Time aspect	Focus
Cohort	Retrospective and prospective (generally prospective)	Exposure
Cross-sectional	One point in time	Population characteristics
Case control	Retrospective and prospective (generally retrospective)	Outcome

Table 10.1 – Types of observational studies

Now that you understand other types of clinical study designs, let's learn about the most important type of clinical study design, the clinical trial.

Learning about the principles of clinical trials

Clinical trials are designed to address many of the problems we have in observational studies and provide medicine with the most trustworthy results possible. For this reason, clinical trials are designed as more or less controlled experiments.

Clinical trials have a specific primary outcome or outcomes of interest to evaluate. These outcomes are related to the main goal of the study. Some examples can be the safety or efficacy of the drug, side effects, or other specific clinical outcomes, such as the survival of patients, which are still in a way related to the efficacy of drugs.

There are four types of clinical trials—*Phase I*, *Phase II*, *Phase III*, and *Phase IV*. All these trials are part of the drug discovery and research pipeline and are closely connected to each other. While the initial phases are meant to mainly evaluate the drug safety profile, later phases, especially after Phase III, are mostly focused on the efficacy of drugs on larger populations. It should be kept in mind that the safety profile is always evaluated and re-evaluated in all phases as either a primary or secondary outcome.

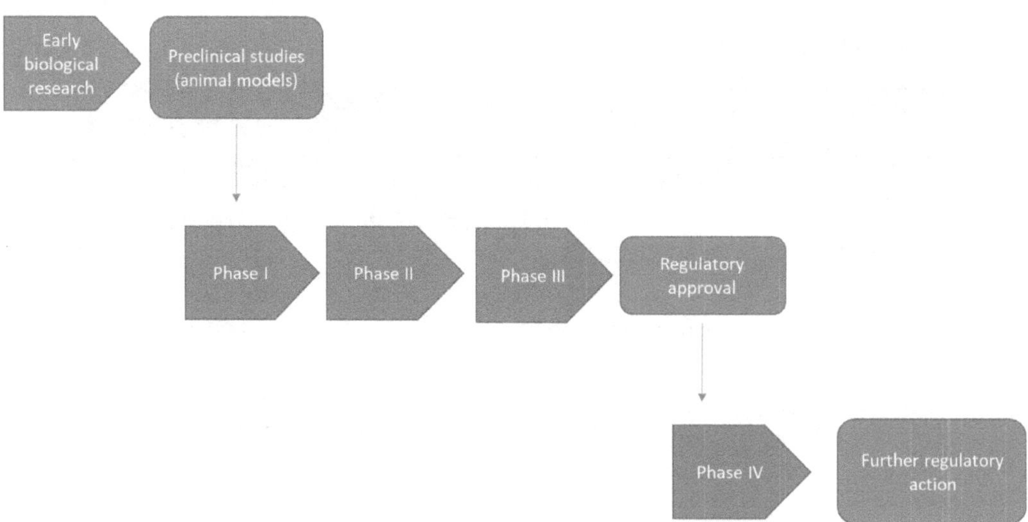

Figure 10.4 – From preclinical research to clinical trials

In *Figure 10.4*, you can see the sequence of phases discussed, and each phase has specific main goals. Early biological and preclinical research was mostly about understanding and discovering biologically active molecules and biological targets. Phase I and II trials are about the initial analysis of drug safety and efficacy. Phases III and IV are about regulatory approval level analysis with the goal of providing the most credible evidence for the safety and efficacy of drugs.

Now, let's discuss reporting in clinical trials.

Reporting in clinical trials

Reporting in clinical trials is a very important aspect of explaining the flow of the clinical study. To achieve clinical trial explainability and clear data reporting, different types of charts are used. Typically, there are different standards established by research and academic institutions around the world. Some of the most known are **Consolidated Standards of Reporting Trials** (**CONSORT**), **Case Reports** (**CARE**), and **Preferred Reporting Items for Systematic Reviews and Meta-Analyses** (**PRISMA**) guidelines, which define how to create simple flowcharts that contain relevant information about the allocation of subjects, randomization, interventions and controls, number of subjects in each group, subjects lost during the follow-up periods, and the final data that is used for the biostatistical analysis.

Here is an example of how a CONSORTIUM flowchart should look with an example of 100 subjects being assigned to the intervention and control groups:

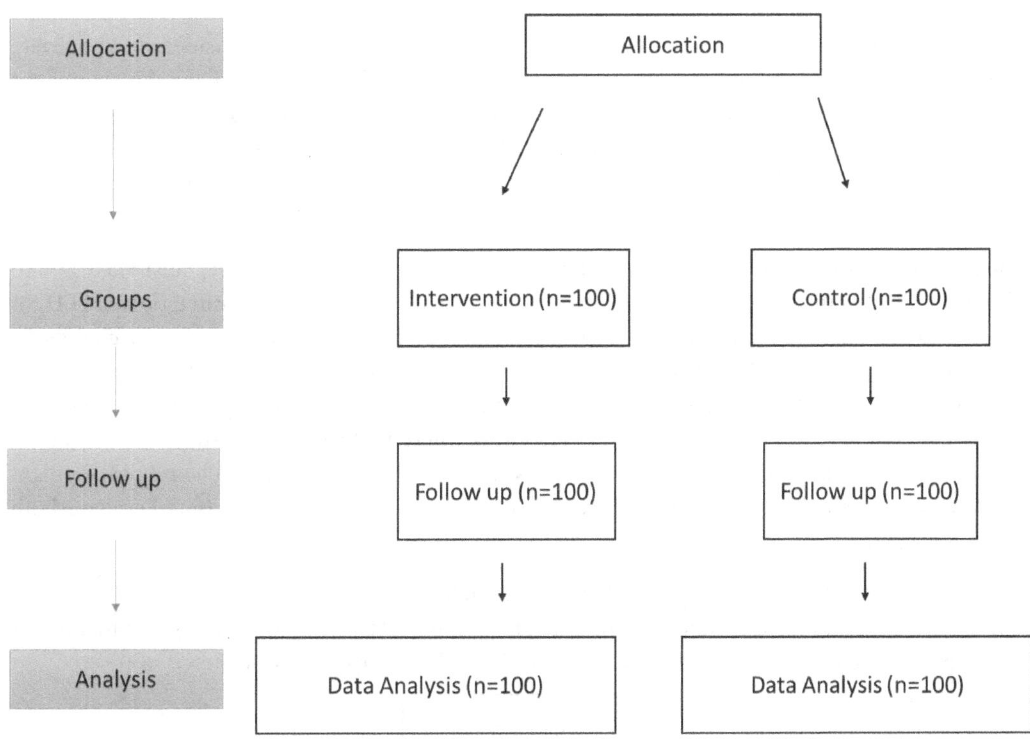

Figure 10.5 – CONSORTIUM flowchart example

In this flowchart, you can see the allocation, groups, intervention versus control groups, each with 100 subjects, and also how many were in the groups during the follow-up. Finally, you can see the data analysis segments in relation to how many subjects were included in the data analysis part.

You can read more about reporting in clinical studies overall here: `https://www.nlm.nih.gov/services/research_report_guide.html`.

Since we now know about clinical trial reporting, let's delve into the specific designs of Phase I clinical trials.

Phase I clinical trials

Phase I clinical trials have the main goal of assessing the safety profile of the specific drug. Since drugs entering Phase I trials are generally the ones not studied well before, fewer subjects are included and careful dose increase is deployed to try to identify the maximal dose that will not cause severe side effects. For this reason, Phase I clinical trials are considered to be dose-escalation studies.

The safety profile is evaluated based on the side effects or toxicity, which is considered to be severe if it might cause serious health-related negative effects. These effects are first defined by clinicians and may differ among different medical areas. Later, these side effects and toxicities are summarized statistically to try to find the dose-dependent phenomenon where these toxicities occur. For that reason, in biostatistics, this phenomenon is called **dose-limiting toxicity** (**DLT**). The majority of Phase I clinical trials have DLT as the main outcome of the study.

When the DLT is identified in a certain number of subjects by increasing the dose, most often 33% or at least one-third of the subjects, the dose identified is called the **maximum tolerated dose** (**MTD**). It should be noted that in Phase I clinical trials, researchers often start with very low doses of a compound administered to subjects, which is most often sub-therapeutic, meaning have no therapeutic effect. Gradually increasing the doses might exhibit some therapeutic effects, but this is not the main outcome for Phase I trials. Instead, safety is the main outcome of interest. Once the maximum dose that does not cause serious side effects or toxicity in subjects is identified, the drug research and development workflow can be redirected to Phase II clinical trials, which will have more subjects and evaluate both safety and efficacy.

Here is an example of a 3+3 Phase I clinical trial where, at first, three subjects are included at the lowest dose, and if the dose is safe without causing the DLT in subjects, a second dose is given to another three subjects, where n=3 (where n is the number of people the dose is administered to):

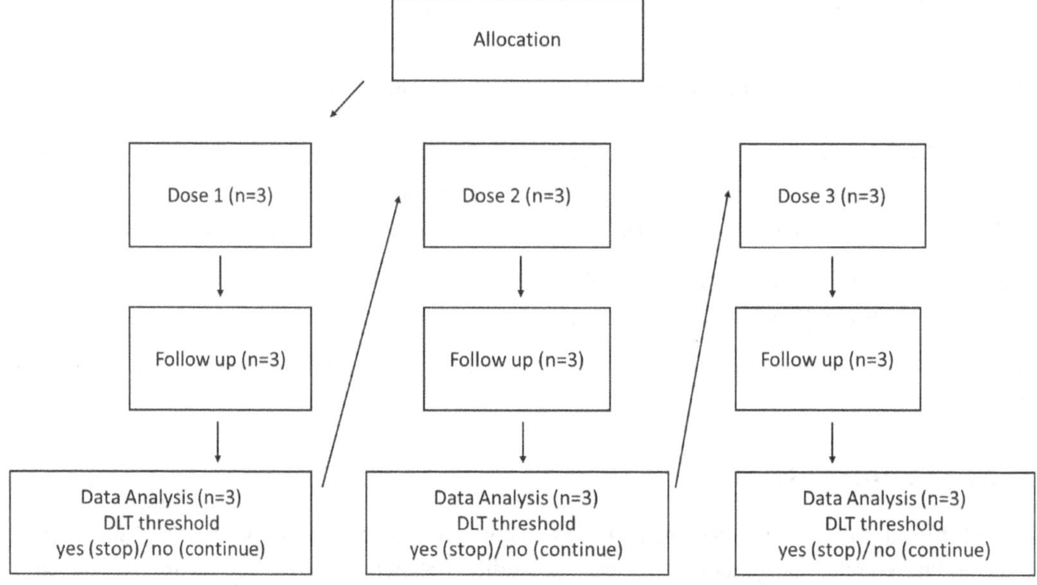

Figure 10.6 – A 3+3 Phase I study design example

As you can see, the decision to stop the study or continue with a higher dose is made based on the DLT frequency in the three subjects analyzed. If there is no DLT, the dose is escalated; if DLT occurs in 1 of 3 subjects or 33%, the same dose is administered again, and if 2 or 3 subjects experience DLT, the study is stopped.

There are other designs, such as the **continuous reassessment method** (**CRM**), which allows for making the predictive dose-response models and using all the available data for all the subjects instead of focusing on the batches of three, and this is a more advanced type of Phase I trial.

Phase II clinical trials

Clinical trials are a part of the connected clinical research chain. This means that Phase II trials can occur only if the Phase I trial was successful. Phase II trials are slowly shifting the balance from evaluating mostly safety to evaluating both the safety and efficacy of drugs and typically involve more subjects from *n=120* or *n=300*.

There are two characteristics of Phase II clinical trials that are essential. First, Phase II clinical trials typically assess both the effectiveness and safety of the studied drug, and for this reason, the control group is included. By comparing intervention against the control, clinical research can infer the potential effectiveness of the drug in addition to the safety profile.

Here is an exemplar flowchart (CONSORT style) for a potential Phase II clinical trial.

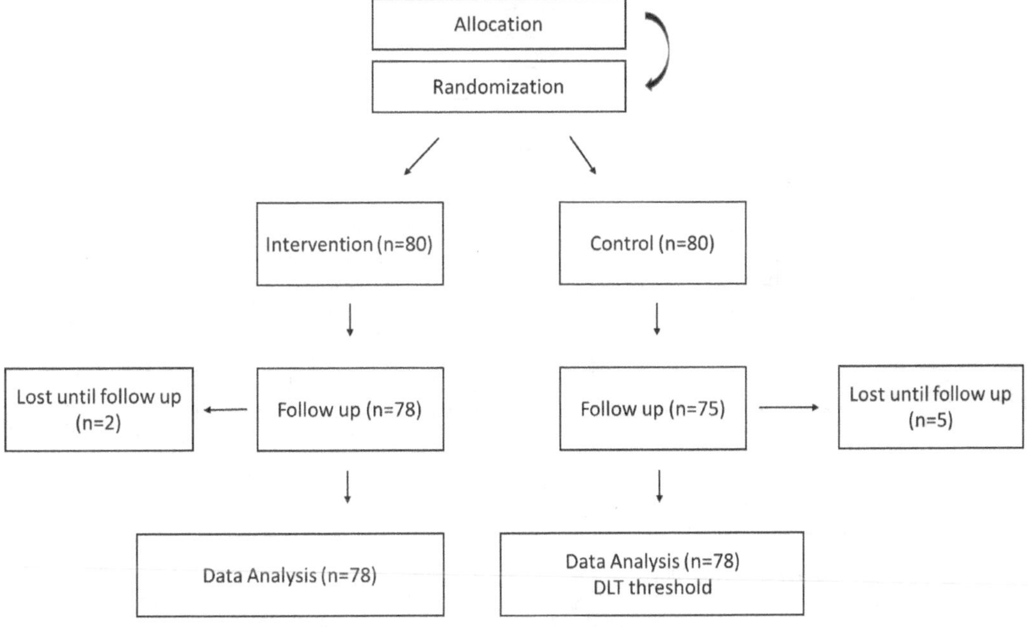

Figure 10.7 – A Phase II clinical trial flowchart example

You may notice that randomization is deployed here (second step, *Figure 10.7*). As mentioned before, randomization can correct for the bias of both known and unknown confounders. Also, randomization allows for causal inference, which is very important for Phase II clinical trials.

Phase III clinical trials

Phase III clinical trials typically involve hundreds to as many as 3,000 subjects, according to the FDA. What is very important for these studies is that they are generally counterfactual and randomized. Part of the causal analysis in **randomized controlled trials** (**RCTs**) is the counterfactual design (some receive an experimental drug, and some receive either a placebo or standard of care), similar to case-control studies. But only with randomization can it achieve causal analysis potential.

This means that subjects receiving experimental drugs and controls are assigned randomly. Why is this important? Well, the randomization principle makes sure that both known and unknown confounder variable bias is corrected. In this way, the experimenters control for both internal and external sources of bias. Controlling for this bias isolates our variables of interest and also allows for causal analysis. Phase III clinical trials are pivotal research tools in the drug evaluation pipeline. They are considered the gold standard in today's clinical research and one of the rare types of study that can actually capture causal information. For this reason, Phase III clinical trials are the key to making decisions by regulatory agencies such as the FDA. RCTs are also the basis for RCT meta-analyses, considered the most credible evidence-based studies in the clinical research world.

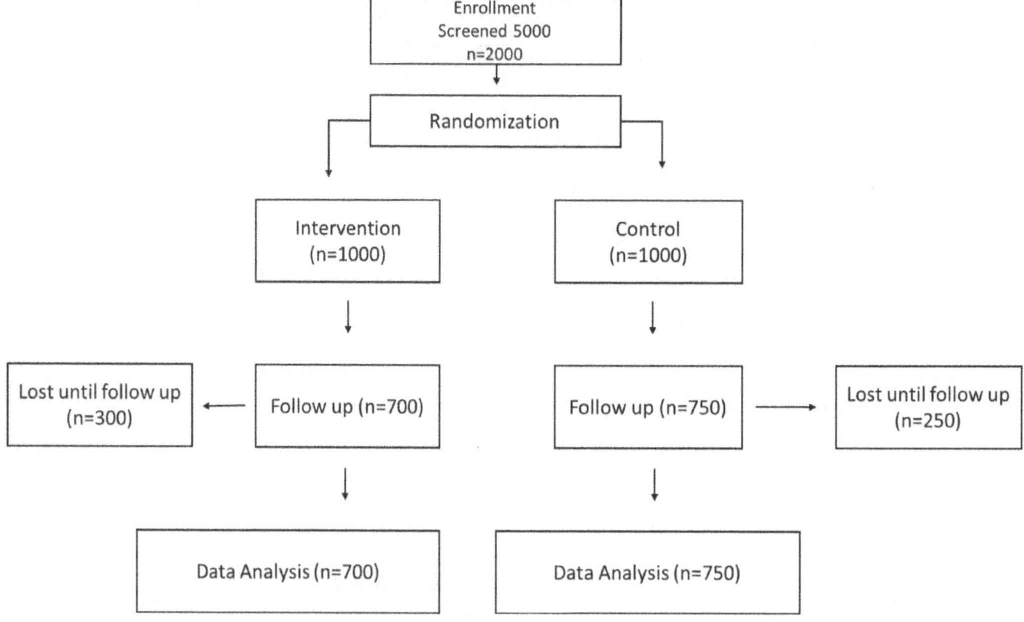

Figure 10.8 – Phase III clinical trial flowchart example

The next type of clinical study design to learn about is the Phase IV clinical trials: those that are deployed once the drug has been approved and are also called post-approval monitoring studies.

Phase IV clinical trials

Phase IV clinical trials are related to post-marketing monitoring of the drug's efficacy, effectiveness, and safety. This means monitoring the drug after it has been approved and administered in the medical systems. Why is this important? Well, once the drug has been adopted for widespread use, more and more data will be available, yielding even more credible results. On the other hand, the new data is from a real-world setting, so the data collected can be considered real-world evidence and a real-world empirical test of what we had in the Phase I to Phase III results. This is especially important because finer details and rare events frequently cannot be assessed without widespread adoption in the real-world setting.

While the real-world setting can be an advantage, it can also be a disadvantage because we cannot control the confounders as we could in the experimental designs. Still, the large sample and real-world setting of the Phase IV clinical trial make it one of the best tools in the assessment of drug performance.

Phase IV clinical trials can involve large populations of subjects receiving already approved therapies, and while there isn't a typical range that could define the number of subjects included, these generally include tens of thousands or more.

Monitoring that happens after the drug has been approved and administered in hospitals is not affected by experimenters and creates data from a real-world situation and actual treatments of subjects. For this reason, Phase IV clinical trials are considered real-world evidence studies.

Real-world evidence provides essential data regarding drugs and further indications for medical professionals on their use.

Once the clinical trials are completed, they can also be included in systematic reviews and meta-analyses, which are at the top of the pyramid of clinical research evidence.

Meta-analyses of clinical trials

Phase IV clinical trials are often combined with meta-analyses to get the most credible evidence possible. Meta-analyses involve including multiple studies, often clinical trials, and combining them into a single large analysis. Meta-analyses are also part of systematic literature reviews on different drugs and can drastically improve the evidence for clinical researchers.

Meta-analyses are used not only to combine the individual studies into the pooled, summarized effect but also to assess the heterogeneity and the quality of individual studies. Observing how individual studies affect the overall effects, which studies could be biased, and whether there is publication bias are some of the most important areas where meta-analyses are essential.

Here is an example of a forest plot from multiple clinical trials and their effects combined into a more trustworthy evidence base:

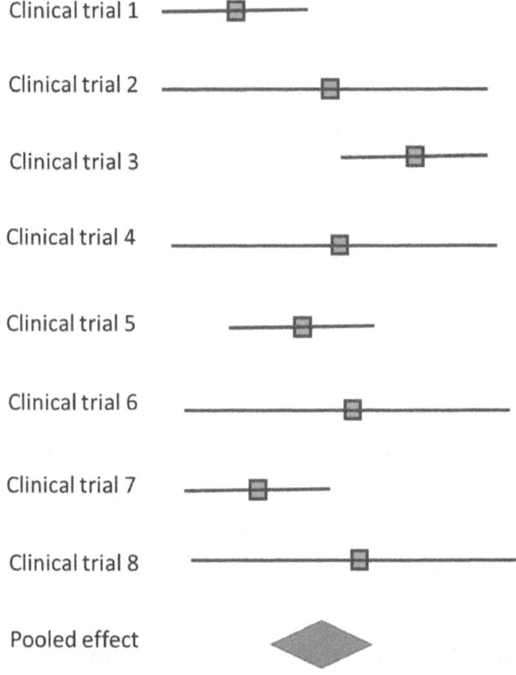

Figure 10.9 – Meta-analysis of clinical trials illustration

As you can see, the pooled effect (diamond) represents the combined evidence from many trials and represents a much more trustworthy estimate in reference to any single trial. Meta-analyses are frequently combined with **systematic literature reviews** (**SLRs**) and are the foundation of clinical research. SLRs and meta-analyses are also used in regulatory frameworks to evaluate the evidence base for already approved drugs and to assess the current status of research in a specific clinical research area. They are especially important in areas such as cardiology and oncology in which a lot of clinical trials are conducted and their systematic reviews and meta-analyses are extremely important.

Calculating sample size for clinical studies

When planning the design for any of the studies we mentioned, it is very important to define the required sample size needed for the study. Why is this important? Well, because the sample size is a very important factor in determining the statistical power.

Let's define the statistical power of the study.

Statistical power is the likelihood of detecting the significant results if they are really present. Smaller studies typically have less chance of detecting true results in comparison to larger ones. This is why it's important to define the required sample size because we need to be confident that the sample we have (size of the sample) will be enough to detect different results and answer the research questions.

Statistical power can be expressed on a scale from 0–1 or using a percentage scale from 0–100%. Typically, at least 0.8 or 80% power is the minimal power set in most studies, but it may be set at an even higher level.

Statistical power is dependent on the sample size but also on the effect size or the magnitude of the difference needed to be detected. Smaller effects are very difficult to detect, and a large sample size is needed to detect them. On the other hand, large effects and large differences are easier to detect and may be detected using smaller samples. Still, it's always a good idea to have a larger sample, if possible.

To learn how to calculate the required sample size, let's explore a few examples in Python.

First, let's load the `statsmodels` power function, which will enable us to calculate the statistical power under different scenarios:

```
#Lets load the statsmodels power function
import statsmodels.stats.power as smp
```

There are typically three parameters to consider when calculating statistical power. Logically, one of them is the `power` level, which is typically `0.8` for most studies. But another very important segment is the effect size. In this case, let's set it to `0.5` (this would correspond to Cohen's $D = 0.5$) and finally, set the statistical significance `alpha` level as `0.05`, the standard value:

```
# Let's specify parameters: effect size, alpha (significance level),
# power, and number of predictors

effect_size = 0.5
alpha = 0.05
power = 0.8

# Perform sample size calculation
sample_size = smp.tt_solve_power(effect_size=effect_size, alpha=alpha,
    power=power, alternative='two-sided')

print("Required Sample Size:", round(sample_size))
```

Here is the output:

```
Required Sample Size: 33
```

As you can see, the required sample size to detect the effect size of `0.5` with a power of 80% is just 33.

Now, let's try to lower the effect size and see what happens:

```
# Assume we want to calculate sample size for a simple linear
regression
# Let's specify parameters: effect size, alpha (significance level),
power, and number of predictors

effect_size = 0.1
alpha = 0.05
power = 0.8

# Perform sample size calculation
sample_size = smp.tt_solve_power(effect_size=effect_size, alpha=alpha,
    power=power, alternative='two-sided')

print("Required Sample Size:", round(sample_size))
```

Here is the output:

```
Required Sample Size: 787
```

When we lower the effect size to `0.1` from `0.5`, the required sample size is now `787`, which is a huge increase. You can see how the effect size actually determines the required sample size. But the power itself also determines the required sample size.

Let's try increasing the power to `0.97`:

```
# Assume we want to calculate sample size for a simple linear
regression
# Let's specify parameters: effect size, alpha (significance level),
power, and number of predictors

effect_size = 0.1
alpha = 0.05
power = 0.97

# Perform sample size calculation
sample_size = smp.tt_solve_power(effect_size=effect_size, alpha=alpha,
    power=power, alternative='two-sided')

print("Required Sample Size:", round(sample_size))
```

Here is the output:

```
Required Sample Size: 1477
```

Now, the required sample size is even larger, `1477`. From these examples, you can see practically how the required sample size increases with smaller effect sizes and larger power settings. The effect size is very important in calculating the required sample size and is typically defined in communication between clinicians and biostatisticians. In clinical biostatistics, there is a term called **MCID** (short for **minimal clinically important difference**), which can also be considered when calculating the effect size for clinical studies. This is another aspect typically defined by clinicians but, again, in communication with statistical experts.

Defining the protocols for clinical studies

One of the most important steps in any study, especially in clinical research, is defining the study protocol.

The study protocol is a document containing a systematic and detailed plan regarding all the steps and procedures to be implemented.

All the details discussed in this chapter are usually in the study protocols.

They typically contain the following:

- Clinical reasoning of the study
- Objectives/research questions
- Study design
- Inclusion/exclusion criteria
- Definition of the biological variables
- Definition of statistical methods
- Data analysis plan
- References

When we define a clear and comprehensive study protocol, it ensures the accuracy, reliability, and reproducibility of clinical research.

Summary

In this chapter, we learned the basics of clinical study designs and how they are related to biostatistics. We also learned about the different types of clinical studies and their use in the pharmaceutical industry and medical research. We also learned about clinical trials as experimental types of clinical studies and their use in the drug discovery and development pipeline. Finally, we learned about the regulatory aspects of approving drugs and the role of biostatistics in them.

In the next chapter, we will be discussing and learning about one specific type of study frequently used in clinical research: survival studies.

11
Survival Analysis in Biomedical Research

Many branches of clinical biostatistics rely on survival analysis. It is the main biostatistical tool in most clinical trials and is used to evaluate drug efficacy and effectiveness in different experimental and observational settings.

In this chapter, we're going to cover the following main topics:

- Understanding survival analysis and how it is used in biomedical research
- Creating Kaplan-Meier curves in Python
- Implementing Cox (proportional hazards) regression in Python

By the end of this chapter, you will gain a deeper understanding of how these methods are applied in biomedical research to analyze and interpret clinical trial data for survival analysis.

Understanding survival analysis and how is it used in biomedical research

Survival analysis is a branch of statistics that focuses on the relationship between time and events. The main aspect of survival analysis is analyzing the duration of time it takes for events to occur. Using this principle, biostatisticians can estimate the probability of occurrence for certain events within a given follow-up period.

Survival analysis was originally developed as a clinical biostatistics discipline and is sometimes used in other areas such as engineering or economy. However, the name *survival analysis* actually originates from clinical research.

Let's explore the main uses of survival analysis in clinical research:

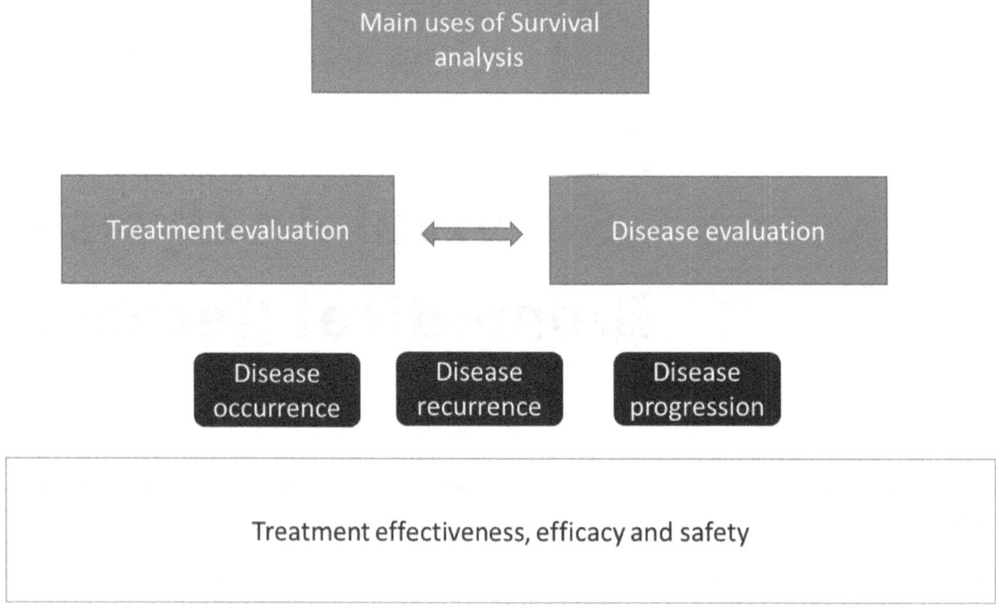

Figure 11.1 – Main uses of survival analysis in clinical research

Since survival analysis and the rates at which events such as disease occur are direct evidence of drug efficacy, survival analysis is one of the main tools of drug evaluation pipelines. Clinical trials in areas such as oncology and cardiology are dominantly survival analyses.

Once combined with a specific design called **counterfactual** and randomization of subjects, survival analysis can be considered a form of causal analysis in which the causal effects of the experimental drug are compared to its counterfactual, or the control group. Control groups are the ones that don't receive the experimental drugs and are considered a comparison reference in clinical trials.

Let's explore the data collection process in survival analysis:

Data collection process in Survival analysis

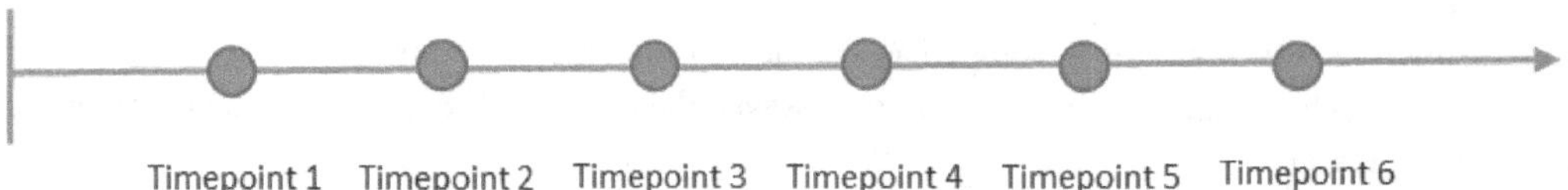

Figure 11.2 – Data collection process in survival analysis

As can be seen, the data collection process is done through a sequence of time points. The time points are equal and, as such, present a time series data. These time points are very important as they add the time aspect to the analysis. We will discuss this further in the rest of the chapter.

Two very important terms in survival analyses are called events and censoring:

- **Event**: An event is an occurrence of interest in survival analysis. It can be an event of mortality, disease progression, recurrence, or any other event that can be used to describe the occurrences used to answer the research question.
- **Censoring**: Censoring is relevant to the data points for which we don't have complete information. For example, once a subject's follow-up is stopped, we have no further information about their survival. This is called **right censoring**. On the other hand, some subjects experience events before entering the study and we have no information on when the event happened. This example is called **left censoring**.

Survival analysis is frequently used also to evaluate the progression and recurrence of diseases such as cancer.

In that sense, we can define different metrics such as the following:

- Disease-free survival
- Progression-free survival
- Recurrence-free survival
- Complication-free survival

In this section, we learned about the basic principles of survival analysis and how survival data is collected. We also learned how to represent different survival metrics. In the next section, we will learn how to visualize survival data using Kaplan-Meier curves

Creating Kaplan-Meier curves in Python

One of the best ways of presenting survival data is by constructing a specific data visualization called the Kaplan-Meier curve. What is the Kaplan-Meier curve? It is a graph that has two axes, X and Y, where X represents time and Y represents the probability of survival. Alternatively, Y can represent the proportion of the events occurring. The time axis (X) can be defined in days, months, or years and can be used to create equal intervals for presenting the survival data. The survival axis is relevant for the outcomes such as survival, mortality, progression of disease, or other clinical events relevant to specific clinical research questions. The Y axis can also be used to indicate whether the data was censored or not. This function most often looks like descending stairs, as shown in the following screenshot, due to the stepwise nature of the function.

This stepwise nature is due to each time point being a step at which the probability of events/survival is shown. Refer to the following figure:

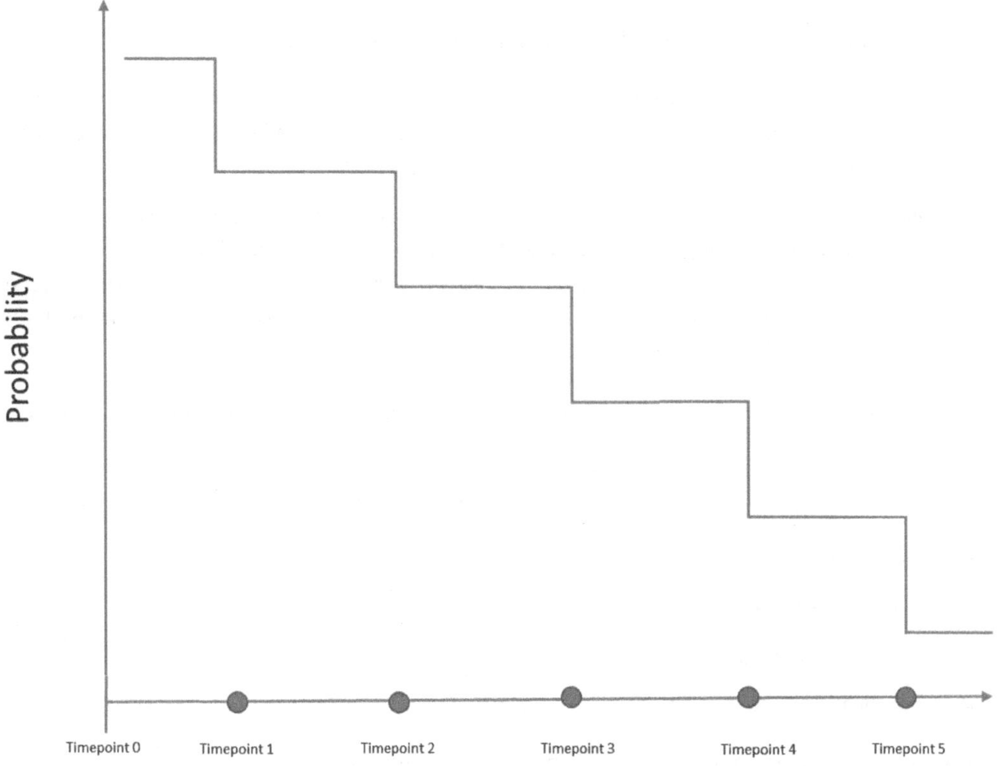

Figure 11.3 – The stepwise nature of the Kaplan-Meier curve

Now let's use Python and the Kaplan-Meier curve in the previous oncology example. This example is related to lung cancer in the lung cancer veterans dataset (`https://github.com/sebp/scikit-survival/blob/v0.22.2/sksurv/datasets/data/veteran.arff`) from the `scikitlearn-survival` package.

We will also be using the `matplotlib` package to plot the Kaplan-Meier curves. Here are the steps for it.

Before running this code, make sure you have installed the `scikit-survival` library by running `pip install scikit-survival`:

1. First, open Python (through Spyder or Jupyter) and load the data:

```
# Import necessary libraries
from sksurv.datasets import load_veterans_lung_cancer
from sksurv.nonparametric import kaplan_meier_estimator
from sksurv.compare import compare_survival
import matplotlib.pyplot as plt

# Load the veterans lung cancer dataset
data_x, data_y = load_veterans_lung_cancer()
```

2. Now let's explore `data_x` (dataframe with independent variables) and `data_y` (series with target dependent-variable `y`), which will contain the relevant data we need to construct the Kaplan-Meier curves and perform further analysis. Refer to the following screenshot:

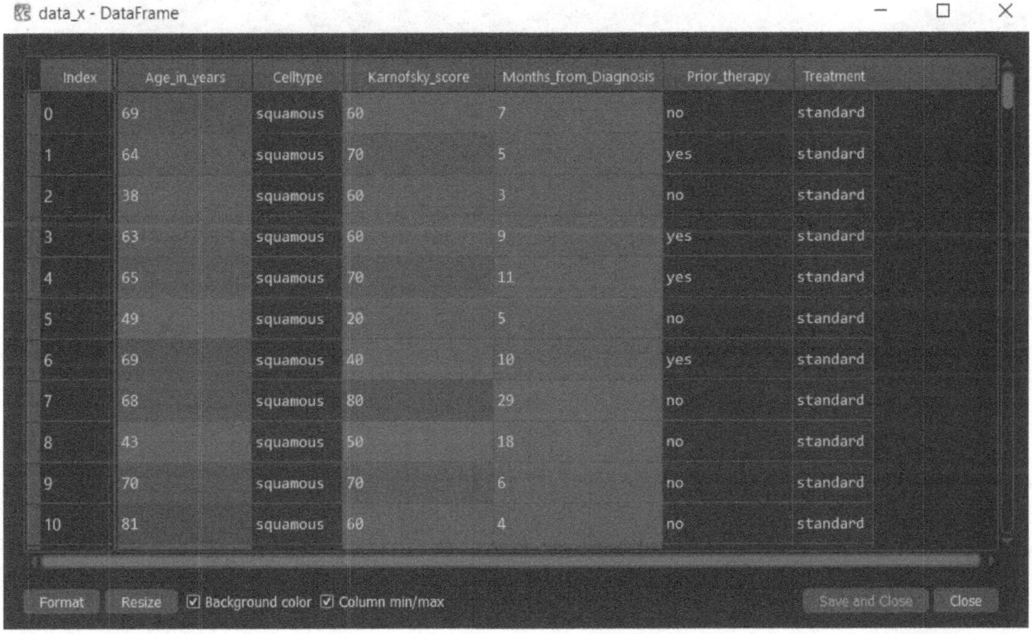

data_x - DataFrame

Index	Age_in_years	Celltype	Karnofsky_score	Months_from_Diagnosis	Prior_therapy	Treatment
0	69	squamous	60	7	no	standard
1	64	squamous	70	5	yes	standard
2	38	squamous	60	3	no	standard
3	63	squamous	60	9	yes	standard
4	65	squamous	70	11	yes	standard
5	49	squamous	20	5	no	standard
6	69	squamous	40	10	yes	standard
7	68	squamous	80	29	no	standard
8	43	squamous	50	18	no	standard
9	70	squamous	70	6	no	standard
10	81	squamous	60	4	no	standard

Figure 11.4 – Exploring the dataframe

As you can see, the `data_x` object we loaded contains descriptive characteristics of the subjects such as age, cell type associated with lung cancer, Karnofsky score, months from diagnosis, prior therapy, and the very important characteristic that is type of treatment. In this study, two types of treatment were used and those are standard of care, noted as **standard**. and experimental, noted as **test**.

3. Now let's explore the `data_y` object as shown in the following screenshot. The *0* column actually contains the `True` and `False` statements, but these are actually determinants of the survival events. `True` means survived and `False` means did not survive. However, we also need the time points for these events. Exploring the **Status** tab will enable this.

Figure 11.5 – Exploring the y variable

The status bar is located below the values in the lower-left corner of the variable explorer. You may change it to the **Survival_in_days** option:

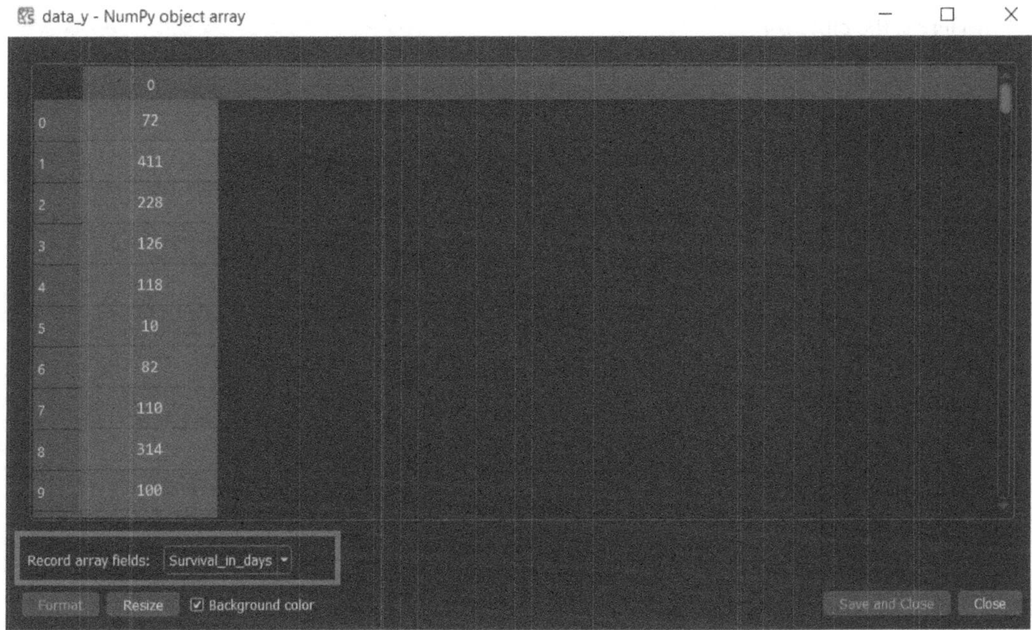

Figure 11.6 – Exploring the time variable

Now you can see the time points for survival.

To summarize, we have now explored `data_x`, which contains descriptive data for the subjects, as well as their disease characteristics. The `data_y` object we loaded contains information about the time points and the events within those time points.

4. The next step is to plot the Kaplan-Meier curve. First, let's use the Kaplan-Meier estimator:

```
# Calculate the Kaplan-Meier survival estimates
time, survival_prob = kaplan_meier_estimator(
    data_y["Status"], data_y["Survival_in_days"])

# Plot the survival curves
plt.step(time, survival_prob, where="post")
plt.ylabel("est. probability of survival $\hat{S}(t)$")
plt.xlabel("time $t$")
plt.show()
```

Here is the plot for it:

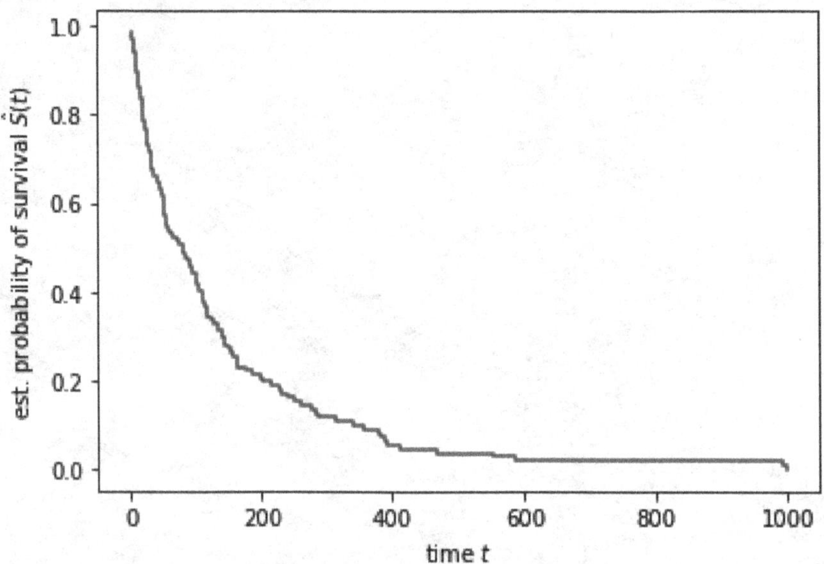

Figure 11.7 – The first Kaplan-Meier plot

The plot shows the time as **time t** and the estimated probability of survival as **S(t)**. As you can see, the survival probability decreases in small steps. In this plot, all the subjects included are the ones using the standard treatment and the ones using the test treatment. We will not differentiate between them for now:

```
# Create a boolean array for the treatment groups
treatment = data_x["Treatment"] == "test"

# Calculate the Kaplan-Meier survival estimates for the first
treatment group
time_treatment, survival_prob_treatment = kaplan_meier_
estimator(
    data_y["Status"][treatment], data_y["Survival_in_days"]
[treatment])

# Calculate the Kaplan-Meier survival estimates for the second
treatment group
time_control, survival_prob_control = kaplan_meier_estimator(
    data_y["Status"][~treatment],
    data_y["Survival_in_days"][~treatment])
```

Notice how the ~ symbol is used in the previous code block. In many regression and survival analyses, we use the ~ symbol to model the relationship between independent and dependent variables:

```
# Plot the survival curves for the first treatment group
plt.step(time_treatment, survival_prob_treatment,
    where="post", label="standard treatment")

# Plot the survival curves for the second treatment group
plt.step(time_control, survival_prob_control,
    where="post", label="test drug")

plt.ylabel("est. probability of survival $\hat{S}(t)$")
plt.xlabel("time $t$")
plt.legend(loc="best")
plt.show()
```

Here is the plot for it:

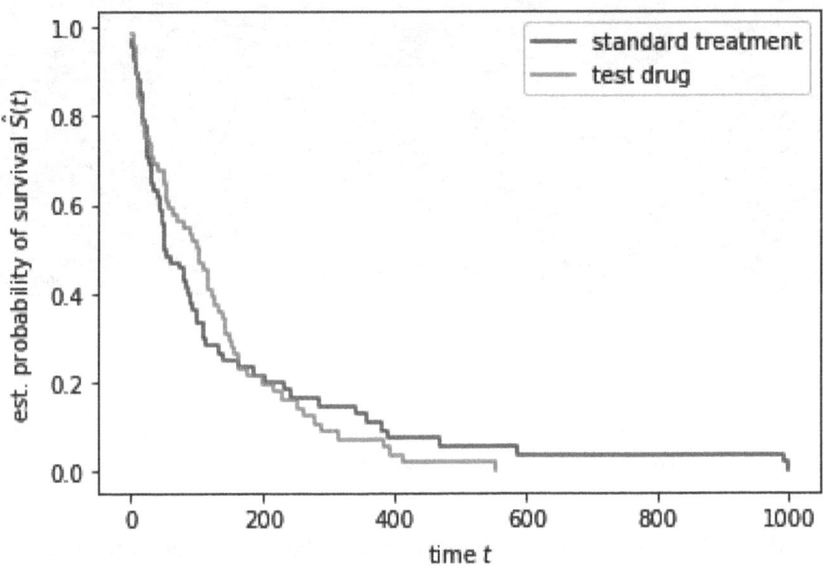

Figure 11.8 – Standard treatment versus test drug Kaplan-Meier curve

Now we can see two curves: one for standard treatment, and one for the test drug. We can see that the curves are very similar by evaluating them visually. The survival probability for both is decreasing in a similar way and is very low, less than 0.1 on a scale from 0 to 1 after 400 days. So, the survival is low for both treatments and neither seem to be very effective.

Now, let's use the logrank test to test for the statistical significance of the phenomena we explored visually so far.

To perform the logrank test, we first need to define the treatment groups and use the `unique` function to perform the logrank test:

```
# Set the group indicator and perform the logrank test
group_indicator = data_x.loc[:, 'Treatment']
groups = group_indicator.unique()
chi2, pvalue= compare_survival(data_y, group_indicator)
chi2
pvalue
```

Here is the output:

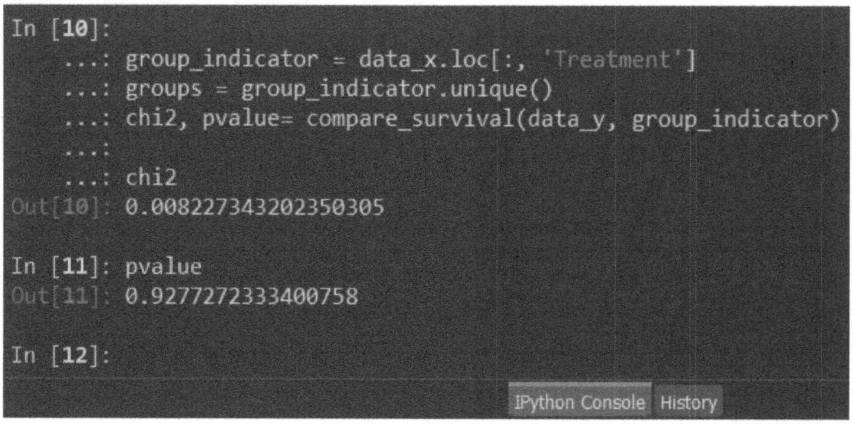

```
In [10]:
    ...: group_indicator = data_x.loc[:, 'Treatment']
    ...: groups = group_indicator.unique()
    ...: chi2, pvalue= compare_survival(data_y, group_indicator)
    ...:
    ...: chi2
Out[10]: 0.008227343202350305

In [11]: pvalue
Out[11]: 0.9277272333400758

In [12]:
                                        IPython Console  History
```

Figure 11.9 – Logrank test output

The `chi2` value is the chi-squared value. The logrank test utilizes the characteristics of the non-parametric chi-squared value to derive the final `pvalue`. `pvalue` is `0.92`, much higher than the threshold of significance (`0.05`).

This is the confirmation of the visual aspects we saw before. There is no significant difference between the experimental drug and standard of care treatment for oncological subjects in the whole sample.

Now, we will use the breast cancer dataset from `sksurv` to compare subgroups of breast cancer subjects based on their tumor size.

First, let's load the data:

```
import matplotlib.pyplot as plt
import numpy as np
from sksurv.datasets import load_breast_cancer
```

```
from sksurv.nonparametric import kaplan_meier_estimator
# Load the breast cancer dataset
data_x, data_y = load_breast_cancer()
```

Here is the output for it:

data_x - DataFrame

Index	21344_at	X221634_at	221816_s_	221882_s_	X221916_at	X221928_at	age	er	grade	size
0	31308	7.8421	10.1326	10.9264	6.47775	5.99189	57	negative	poorly differentiated	3
1	12677	8.78033	10.2135	9.55509	4.96805	7.05113	57	positive	poorly differentiated	3
2	93602	7.85565	10.1645	9.30805	4.28378	6.82899	48	negative	poorly differentiated	2.5
3	7873	6.67838	10.6601	10.2082	5.7134	6.92725	42	positive	poorly differentiated	1.8
4	31229	7.35856	11.5703	10.9318	5.81727	6.65545	46	positive	intermediate	3
5	7949	7.42558	10.3911	8.52119	6.06704	7.38362	58	positive	intermediate	2
6	87718	8.65571	10.4857	9.12129	6.11589	4.53965	44	negative	poorly differentiated	2
7	89667	7.77929	10.857	8.12652	7.17802	8.48244	58	positive	well differentiated	2.5
8	50944	7.86945	11.5844	8.08564	6.18076	6.65666	47	positive	poorly differentiated	3
9	31437	7.36249	10.7831	8.98696	3.14465	7.00675	38	positive	intermediate	2.5
10	1048	6.70043	11.1837	8.59565	2.58012	6.31818	59	positive	poorly differentiated	2
11	34401	7.89692	11.2944	8.45861	7.04926	5.25835	50	positive	well differentiated	0.8
12	73383	7.78631	9.79311	9.51593	6.82037	6.86561	46	negative	poorly differentiated	4.5

Format Resize ☑ Background color ☑ Column min/max Save and Close Close

Figure 11.10 – Comparison of the subgroups of the breast cancer dataset

As you can see, there is a column with the **X221928_at** marking. These are the gene expressions, which we don't need for this analysis. So just scroll to the right and you will find the column size, adjacent to the **age**, **er**, and **grade** columns.

Now let's compare the subgroups based on the tumor size where 2 cm will be the threshold.

Here is the code:

```
# Print column names
print("Feature columns:", data_x.columns)
print("Target columns:", data_y.dtype.names)
event_col = 'e.tdm'
time_col = 't.tdm'
# Split the data based on the 'size' column
```

```
group_a = data_y[data_x["size"] >= 2]
group_b = data_y[data_x["size"] < 2]

# Calculate the survival for each group
time_a, survival_prob_a = kaplan_meier_estimator(
    group_a[event_col], group_a[time_col])
time_b, survival_prob_b = kaplan_meier_estimator(
    group_b[event_col], group_b[time_col])

# Plot the Kaplan-Meier curve
plt.step(time_a, survival_prob_a, where="post", label="Size >= 2")
plt.step(time_b, survival_prob_b, where="post", label="Size < 2")
plt.ylabel("est. probability of survival $\hat{S}(t)$")
plt.xlabel("time $t$")
plt.legend(loc="best")
plt.show()
```

Here is the plot for it:

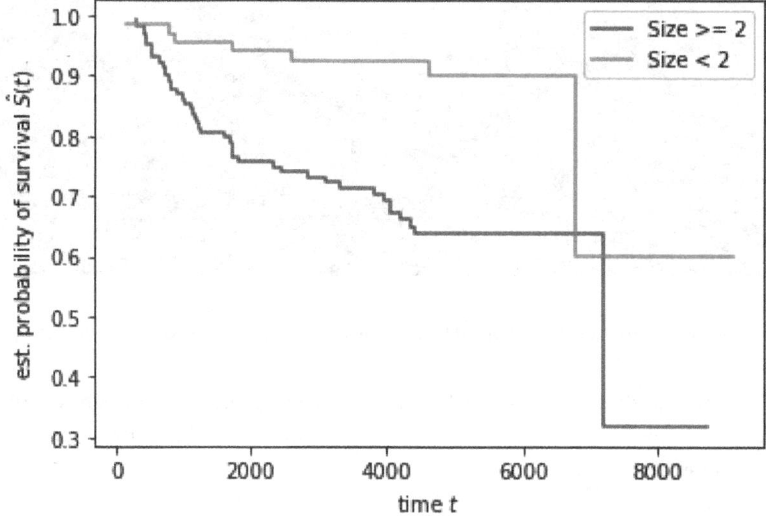

Figure 11.11 – Kaplan-Meier curve showing comparison of survival based on tumor size categories

It is intuitive to visually conclude that the subgroup with a tumor size smaller than 2 cm has better survival probabilities across time. You can notice how this group's probability is descending much more slowly compared to the subgroup with a tumor size greater than or equal to 2 cm.

We can now perform the logrank test to confirm this conclusion using a non-parametric hypothesis test the same way as for the veteran lung cancer dataset. Refer to the following code block:

```
# Perform the log-rank test
group_indicator = (data_x['size'] >= 2).astype(int)
chi2, pval = compare_survival(data_y, group_indicator)
chi2
pval
```

This is what we get:

```
In [21]: chi2, pval = compare_survival(data_y, group_indicator)

In [22]: chi2
Out[22]: 12.548757936664838

In [23]: pval
Out[23]: 0.00039646974061676056

In [24]:
```

Figure 11.12 – Logrank test for tumor size versus survival comparison

We can see that the result is statistically significant, but our interpretation focus should be placed more on the visual aspects of the magnitude of difference as seen on the Kaplan-Meier curve. Now we can move to the next step in survival analysis, which is implementing the Cox proportional hazards model.

Implementing Cox (proportional gazards) regression in Python

Now we will use another package to perform a specific method called Cox proportional hazards regression. First, let's discuss the method itself.

Cox proportional hazard regression is a statistical method used in survival analysis to evaluate the relationship between the time until an event of interest occurs and predictor variables. This method was developed by Sir David Cox in 1972.

What is very good about this type of regression is that it can handle censoring of the data and allow us to create multivariate models. This is very useful when we want to add other covariates in the survival analysis, or adjust for other covariates such as age or type of tumors in oncology.

The Cox model is based on the hazard function, which describes the instantaneous rate of failure at a given time. The hazard is assumed to be proportional across different levels of the predictor variables. One of the key assumptions of the Cox model is that the **hazard ratio** (**HR**) is constant over time, meaning that the ratio of the hazard rates for any two individuals is constant over time. The model can handle censored data, where the event of interest has not occurred by the end of the study. Censoring occurs when the survival time for some subjects is not fully observed.

The output of Cox regression includes HRs, which quantify the relative risk of experiencing the event for different levels of a predictor variable. A HR greater than 1 indicates an increased risk, while a HR less than 1 indicates a decreased risk.

One of the strengths of Cox regression is that it does not require assumptions about the shape of the baseline hazard function.

Cox regression is widely used in medical research, epidemiology, and other fields where the time until an event is of interest.

Before running the code, we will need to install the package we haven't installed so far, `lifelines`, by using `pip install lifelines`.

Now let's see how to implement the proportional hazards regression in Python:

```
import matplotlib.pyplot as plt
import numpy as np
import pandas as pd
from lifelines import CoxPHFitter
from sksurv.datasets import load_breast_cancer
from sksurv.nonparametric import kaplan_meier_estimator
from sksurv.compare import compare_survival

# Load the breast cancer dataset
data_x, data_y = load_breast_cancer()

# Print column names
print("Feature columns:", data_x.columns)
print("Target columns:", data_y.dtype.names)

# Replace 'cens' and 'time' with your column names
event_col = 'e.tdm'  # replace with your column name
time_col = 't.tdm'  # replace with your column name

# Convert 'er' to categorical variables
data_x['er'] = pd.Categorical(data_x['er'])

# Create dummy variables
```

```
data_x = pd.get_dummies(data_x, drop_first=True)

# Add a new covariate based on the 'size' column
data_x['size_group'] = (data_x['size'] >= 2).astype(int)

# Select only 'age', 'er_positive', and 'size_group' columns
data_x = data_x[['age', 'er_positive', 'size_group']]

# Prepare the data for CoxPHFitter
df = data_x.copy()
df[time_col] = data_y[time_col]
df[event_col] = data_y[event_col]

# Fit the CoxPHFitter model
cph = CoxPHFitter()
cph.fit(df, duration_col=time_col, event_col=event_col)

# Print the summary
cph.print_summary()
```

This is what we get as the output:

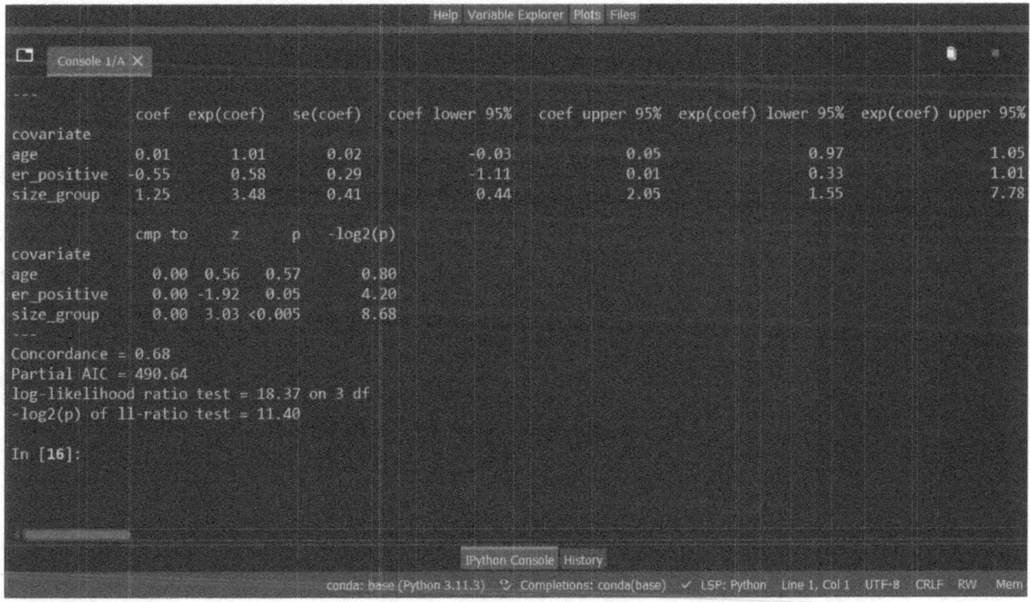

Help
Variable Explorer
Plots
Files
Console 1/A
---
coef exp(coef) se(coef) coef lower 95% coef upper 95% exp(coef) lower 95% exp(coef) upper 95%
covariate
age 0.01 1.01 0.02 -0.03 0.05 0.97 1.05
er_positive -0.55 0.58 0.29 -1.11 0.01 0.33 1.01
size_group 1.25 3.48 0.41 0.44 2.05 1.55 7.78
cmp to z p -log2(p)
covariate
age 0.00 0.56 0.57 0.80
er_positive 0.00 -1.92 0.05 4.20
size_group 0.00 3.03 <0.005 8.68

Concordance = 0.68
Partial AIC = 490.64
log-likelihood ratio test = 18.37 on 3 df
-log2(p) of ll-ratio test = 11.40
In [16]:
IPython Console
History
conda: base (Python 3.11.3)
Completions: conda(base)
LSP: Python
Line 1, Col 1
UTF-8
CRLF
RW
Mem

Figure 11.13 – Cox regression output

This data contains the most relevant number in relation to the interpretation of results for this breast cancer project. The coefficients (**coef**) are actually the slopes on the regression line and are best interpreted when converted to HRs. Their conversion to HRs is very simple. We just need to exponentiate the coefficients and their confidence interval. The HR for the variable is under the column name **exp(coef)** and it is **1.01** for **age, 0.58** for **er_positive** tumors, and **3.48** for **size_group**. However, how do we interpret these? Well, we do this by considering **1** as the reference for comparisons. The value of 1 actually means no difference. The value of 2 means twice the increased hazard or risk. A value of 0.5 means half as much hazard or risk and so on.

Let's interpret the results of the preceding output using this principle. The HR for age is 1.01, which is just a 0.01 increase compared to 1. This is a 1% increase in risk, which is a very small magnitude of difference and a non-significant result of **p=0.57**. The HR for er_positive is 0.58, which would be an almost double decrease in risk, and it's on the threshold of statistical significance if we use the confidence interval for interpretation instead of p-values. When some result is on the threshold of significance level, we cannot be sure whether it's significant but it's an interesting result and should be investigated further, especially as the magnitude of difference is large in this case.

Now, let's look at the most interesting result. Since our `size_group` variable is coded as ≥ 2 cm as 1 and <2 cm as 0, what we see as results is associated with the >=2cm group. The HR for >=2cm in tumor size is 3.48, an increase of more than three times in the magnitude of risk increase. Also, this result is statistically significant and quite credible at $p<0.05$.

Finally, we can interpret the lower and upper confidence intervals. These are under the **exp(coef) lower** and **exp(coef)upper** columns. The principle is very simple. If the whole interval is above 1, we can consider it as a significant increase in risk and if it's below 1 in its whole range, we can consider it as a significant decrease in risk or hazard level. We can also use the confidence intervals to evaluate uncertainty around the HRs. It should be kept in mind that it is within the 95% confidence interval that we can expect that 95% of HRs measured would be if we repeated the analysis many times.

OK, now let's plot the results using the Kaplan-Meier curve, implementing this with the `lifelines` package:

```
from lifelines import KaplanMeierFitter
# Create a KaplanMeierFitter object
kmf = KaplanMeierFitter()

# Fit the KaplanMeierFitter model and plot the survival function with
# confidence intervals
for name, grouped_df in df.groupby('size_group'):
    kmf.fit(grouped_df[time_col], grouped_df[event_col],
        label='Size group ' + str(name))
    kmf.plot(ci_show=True)
```

```
plt.ylabel("est. probability of survival $\hat{S}(t)$")
plt.xlabel("time $t$")
plt.show()
```

Here is the plot for it:

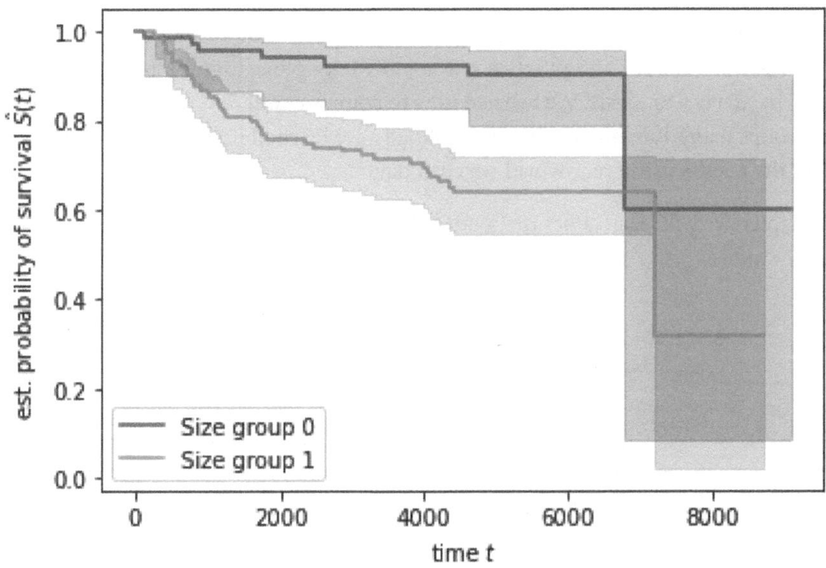

Figure 11.14 – Kaplan Meier curves of different size groups

You can now see that the groups are coded as 0 and 1. The 0 group is the one with a <2cm tumor size and group 1 is the group with a tumor size of >=2 cm. Again, it is clearly visible that the group with <2cm has better survival probabilities over time points and that its decline is also slower. In addition to the line curves, we can also see the confidence intervals shaded around the Kaplan-Meier curve lines. These are the 95% confidence intervals and we can see that for a significant part of the curves, there is a gap between their confidence intervals and this is another indication of a statistically significant difference.

In this section, we learned how to implement the Cox proportional hazards model in Python, as well as visualize and interpret the results. Now let's summarize what we learned in the chapter.

Summary

In this chapter, we learned about the basics of survival analysis, including where it is used and how we can improve drug evaluation. We also learned about its use in biomedical research through the use of time series data. Also, we learned how to implement survival analysis using two Python packages, `scikit-survival` and `lifelines`. We evaluated different exemplar data in the oncology area, the veteran lung cancer and breast cancer datasets, from the `scikit-survival` package. Finally, we covered the theory and practice of Kaplan-Meier plots, logrank tests, and Cox regression typically used in survival analysis. We learned how to compare different treatments and survival rates of different groups using Kaplan-Meier curves. Finally, we learned how to implement these methods and interpret the results using real-world survival data.

In the next chapter, we will learn about meta-analysis. We will also learn how to synthesize the evidence from multiple studies.

12

Meta-Analysis – Synthesizing Evidence from Multiple Studies

Most biostatistical analyses are conducted in different studies and focus on a specific sample drawn from a population. The larger the sample, the greater the credibility and statistical power of the studies. This is the basic principle of biostatistics.

There is another way to improve the credibility of evidence and statistical power. **x** is a statistical technique used in biostatistics to combine the results of multiple independent studies on a specific topic to derive a better more credible and trustworthy result. Further meta-analyses are components of systematic literature reviews that summarize results and assess the quality of studies on specific biological or biomedical topics.

In this chapter, we're going to cover the following main topics:

- Understanding meta-analysis and synthesizing evidence from multiple studies
- Understanding random effects meta-analysis and fixed effects meta-analysis
- Exploring and learning meta-regression and which packages to use for its implementation in Python
- Learning how to interpret meta-analysis

The main learning outcome of this chapter is that you will understand the principles of meta-analysis. You will learn how to rate evidence quality and how to summarize evidence. Also, you will learn about meta-regression as a specific form of meta-analysis. Finally, you will learn how to interpret the different plots typically used in meta-analysis and prepare for the example project that will be implemented in *Chapter 13*.

Understanding meta-analysis and synthesizing evidence from multiple studies

When individual studies are performed, they usually involve collecting data from a small portion of the population and an inference is made on that portion or sample. Every individual study has a certain amount of evidence with a level of credibility and uncertainty around that evidence, depending on the size of the study.

When combined, multiple studies will have more credibility and a stronger evidence base than individual studies. For this reason, meta-analysis can be used to combine studies into a more credible synthesis of evidence. The term *meta* means *high level(above)*, so meta-analysis is a type of analysis where multiple individual studies are combined to create a construct of a new synthesis analysis which is 'above' the individual studies due to more evidence.

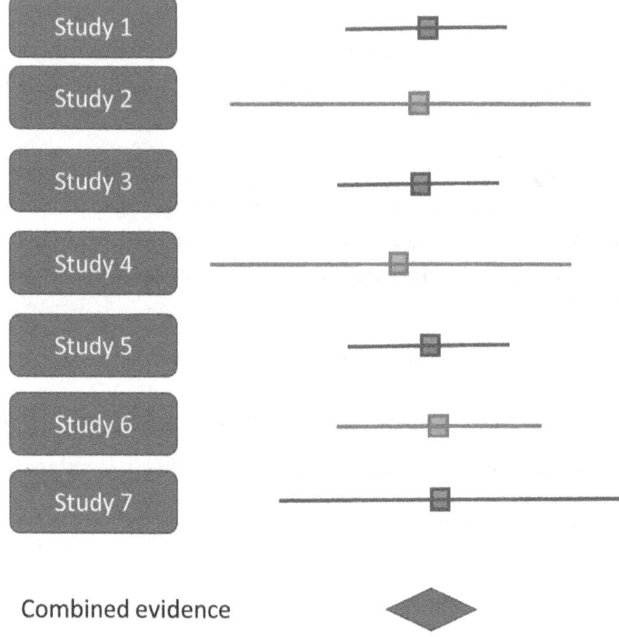

Figure 12.1 – Meta-analysis example

Meta-analysis is a collection of statistical and other scientific methods used to combine individual studies; however, it is also much more. We use meta-analysis as a part of performing systematic reviews of literature in specific areas of life science. We can assess the quality of studies, heterogeneity, how they relate to each other, and summarize the knowledge of whole scientific topics of interest.

For example, we can use meta-analysis to summarize how specific types of chemotherapy work for oncology patients across many clinical trials and get the most credible evidence possible for this topic. For this reason, meta-analysis is considered to be at the top of the pyramid of evidence in biostatistics and clinical research:

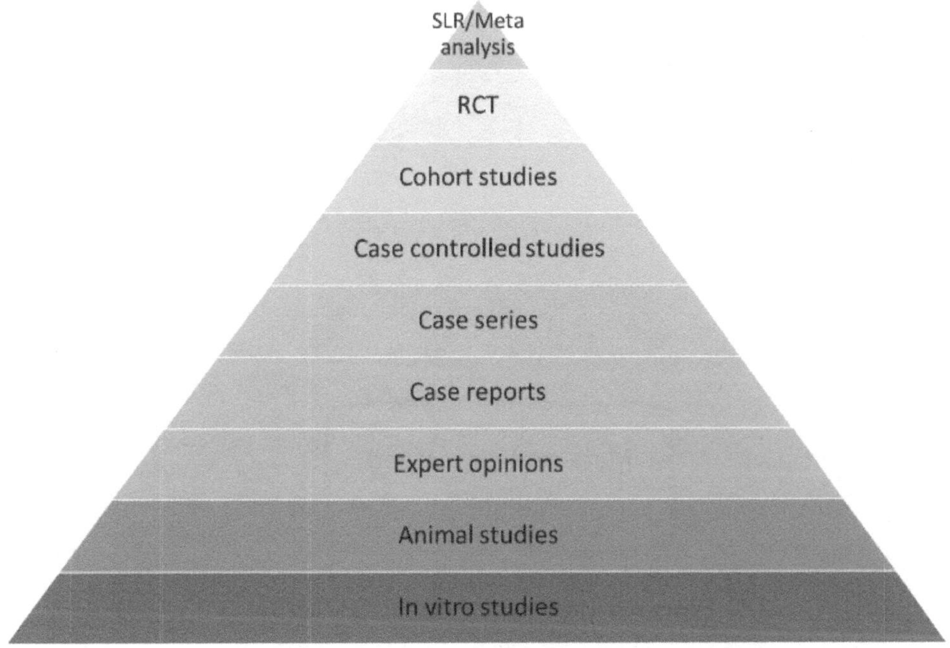

Figure 12.2 – Evidence pyramid

We can see that **Systematic Literature Review** (**SLR**) and **meta-analysis** are at the top of the pyramid of evidence. Most of the evidence is contained in them, and then the pyramid descends towards the next most credible category, which is **Randomized Controlled Trials** (**RCTs**). We can see other forms of evidence in the lower parts of the pyramid, such as cohort studies, case-controlled studies, case series, reports, and expert opinion, and the pyramid ends with animal studies and preclinical in vitro research.

The main question is "how can we compare multiple studies that generally differ in sample size, results, and even the conclusion drawn from them?"

The basic principle of meta-analysis is to assign weights to studies based on their credibility and level of evidence. If the study has a high level of evidence, we can assign it a larger weight and vice versa.

In the next chapter, we will use Python to implement this weighting and construct meta-analyses in a real-world example project. But before that, learning about the theoretical and mathematical segments of meta-analysis is vital.

Meta-analysis method structure

Meta-analysis methods tend to be complex. To simplify them, we can separate each meta-analysis method into two important segments: the effects segment, which deals with different effects that may occur in the data, and the estimator segment, which deals with the overall effect pooling.

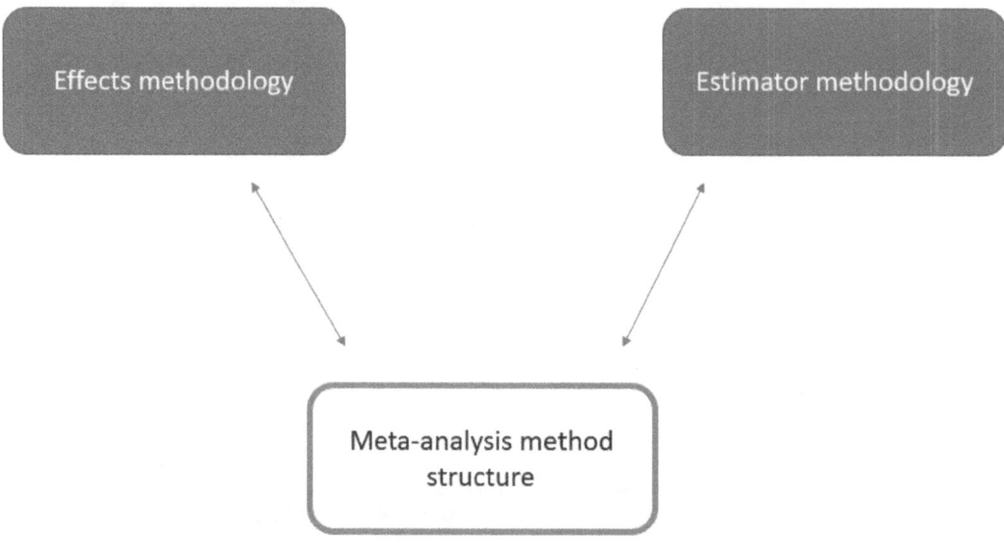

Figure 12.3 – Method structure for meta-analysis

Why is this important? Because different types of data, different types of studies, and the different effects we identify within them require the use of different meta-analysis methods. To simplify the process of learning when to apply different types of meta-analysis methods, let's learn about the fixed and random effects models.

Understanding random effects meta-analysis and fixed effects in meta-analysis

Meta-analysis in its basic form is meant to evaluate a singular summarized parameter from two or more studies and construct an overall majority effect estimate. But in practice, different individual studies, different experiment or study designs, different methodologies, and different populations are all different factors that can influence the final outcomes of the studies. This means that the majority of meta-analyses are actually composed of a number of different effects instead of one singular fixed effect. A solution to this problem is a meta-analysis method called random effects meta-analysis. This type of evidence synthesis allows biostatisticians to adjust for the potential random effects originating from the sources mentioned.

This is achieved statistically by considering every individual study as a random variable with potential different effects and summarizing them into one effect as a product of multiple potentially different effects.

Refer to the following figure:

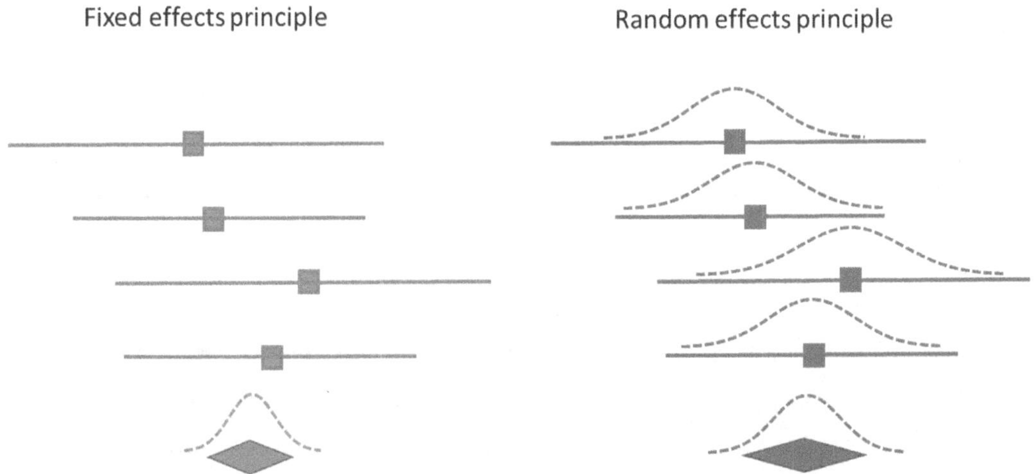

Figure 12.4 – Fixed effects versus random effects in meta-analysis models

Random effects meta-analyses are typically more conservative than fixed effects meta-analyses, and meta-analysis with the random effects model will have a larger confidence interval than the same fixed effects version.

You can see this in the preceding figure by observing the diamond on the right side of the image and the distribution around the model based on the random effects principle, which is larger than the diamond on the left and its distribution, which is based on the fixed effects principle.

Many meta-analysis practitioners often include both random effects and fixed effects analysis into the meta-analysis and use them to estimate the overall effects in both scenarios.

Let's now take a look at meta-analysis estimators.

Meta-analysis estimators

There are many different ways to estimate the overall pooled effects in meta-analysis, and to do this, we use estimators. Depending on the data and meta-analysis design, different estimators may be used, and they have different characteristics. Let's explore some of the most widely used estimators in meta-analysis.

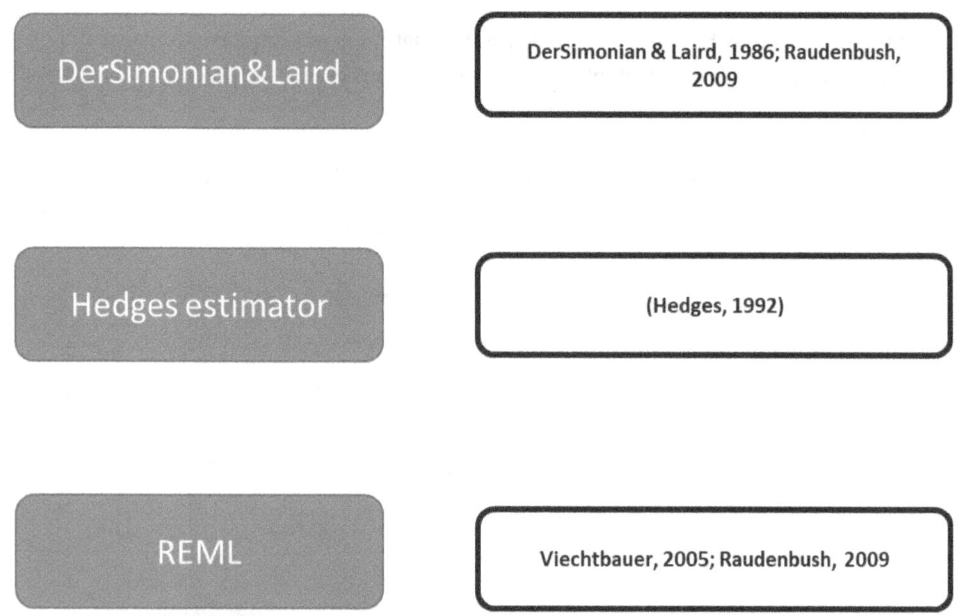

Figure 12.5 – Different meta-analysis estimators

The first in the list is DerSimonian&Laird, one of the most widely used implementations. This method is based on inverse variance, meaning that a larger variability in the effect data will assign less weight and a smaller variability will assign more weight to any individual study in a meta-analysis. For this reason, for the DerSimonian&Laird method to work, the assumption of reverse proportionality between variability and the study effect weights must hold.

This assumption indeed holds for the majority of clinical research frameworks and, for this reason, DerSimonian&Laird is one of the most frequently used estimators in biostatistics.

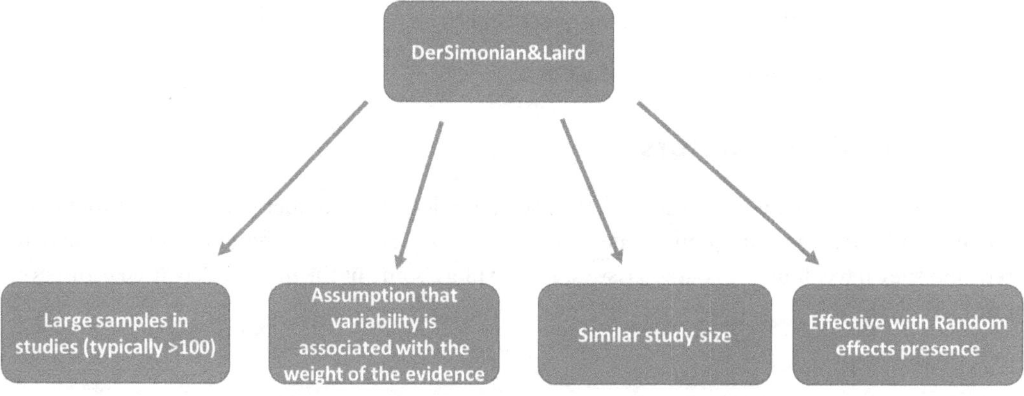

Figure 12.6 – DerSimonian&Laird estimator characteristics

The DerSimonian&Laird estimator is a standard in most clinical research meta-analysis, especially if the sample size is above 100, so most phase II and phase III trials can be run using this estimator. Also, if the study size is not very different between individual studies, the DerSimonian&Laird estimator will also be very flexible and will work well. Finally, the DerSimonian&Laird estimator is one of the best implementations for random effects models and is one of the most widely used estimators in the field of meta-analysis.

To read more about the DerSimonian&Laird estimator, you may use this publication as a reference: DerSimonian R and Laird N. Meta-analysis in clinical trials. Controlled Clinical Trials 1986, which may be accessed via `https://pubmed.ncbi.nlm.nih.gov/3802833/`

Now, let's discuss Hedges' estimator.

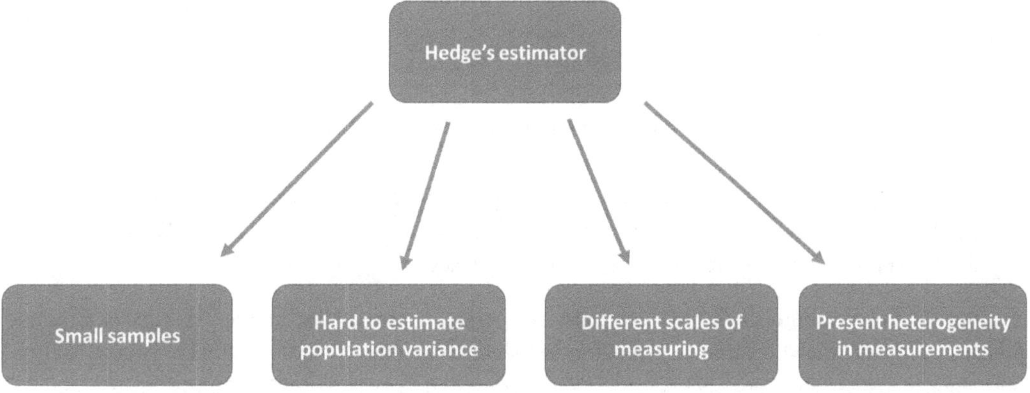

Figure 12.7 – Hedge's estimator characteristics

Hedges' estimator is well suited for smaller sample studies in which the population variance is difficult to estimate. Further, if the measurements are on different scales, this method is one of the best options and it will also work moderately well in the presence of heterogeneity in measurements and effects.

You can read more about this method here: Hedges, L. V., & Olkin, I. (1985). Statistical methods for meta-analysis. San Diego, CA: Academic Press. `https://www.researchgate.net/publication/216811655_Statistical_Methods_in_Meta-Analysis`

The **restricted maximum likelihood** (**REML**) method is more flexible than most other methods. Let's explore the situations in which can we use the REML estimator for meta-analysis:

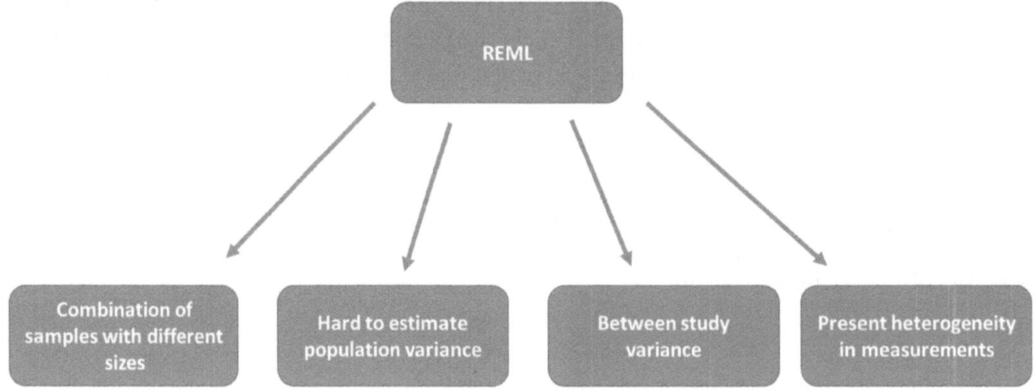

Figure 12.8 – REML estimator characteristics

REML is an implementation in meta-analysis that is well suited for situations with samples of different sizes, different variances, and presence of between-study variance. As such, it is also another very popular implementation for random effects meta-analyses.

You can read more about the REML estimator in this publication: Viechtbauer, W. (2005). Bias and Efficiency of Meta-Analytic Variance Estimators in the Random-Effects Model. Journal of Educational and Behavioral Statistics, 30(3), 261-293 `https://doi.org/10.3102/10769986030003261`, and you may access it here: `https://journals.sagepub.com/doi/10.3102/10769986030003261`.

Now, let's explore the package that is optimal for Python meta-analysis.

Exploring and learning meta-regression and which packages to use for its implementation in Python

In meta-analysis we are not interested just in comparing one overall effect (size or outcome), but also seeing how different covariates affect that overall effect. Why is this important? Because, in many biological settings, the overall effect is summarized and there could be biological covariates affecting the way the overall effect is behaving. For example, if a vaccine treatment is the main goal of the analysis, the subjects receiving the vaccine could be compared against those not receiving it. But, among biology researchers and clinicians, it is known that age affects the effectiveness of the vaccine. So, the meta-analysis would need to evaluate how age affects the vaccine too in a process called **meta-regression**.

Meta-regression is an extension of linear regression into the meta-analysis framework and typically involves comparing the association between the main outcome effect and a certain covariate, such as age in the vaccine example. Meta-regression typically involves creating a meta-regression plot, where the *x* axis would be the covariate of interest (e.g., age) and the *y* axis would be the treatment effect of interest (e.g., the vaccine effect in the form of a hazard ratio).

One approach that is frequently used in meta-regression is to show individual studies' trustworthiness or weight in the overall meta-regression by presenting each study with a bubble or interval. Frequently, the term *bubble plot* may be used to describe how meta-regression plots look, and they may have a combination of bubbles for individual studies and a trend line that explains the association between *x* and *y*.

Here is what a meta-regression plot looks like:

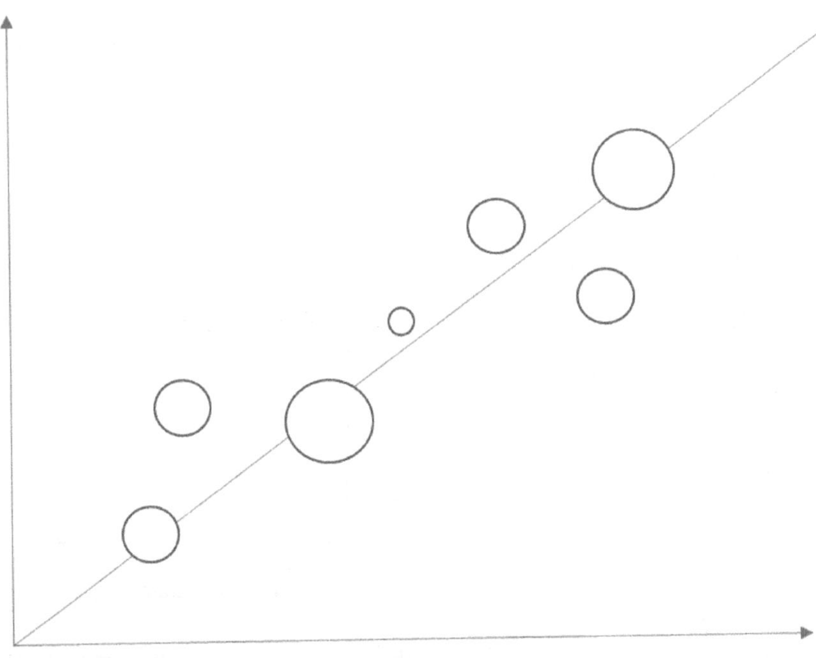

Figure 12.9 – A meta-regression plot

Two Python packages are specifically designed for high-quality meta-regression, and we will be using the, in *Chapter 13*:

- Package 1: `PythonMeta`
- Package 2: `PyMARE`

Now, let's explore how to interpret the results that are typically seen in meta-analysis. We will need this for the implementation.

Learning how to interpret meta-analysis

Meta-analysis is interpreted using multiple types of data and the associated plots. By far the most important part of the meta-analysis workflow is interpreting forest plots.

Interpreting forest plots

Have a look at the following figure:

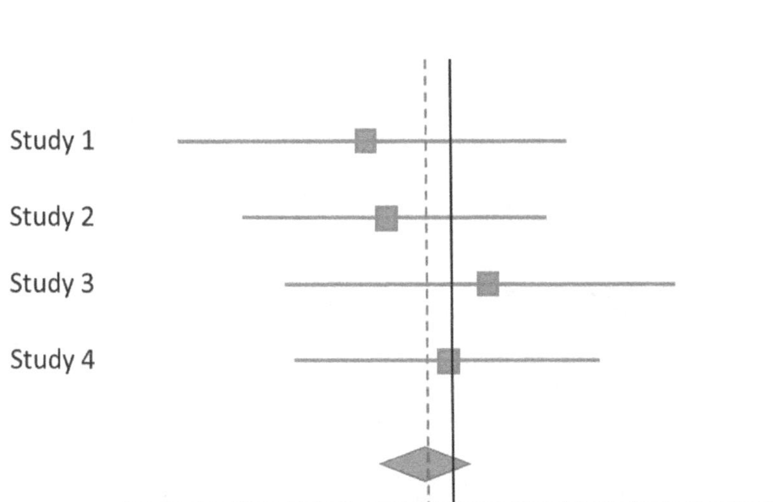

Figure 12.10 – Typical representation of a forest plot in a meta-analysis

As you can see, the forest plot looks like a tree, and that is where its name came from. The tree consists of individual studies and the overall effect. When interpreting forest plots, it is never a good idea to jump to the final overall effect conclusions. Start by looking at the individual studies and see how they relate to each other. Are they similar in the point estimates, central points, and confidence intervals?

Then, look at the heterogeneity, which is a number below the plot telling us how heterogeneous the results of the studies are. This can be seen by observing the boxes in reference to the dotted line.

Forest plots typically contain different information of a clinical nature, such as compounds used in the analysis or the clinical subgroups of subjects. Only after taking all these into account should you make a conclusion based on the overall effect. This effect is presented typically via the diamond below the plot, but may also be a box with error bars.

Meta-analysis plots always have a reference point, which is considered the 0 point or the point where no effect is observed. For most meta-analyses, this point is either 0 (for differences) or 1 (for ratios). For mean difference and correlation meta-analyses, it would be 0. For estimands such as **odds ratio (OR)**, **risk ratio (RR)**, and **hazard ratio (HR)**, the reference is 1. This means that each value below 1 is a decrease in OR, RR, and HR, and each value above 1 is an increase in these metrics.

How to interpret publication bias analysis

Publication bias is one of the most important segments of any meta-analysis and SLR. There are two main ways we can assess publication bias statistically.

The first way is to construct a plot called a **funnel** plot. This type of plot has a triangular shape, and all the studies are presented as points in reference to the triangular shape. Refer to the following figure:

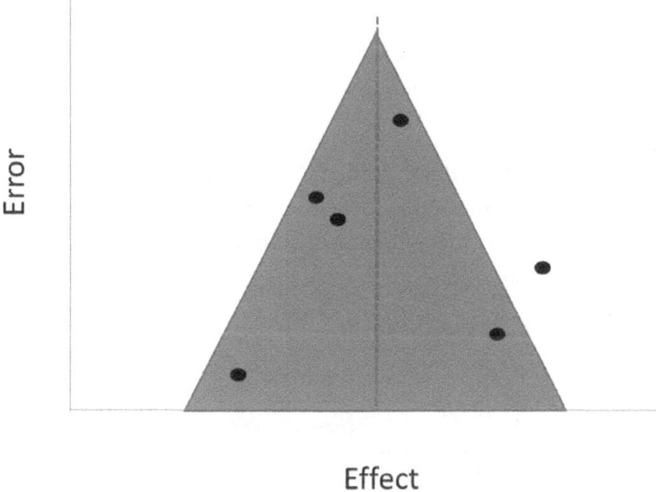

Figure 12.11 – Typical representation of a funnel plot in a meta-analysis

Funnel plots represent effect metrics such as HR or RR on the x axis and error terms on the y axis. The triangle is formed around the effect point estimate, and the triangle represents the 95% confidence interval area. If all the studies are within the triangle, we can conclude statistically that there is no detectable publication bias. If there are studies outside the confidence interval, than it's a good idea to perform Egger's test, which will show numerically if publication bias testing will provide significant or non-significant results. Egger's test is usually combined with a funnel plot, even if there aren't any studies outside the range, to get a more meaningful set of results in a meta-analysis.

How to interpret the sensitivity analysis

Sensitivity analysis is often conducted to see how each study affects the results. Why is this important? We can validate the meta-analysis design by making sure that no individual study is affecting the results too much and that the overall effect is indeed a combined result of multiple studies and not a single study with a large weight.

If the results are stable when omitting any individual study, then we can say it is validated in a process called **leave-one-out analysis**, where overall effects are computed for every single omitted study individually.

Assessing the quality of the studies

All studies in a meta-analysis should be interpreted individually, along with an interpretation of the overall effect. For each study, a meta-analysis practitioner should evaluate the credibility and quality of the study. Making assessments about the sample size, statistical power, inclusion criteria, subject characteristics, methodology, and potential limitations of individual studies should be an essential part of the meta-analysis interpretation.

Meta-analysis is not meant just to evaluate pooled overall effects from multiple studies, but also to summarize the quality of evidence originating from studies related to a specific biological or biomedical topic. This is one of the reasons why meta-analysis is an integral part of SLRs, which are at the top of the evidence pyramid in research today.

Here is an example of a study quality assessment:

Study quality assessment

	Random sequence generation	Allocation concealment	Selective reporting	Incomplete outcome data
Study 1	+	+	-	-
Study 2	+	-	-	?
Study 3	?	+	?	+
Study 4	+	?	+	+

Figure 12.12 – The publication bias graph

As you can see, there are four studies in this example, and each of them is assessed for:

- Random sequence generation
- Allocation concealment
- Selective reporting and
- Incomplete outcome data

Some other quality assessment areas may be covariate data completeness, methodology robustness, control group quality, and population coverage.

Making final conclusions in a meta-analysis

Making final conclusions is focused not just on the overall effect, but also on all the nuances related to the overall effect. As mentioned before, the final interpretation includes assessing individual study quality, the overall quality of studies in a certain field, and their advantages and limitations. Interpreting potential reasons for heterogeneity in the studies, their methodologies, and other differences are important aspects to consider when making the final conclusion in a meta-analysis.

Heterogeneity tells about potential differences in the designs of the studies, and also about potential differences in the populations studied. This brings us to making conclusions about the potential subgroups in a meta-analysis. Subgroup evaluation can be as important as the overall evolution of the study. While overall estimates will enable robust and evidence-rich analysis from the perspective of sample size and the number of studies, subgroup analysis will provide more specific information about the groups of subjects evaluated.

The following are the most important segments to interpret in a meta-analysis:

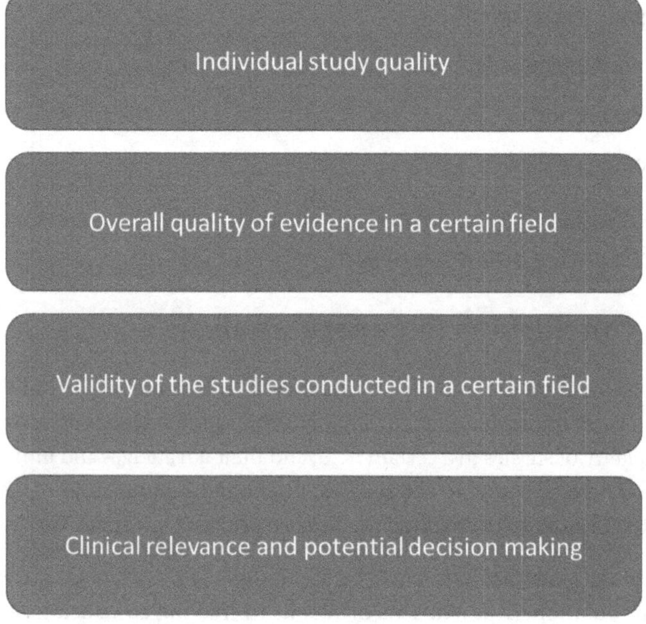

Figure 12.13 – Main aspects to interpret in any meta-analysis

Subgroup analysis is especially important for making final conclusions, because subgroup analysis can reveal whether certain parts of the population would have better or worse effects from different treatments/drugs.

Since meta-analysis is frequently part of the SLRs and is intended to summarize the current state of knowledge in a certain area, interpretation should be an objective assessment in terms of the quality of evidence presented. For this reason, it is important that conclusions originating from quality assessments of individual studies, and also from bias analysis, are accurate. Finally, since meta-analysis is at the top of the evidence pyramid, it is often used to make conclusions that will lead to future decisions in regulatory frameworks in different pharmaceutical and biomedical areas.

Summary

In this chapter, we learned the basic principles of meta-analysis and its role at the top of the evidence pyramid in both biostatistics and clinical research. We learned how to summarize data from multiple studies and how to evaluate the quality of individual studies. Finally, we learned how to interpret different meta-analysis plots and prepare for the example project, which will be the main topic of the next chapter.

In the next chapter, we will learn about the Python implementations of meta-analysis and interpret the results using a real-world dataset of oncology survival study examples.

13
Survival Predictive Analysis and Meta-Analysis Practice

Survival analysis is one of the main tools for assessing evidence in clinical research. Frequently, multiple estimates of effects from multiple survival studies are used to create a stronger evidence base by conducting a meta-analysis. Survival data has specific metrics that need to be complementary to each other to be combined in a meta-analysis. Also, the context surrounding the data needs to be evaluated carefully before deciding whether the data is appropriate for meta-analysis. The code for implementing the meta-analysis in Python is very specific, and we will be discussing it in this chapter.

In this chapter, we're going to cover the following main topics:

- Understanding survival and meta-analysis data
- Implementing the DerSimonian and Laird inverse variance method and investigating heterogeneity in meta-analysis
- Plotting the forest plots for meta-analysis
- Mastering meta-regression

Understanding survival and meta-analysis data

When performing meta-analysis, it is always important to start with the research question. For survival data, most research questions are associated with time as a central aspect. Here are three typical time-dependent research questions:

- What is the increase or decrease in a disease or event probability for subjects in a specific population or groups at given time points (for example, expressed as hazard ratio)?
- What is the median survival time for a specific population or groups in a given period (for example, 5 years, 10 years, or 15 years)?

- What is the rate at which a certain event associated with a disease occurs (for example, cancer-related mortality, the progression of a disease, or other complications) in a population or in specific treatment groups?

As you can see, some statistical metrics are mentioned in those research questions.

Survival metrics are key when it comes to relating research questions to research answers. These metrics are used to relate time to events, explaining at what rate the event of interest occurs in relation to time.

Let's start with basic survival metrics.

- **Overall survival**: **Overall survival** (**OS**) is the simplest form of survival metric. It is simply the time between a specific starting point and a mortality event (death). In clinical biostatistics, it is typically defined as the time between the initiation of treatment and a mortality event. At each time point, researchers can measure the survival and use such data for the prognosis of future subjects.

  It is calculated as *OS = % survived / n*. In this formula, *n* is the number of subjects.

- **Disease-free survival**: The **disease-free survival** (**DFS**) metric is another very important survival metric. In clinical biostatistics, the DFS metric tells us the amount of time between the initiation of the study and either disease recurrence or a mortality event. The DFS metric is slightly more complex than the OS metric because there are three time points involved: treatment starting, recurrence of the disease, and mortality event. The DFS metric is essential in clinical research as resolving a clinical research problem (usually a disease itself) usually means removing any signs of disease.

  It is calculated as *DFS = % disease free&death / n*.

- **Progression-free survival**: The **progression-free survival** (**PFS**) metric has a similar structure to disease-free survival, except for one detail: instead of focusing on disease recurrence, it focuses on disease progression. Disease progression, in this case, tells an experimenter about the dynamics of a disease, such as when a disease worsens.

  It is calculated as *PFS = % progression free&death / n*.

- **Recurrence-free survival**: The **recurrence-free survival** (**RFS**) metric has a similar estimate structure to the previous ones, but it focuses on the recurrence of a disease. In other words, it is the survival time until the disease occurs again or there is a mortality event.

  It is calculated as *RFS = % recurrence free&death / n*.

- **Median survival**: Survival time can also be expressed simply as a median survival metric. A logical question to ask is, why median survival rather than mean survival? Well, survival data can be unpredictable and is frequently non-linear. That is why survival is typically better expressed as a median.

Any of the previously mentioned metrics can be expressed also as median survival metric:

- Median OS (e.g., 18 months)
- Median DFS (e.g., 15 months)
- Median PFS (e.g., 12 months)
- Median RFS (e.g., 11 months)

Notice how none of the types of median survival can be larger than the OS metric. This is because the OS is dependent only on the mortality event, while others have additional criteria but also contain the mortality event.

These survival metrics can be combined to form a larger evidence base using the meta-analysis approach.

Meta-analysis and survival data

In theory, any of the survival metrics mentioned in the previous section could be used to perform a meta-analysis. However, there are two main approaches that an experimenter or a researcher can take in relation to meta-analytic thinking.

- **Summarizing the evidence of multiple studies without actually comparing groups**: This approach focuses on just summarizing the evidence and achieving a higher metric, whether OS, DFS, or PFS, in relation to survival data or any other non-survival metric.

  Let's look at a simple form of evidence synthesis.

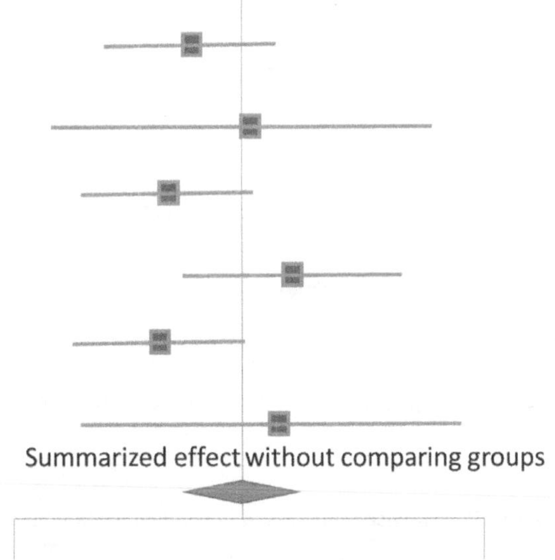

Figure 13.1 – Simple evidence synthesis using meta-analysis

In *Figure 13.1*, different studies' estimates of treatment effect are presented as boxes and confidence intervals as error bars (the lines to the left and right of the boxes). When multiple studies' estimates are combined, the product is the overall effect (the diamond at the bottom of the figure). The confidence interval tells about the uncertainty of the results. So, the larger the interval, the less confident we are in the result. Typically, the diamonds, which represent the overall effects, have smaller confidence intervals compared to the individual studies, which means there is more certainty in the overall evidence than in any of the individual studies. In this case, just summarizing the studies' effects has its own specific scientific purpose of increasing the evidence level for a specific topic by using meta-analysis.

- **Summarizing the evidence while comparing different groups, such as treatment and placebo groups**: In this approach, the focus is on comparing the treatment effects to a placebo group, or a group receiving no treatment. These groups are called control groups. Keep in mind that treatment can be compared against what we call the **standard of care** (**SC**). Having a reference group to compare against is very important to making comparative claims about the treatments studied. We call such studies controlled studies.

Survival meta-analyses are most frequently pairwise; that is, they compare two groups or two treatments, as shown here:

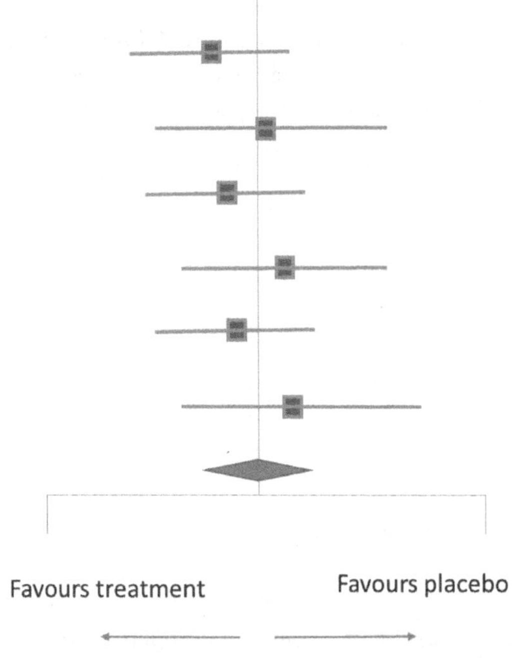

Figure 13.2 – Pairwise comparison in survival meta-analysis

In this example, you can see arrows showing which direction relative to the diamond favors treatment and which favors the placebo. The diamond is in the center, so the overall effect is neutral in this case. If it was to the left, it would favor the treatment, and if it was to the right, it would favor the placebo. This is the most common setting for a survival data meta-analysis, and the metric typically used for it is the **hazard ratio** (**HR**). Generally, a lower HR favors treatments and a larger HR favors controls. Note that controls may be placebo-based but they also may be based on the SC.

Now let's talk about HR as a metric in more detail. The HR is as follows:

HR = hazard rate treatment/hazard rate (controls),

where the hazard rate is the rate of the event relevant to the survival analysis. The HR itself is a metric that tells us about the difference in rates for different biological groups or treatment groups in life science research. It is frequently presented with a 95% confidence interval. The confidence interval is a variability metric that tells us about the statistical trustworthiness of the result we are observing and calculating. Without confidence intervals, HRs as point estimates would have much less scientific value.

Usually, a table like the following is needed to start inputting the data for any meta-analysis.

Study name	Metric (point estimate)	Metric (variability)	Sample information (eg. N subjects)	Clinical information (eg. clinicalsubgroup)

Figure 13.3 – Typical survival data template

As you can see, the table contains study names. When performing meta-analysis and creating forest plots for interpretation, we need to specify individual study names. They could be placed in a specific order, such as alphabetic order, or they could be ordered by year or whatever other criterion makes the most sense for the specific meta-analysis.

We need to add the metric information that will be used to perform the meta-analysis. While most other information from this table is descriptive in nature, the metric information is what goes into the meta-analysis calculation models.

It should be noted that variance, standard errors, and confidence intervals can all be converted to SDs for the purpose of calculating the standardized mean difference and performing meta-analysis.

Study name	Metric (point estimate)	Metric (variability)	Sample information (eg. N subjects)	Clinical information (eg. clinicalsubgroup)
Study 1	Hazard ratio 1	95% Confidence interval 1	N1	Specific clinical characteristics
Study 2	Hazard ratio 2	95% Confidence interval 1	N1	Specific clinical characteristics
Study 3	Hazard ratio 3	95% Confidence interval 1	N1	Specific clinical characteristics
Study 4	Hazard ratio 4	95% Confidence interval 1	N1	Specific clinical characteristics
Study 5	Hazard ratio 5	95% Confidence interval 1	N1	Specific clinical characteristics
Study 6	Hazard ratio 6	95% Confidence interval 1	N1	Specific clinical characteristics
Study 7	Hazard ratio 7	95% Confidence interval 1	N1	Specific clinical characteristics
Study 8	Hazard ratio 8	95% Confidence interval 1	N1	Specific clinical characteristics

Figure 13.4 – Example of survival data

As you can see, each cell in the table should contain inputs specific to a certain study. We have eight hazard ratios, in this case, and each of them is bound to a 95% confidence interval. Confidence intervals are not the only type of variability metric. Instead of these, standard deviations, variances, standard errors, and even t-values can be used. What is interesting in statistics is that all these can be converted to each other, so if we know any one of these for the HRs, we can convert all of them to another variability metric. Typically, meta-analysis models use either confidence intervals, variances, or standard errors in their methods to construct a meta-analysis model.

An HR-based meta-analysis structure would look like a forest plot. A number of variability metrics can be used to derive the confidence interval, so each of them can be used to represent data variability in survival analysis. These metrics include variance, standard deviation, standard error, and the t-value.

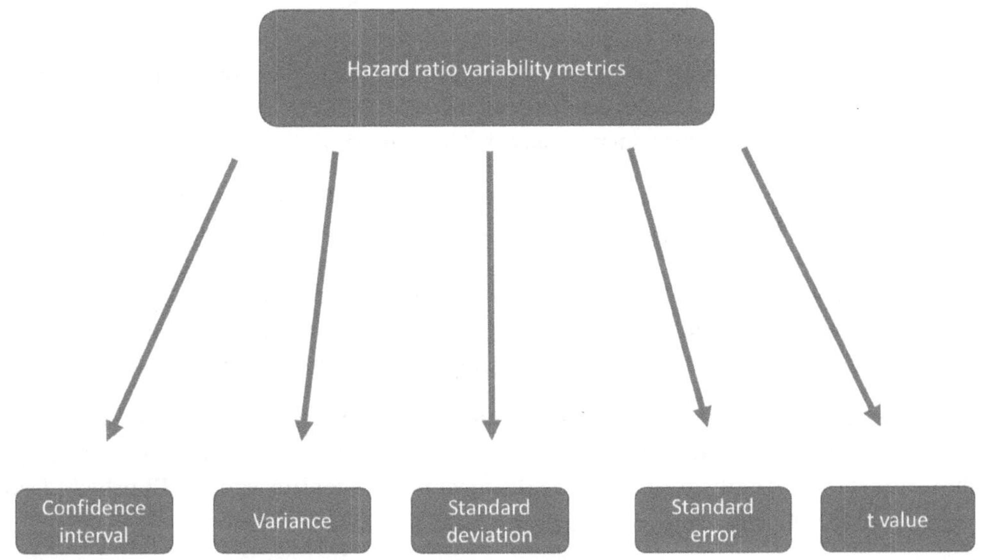

Figure 13.5 – HR variability metrics

These are the metrics typically found throughout the literature, and at least one of them is needed to perform meta-analysis using HRs. As mentioned previously, any one of these metrics can be converted to any other, and all can be converted to confidence intervals and used in a survival meta-analysis.

Now let's explore point estimates (HRs) and uncertainty estimates (confidence intervals) using an example forest plot.

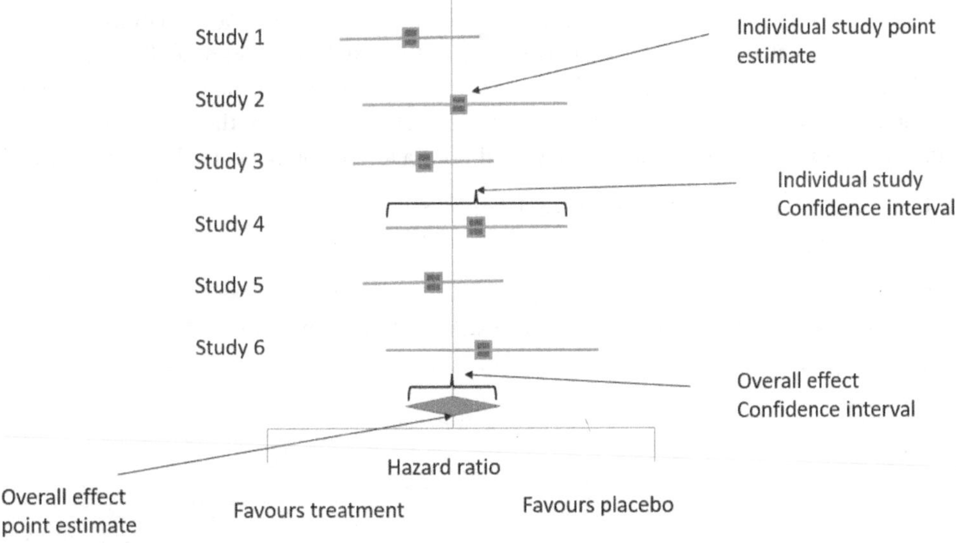

Figure 13.6 – Exploring point estimates and confidence intervals in a forest plot

Using *Figure 13.7*, you can observe specific details that are very important for any meta-analysis. The forest plot shows multiple aspects that are used to calculate and interpret any meta-analysis model. First, we can see individual study point estimates, in this case, HRs. For each individual HR, you can see the individual study confidence interval, specified by error bars either side of the HRs.

For the overall effect, the point estimate and its confidence interval can be found using the diamond, which is usually at the bottom of the forest plot but may be in other positions (in subgroup analysis). The center of the diamond is the point estimate (the HR), and the sides (edges) of the diamond represent the confidence interval of the overall effect. The confidence interval typically used in a meta-analysis is set to 95%. Note that this may change in specific studies. Some researchers may need a higher confidence level, such as 99%, and this is typically set in the software used in meta-analysis.

Finally, a vertical line is the reference line for HRs, and the interpretation of the forest plot is dependent on specific reference values. For HRs, the reference line is always set at 1. An HR of 1 means there is no effect of the treatment, an HR above 1 favors control, and an HR below 1 favors the experimental treatment.

In this section, we learned about HRs as survival metrics and how they relate to confidence intervals and other variability metrics. We also learned how to visualize them using forest plots.

Implementing the DerSimonian and Laird inverse variance method and investigating heterogeneity in meta-analysis

Let's look at why we would use the DerSimonian and Laird inverse variance method and how it relates to survival data meta-analysis. The DerSimonian and Laird method makes use of HRs and different variability metrics, such as variances and confidence intervals, to create a meta-analysis model based on inverse variances, that is, inverse variability metrics of the HRs. This works by adding weight to each study based on the inverse variances of the studies. In most meta-analysis approaches, if a study has less variability, it will be assigned more weight in the overall result. Finally, the overall HR and overall confidence interval will be calculated and presented using a forest plot that plots the individual studies.

We will be using the `PythonMeta` package. Load it as PMA:

> **Note**
>
> Make sure to run `pip install PythonMeta` before running the code.

```
import PythonMeta as PMA
```

First, we need to load three classes of functions for loading the data, creating the model, and creating the forest plot:

```
# Load classes
data = PMA.Data()
model = PMA.Meta()
figure = PMA.Fig()
```

Now, let's create a dictionary that contains the settings and set the data type as follows:

```
# Define settings
settings = {
    "datatype": "CONT",   # 'CATE' for binary data or 'CONT' for
continuous data
    "models": "Fixed",    # 'Fixed' or 'Random'
    "algorithm": "IV",    # based on datatype and effect size
    "effect": "MD"   # RR/OR/RD for binary data; SMD/MD for continuous
data
}
# Set data type
data.datatype = settings["datatype"]
```

In this case, we used `"models": "Fixed"`, which entails a fixed-effects model. This model is used when we can assume that all studies estimate the same underlying treatment effects and that observed differences are mostly due to sampling error and natural variability. Studies may come from different hospitals, different populations, different doctors, and different investigators; even different study methodologies may contribute to variability between treatment effects. A lot of different aspects may contribute to variability between effect sizes, but they are still related to the same outcome and type of treatment. In that case, it's best to use a random effects model, which allows for the difference observed earlier to be a part of the overall effect calculation.

Next, we can create a practice dummy dataset. We need to specify the study name, HR1, its variance and *n* subjects, HR2, its variance, and its number of subjects in the form of an array:

```
# Dummy data
samp_cont = [
    "Study1 (2005), 22.9,  6.0,   15,   27.4,  6.5,   24",
    "Study2 (2015),    7.8,   5.2,   51,   11.9,   5.3,   53",
    "Study3 (2009),    20.38, 5.26, 35,   24.32, 5.23, 35",
    "Study4 (2012),   6.67, 8.28, 43,   12.67, 9.87, 43",
    "Study5 (2001),     18.49, 7.16, 50,   20.72, 8.67, 60",
    "Study6 (2010),   11.8,   5.7,   40,   13.0,   5.2,   40",
    "Study7 (2016),   6.8,   4.7,   40,   8.0,   4.2,   45",
    "Study8 (2011),   12.9,   2.7,   40,   9.2,   5.2,   40"
]
```

Here is the output:

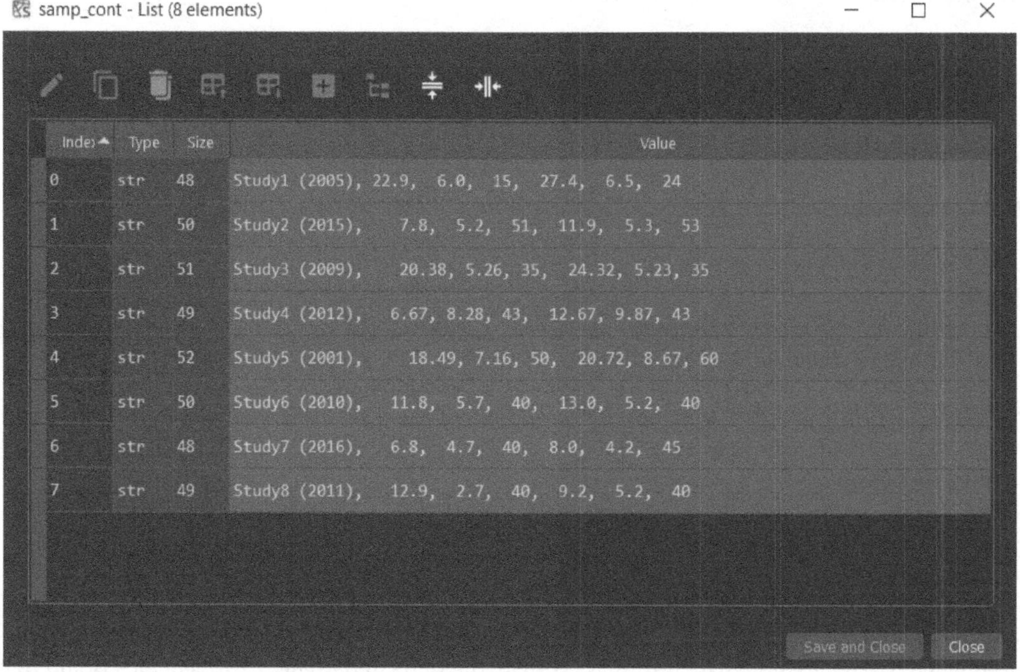

Index	Type	Size	Value
0	str	48	Study1 (2005), 22.9, 6.0, 15, 27.4, 6.5, 24
1	str	50	Study2 (2015), 7.8, 5.2, 51, 11.9, 5.3, 53
2	str	51	Study3 (2009), 20.38, 5.26, 35, 24.32, 5.23, 35
3	str	49	Study4 (2012), 6.67, 8.28, 43, 12.67, 9.87, 43
4	str	52	Study5 (2001), 18.49, 7.16, 50, 20.72, 8.67, 60
5	str	50	Study6 (2010), 11.8, 5.7, 40, 13.0, 5.2, 40
6	str	48	Study7 (2016), 6.8, 4.7, 40, 8.0, 4.2, 45
7	str	49	Study8 (2011), 12.9, 2.7, 40, 9.2, 5.2, 40

Figure 13.7 – Data sample for meta-analysis

The data has specific values as numbers but it is initially organized as strings; the `getdata()` function will be used next to extract the numerical values needed for the meta-analysis.

We cannot use the object; we need to apply the `getdata()` function to create a data object for further processing.

The first step is to load the data. For this, we can use the `getdata()` function:

```
# Load data
studies = data.getdata(samp_cont)
```

At this stage, we need to use model functions and specify the potential subgroups, datatype, model settings, algorithm (method), and effect. For now, we will not be adding the subgroup, but let's leave it in the code as we will need the same workflow later:

```
# Set the subgroup, data type, models, algorithm, and effect for meta-
analysis
model.subgroup = data.subgroup
model.datatype = data.datatype
model.models = settings["models"]
model.algorithm = settings["algorithm"]
model.effect = settings["effect"]
```

Perform the analysis but using the `model.meta()` function. This function will implement the DerSimonian and Laird inverse variance method:

```
# Perform the analysis
results = model.meta(studies)
```

This function is used to perform the meta-analysis. The `results` object will be created as output and used for the next step, which is the data visualization of the meta-analysis results.

In this section, we learned how to perform a simple meta-analysis implementation and create a meta-analysis `results` object. Next, we need to create the forest plots.

Plotting the forest plots and funnel plots for meta-analysis

The forest plot is the most important data visualization in a meta-analysis, as discussed in the previous chapter. So, when choosing which Python package to use for meta-analysis, one of the first aspects to observe (of course, in addition to the validity of the package) is its forest plot data visualization potential.

The coding part is very simple: we will use the `figure.forest()` and `figure.funnel()` functions to create the forest and funnel plots. Keep in mind that we previously defined `figure = PMA.Fig()` as the function to be called:

```
# Show forest plot (as seen in the Figure 13.9)
figure.forest(results).show()
# Show funnel plot
figure.funnel(results).show()
```

Here is the resulting forest plot for the overall evidence synthesis.

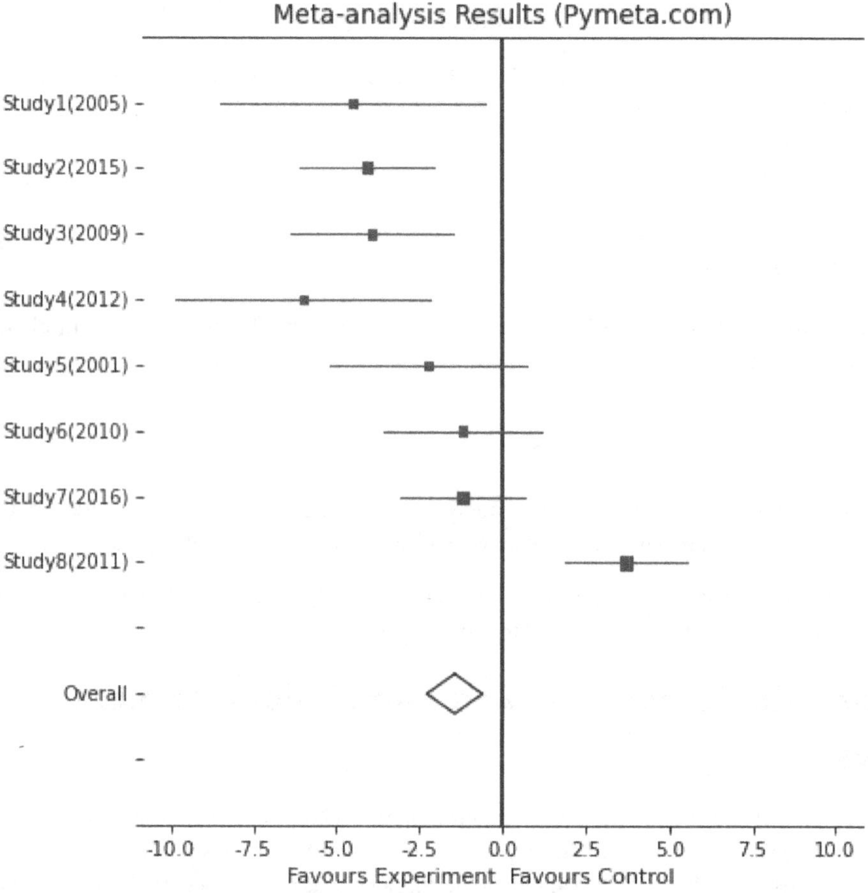

Figure 13.8 – The forest plot

By observing the forest plot, we can evaluate both the individual studies' difference and the overall effect, represented by the diamond.

As you can see in the following figure, the diamond is to the left of the reference line, which means it favors the experimental treatment rather than the control group. The left and right sides of the diamond represent the confidence interval, and we can see that the whole of the confidence interval is to the left of the reference line, indicating a statistically significant result.

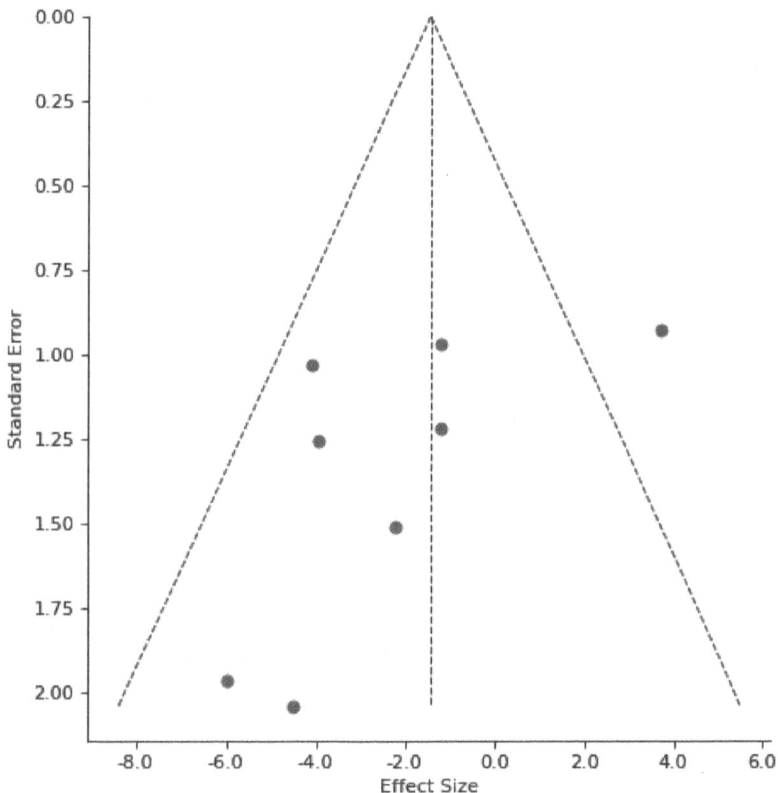

Figure 13.10 – The funnel plot

The second plot created is the funnel plot, and it's used to assess the publication bias. Using a triangle as seen in the preceding figure and the central reference line, a funnel plot can be used to evaluate the symmetry of the studies' effects relative to the reference line. Typically, a high level of symmetry is expected in the area where both negative and positive effects are published. If the studies lean toward each of the sides, there is a possibility of publication bias. We will talk more about the publication bias when evaluating real-world data in the next chapter.

Since we implemented the meta-analysis using studies, the next step is to perform a subgroup meta-analysis.

The subgroup analysis

Next, we need to assess whether there are any subgroups in the data. What are subgroups in studies, biologically speaking? Subgroups are specific categories of patients/subjects who share a specific common characteristic that differentiates them biologically and biostatistically from the rest of the population. Examples include subgroups that share a common mutation, the same blood type, or a specific gene expression.

For this example, we will be setting the subgroups based on hypothetical mutations. A mutation is a change in a genetic sequence that can alter gene function and lead to variations in how an organism responds to treatments. Let's name the hypothetical mutations `Mutation_A` and `Mutation_B` and explore in this practice example whether they affect the treatment outcome.

Let's take the same steps as before but with this slight subgroup-based modification of the code:

```
import PythonMeta as PMA

# Load classes
data = PMA.Data()
model = PMA.Meta()
figure = PMA.Fig()

# Define settings
settings = {
    "datatype": "CATE",    # 'CATE' for binary data or 'CONT' for
continuous data
    "models": "Random",    # 'Fixed' or 'Random'
    "algorithm": "IV",   # based on datatype and effect size
    "effect": "RR"   # RR/OR/RD for binary data; SMD/MD for continuous
data
}

# Set data type
data.datatype = settings["datatype"]

# Provided data with subgroup
```

Now let's define the subgroups as specified (one subgroup with `Mutation_A` and one subgroup with `Mutation_B`):

```
a samp_cate = [
    "Study1 (2005), 15, 30, 8, 30",
    "Study2 (2015), 20, 50, 10, 50",
    "Study3 (2009), 25, 70, 15, 70",
    "Study4 (2012), 10, 50, 20, 50",
    "<subgroup>name=Mutation_A",
    "Study5 (2001), 30, 110, 25, 110",
    "Study6 (2010), 15, 80, 80, 80",
    "Study7 (2016), 15, 85, 60, 90",
    "Study8 (2011), 20, 90, 30, 90",
    "<subgroup>name=Mutation_B"
]
```

Notice how I used `<subgroup>name=Mutation_A` to define the first subgroup and `<subgroup>name=Mutation_B` to define the second subgroup. In this dataset, we consider these two subgroups as biologically different; we want to compare them statistically using subgroup analysis.

Now, using the variable explorer, you can see what the dataset looks like and what types of data were just created:

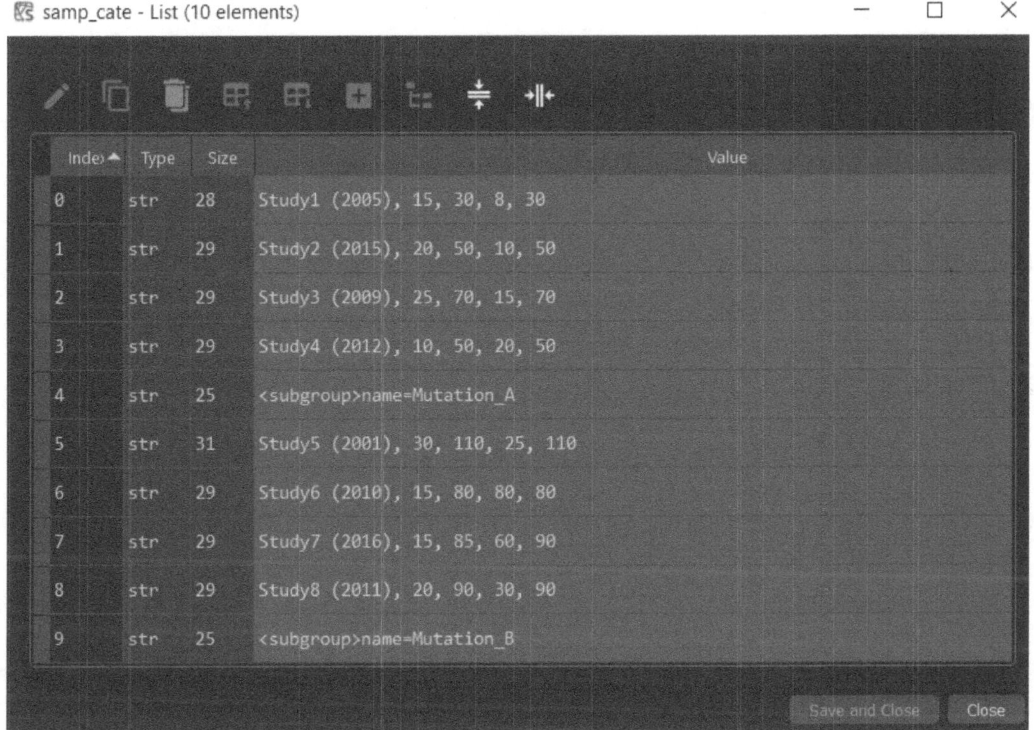

samp_cate - List (10 elements)
Index Type Size Value
0 str 28 Study1 (2005), 15, 30, 8, 30
1 str 29 Study2 (2015), 20, 50, 10, 50
2 str 29 Study3 (2009), 25, 70, 15, 70
3 str 29 Study4 (2012), 10, 50, 20, 50
4 str 25 <subgroup>name=Mutation_A
5 str 31 Study5 (2001), 30, 110, 25, 110
6 str 29 Study6 (2010), 15, 80, 80, 80
7 str 29 Study7 (2016), 15, 85, 60, 90
8 str 29 Study8 (2011), 20, 90, 30, 90
9 str 25 <subgroup>name=Mutation_B
Save and Close
Close

Figure 13.11 – Variable explorer (the data for subgroup analysis)

You can see again that the data has specific values as numbers, but it's essentially organized as strings; the `getdata()` function will be used next to extract the data.

The rest of the code is the same as in the first example; we need to plot the forest plot and the funnel plot in the end:

```
# Load data
studies = data.getdata(samp_cate)
```

```
# Set the subgroup, data type, models, algorithm, and effect for meta-
analysis
model.subgroup = data.subgroup
model.datatype = data.datatype
```

```
model.models = settings["models"]
model.algorithm = settings["algorithm"]
model.effect = settings["effect"]

# Perform the analysis
results = model.meta(studies)

# Show forest plot as in Figure 13.13
figure.forest(results).show()

# Show funnel plot
figure.funnel(results).show()
```

Here is the plot for it:

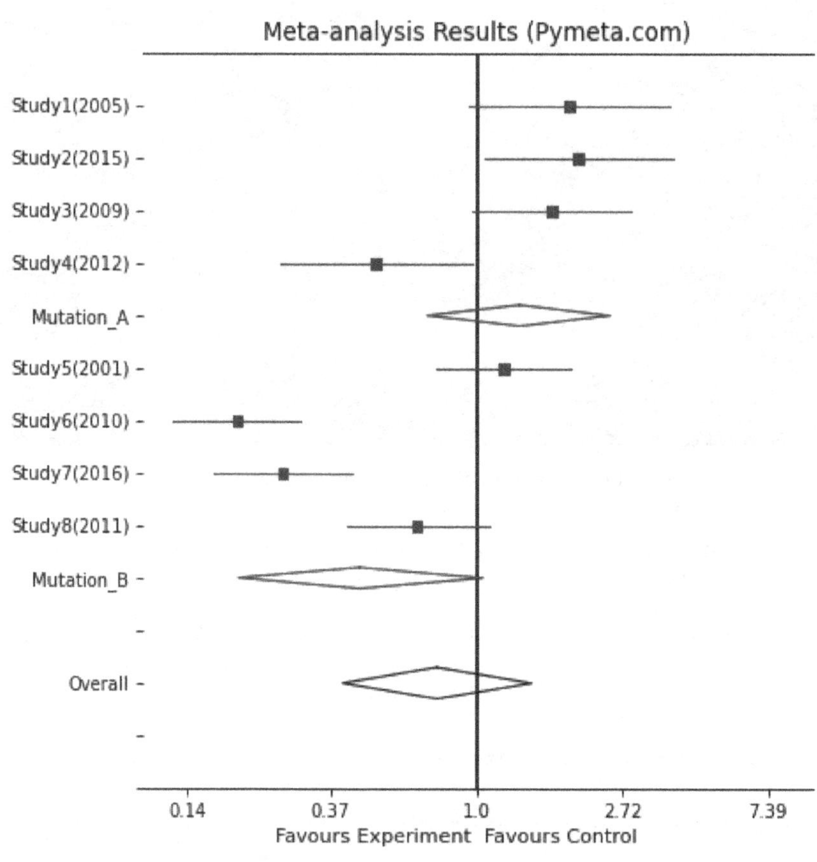

Figure 13.12 – Subgroup meta-analysis forest plot

The resulting forest plot not only shows individual studies' effects but also the effects for subgroups and the overall effect. We can see that there are three diamonds here. They represent the mutations and the overall effect. By comparing these diamonds, we can evaluate both the subgroup difference and the overall effect. For example, we can see that the overall effect is slightly below 1, which means the treatment effect is present. But the important aspect here is that the treatment effect is stronger in a subgroup with `Mutation_B`. We can see that the diamond for `Mutation_B` is actually much below 1 (the central edges of the diamond). The diamond for `Mutation_A` is above 1, which means that the treatment only worked for subjects with `Mutation_B`. (Note that the reference line is 1.)

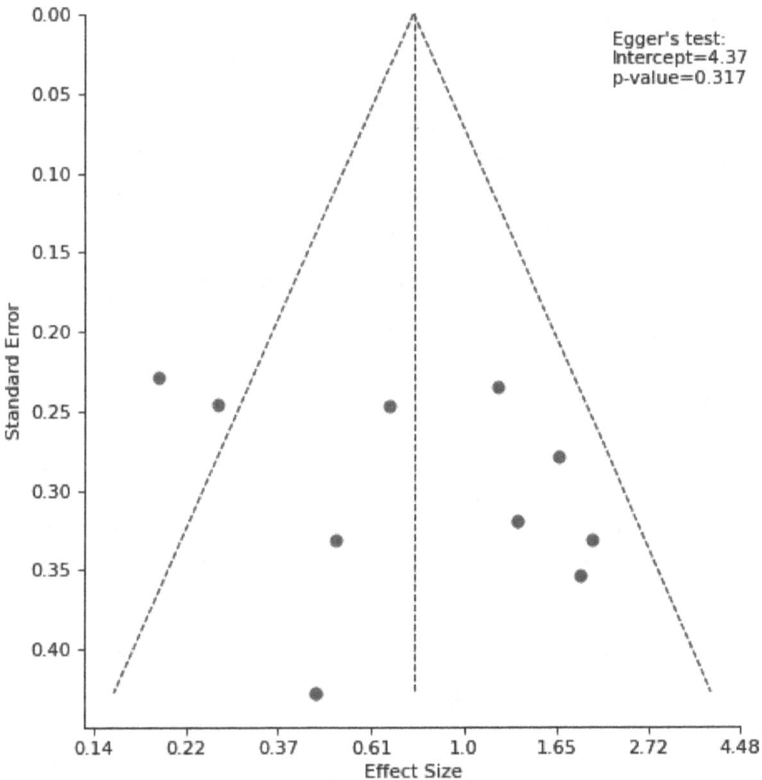

Figure 13.13 – The funnel plot for subgroup meta-analysis

This is the funnel plot as output. How do we interpret the funnel plot results? First, we look at the individual studies' symmetry. We look to see whether the studies (points) have a symmetrical distribution on the left and right of the reference line (central line).

So far, we have learned how to implement meta-analysis for overall evidence synthesis and perform the subgroup analysis. Now we can move to the next step, which is meta-regression.

Mastering meta-regression

In a meta-analysis, it is often important to assess how different variables affect the overall meta-analysis results. For example, sometimes we want to assess how age affects the effect of therapy or is a specific biomarker associated with the overall results that we see on the forest plot. Instead of just summarizing the results into the overall pooled results, we will now consider a specific variable called the moderator to see whether any specific moderators affect the results. In meta-regression, a moderator is a study-level variable that influences the overall effect variability. One such example is age. Age can influence how a treatment works across different studies, if the patients differ in age from study to study. When including moderators, we can explain the heterogeneity and sensitivity to different variables in the meta-analysis, identifying factors that may affect the overall outcomes. Meta-regression is a technique used for this purpose, and we will be implementing it using the `statsmodels` package. Finally, we can visualize the association between the moderator (or any other covariate) and the outcome using a specific plot type called a **bubble plot**, where bubbles will represent individual studies, the level of confidence will be reflected in the size of the bubbles, and the trendline will represent the association between the moderator variable and the outcome.

Before starting with meta-regression, let's load the required libraries:

```
# Import necessary libraries
import numpy as np
import pandas as pd
import statsmodels.api as sm
import matplotlib.pyplot as plt
```

Let's create another random example for a data construct where meta-regression can be implemented. We will use the `np.random()` function to create effects for 10 studies and their variances.

We will be using the `random.normal()` function from numpy. This function is based on a random number generator that generates data with a normal distribution (a bell-shaped distribution):

```
# Create a dummy dataset, random seed for reproducibility
np.random.seed(0)
n_studies = 10
effect_sizes = np.random.normal(
    loc=0.2, scale=0.1, size=n_studies)
variances = np.random.uniform(
    low=0.01, high=0.05, size=n_studies)
weights = 1 / variances
covariates = np.random.normal(loc=0, scale=1, size=n_studies)
```

We can use the `pd.Dataframe()` function to create a DataFrame, which is easier to explore. In essence, many meta-regressions are actually weighted regressions, and that is the reason we need them in the dataset:

```
# Create a DataFrame
df = pd.DataFrame({
    'effect_sizes': effect_sizes,
    'variances': variances,
    'weights': weights,
    'covariates': covariates
})
#Add an intercept. This is needed in modeling the regression
df['intercept'] = 1
```

Here is the output:

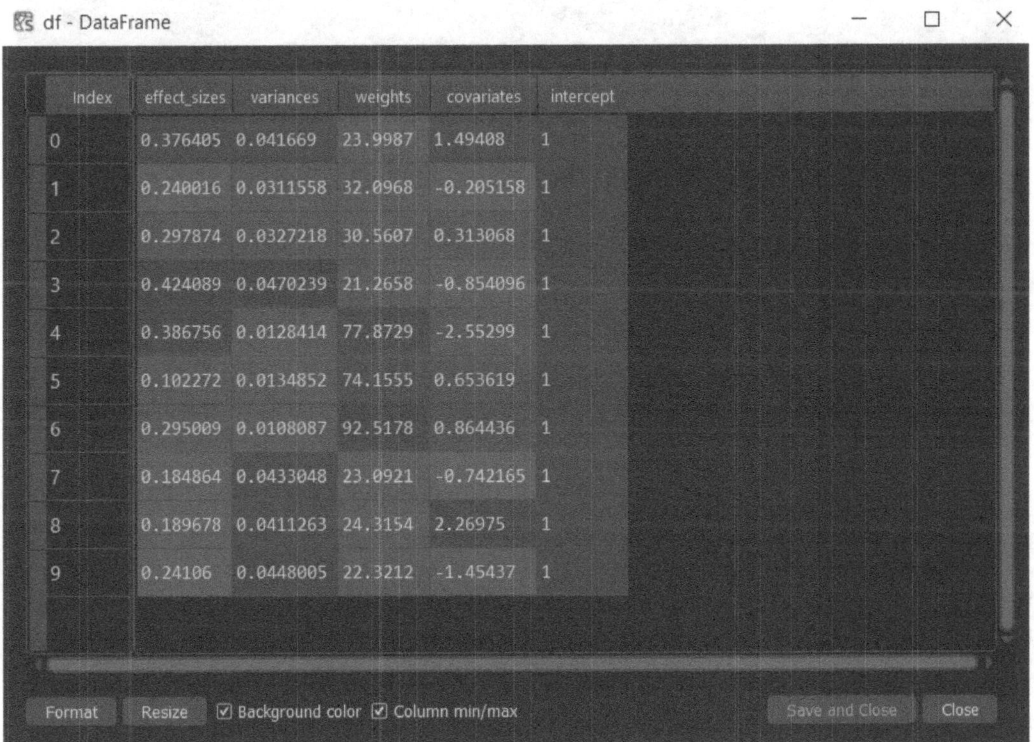

Index	effect_sizes	variances	weights	covariates	intercept
0	0.376405	0.041669	23.9987	1.49408	1
1	0.240016	0.0311558	32.0968	-0.205158	1
2	0.297874	0.0327218	30.5607	0.313068	1
3	0.424089	0.0470239	21.2658	-0.854096	1
4	0.386756	0.0128414	77.8729	-2.55299	1
5	0.102272	0.0134852	74.1555	0.653619	1
6	0.295009	0.0108087	92.5178	0.864436	1
7	0.184864	0.0433048	23.0921	-0.742165	1
8	0.189678	0.0411263	24.3154	2.26975	1
9	0.24106	0.0448005	22.3212	-1.45437	1

Figure 13.14 – Effect sizes DataFrame

The next step is to actually perform the weighted meta-regression. We will be using the `WLS()` function to implement the weighted least-squares regression and fit the model to the data:

```
# Perform weighted least squares regression
model = sm.WLS(df['effect_sizes'],
    df[['intercept', 'covariates']],
    weights=df['weights'])
results = model.fit()
```

Here is the output:

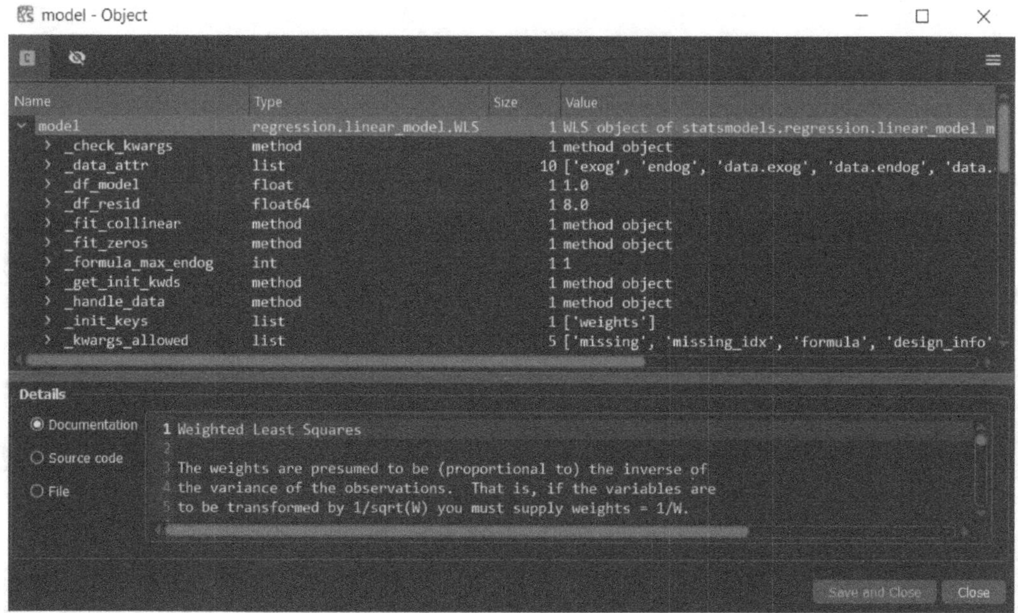

Figure 13.15 – Exploring the created model

By opening the model in the variable explorer, you can see various details about the created model as well as its documentation. When creating meta-regression models, always read the documentation to verify whether the right model was used for the data.

Let us print the results in the form of coefficients and their corresponding confidence intervals and statistical significance (the p-values):

```
# Print the results
print("Coefficients:")
print(results.params)

print("\nConfidence Intervals:")
```

```
print(results.conf_int())

print("\nP-values:")
print(results.pvalues)

print("\nStandard Errors:")
print(results.bse)
```

Here are the results:

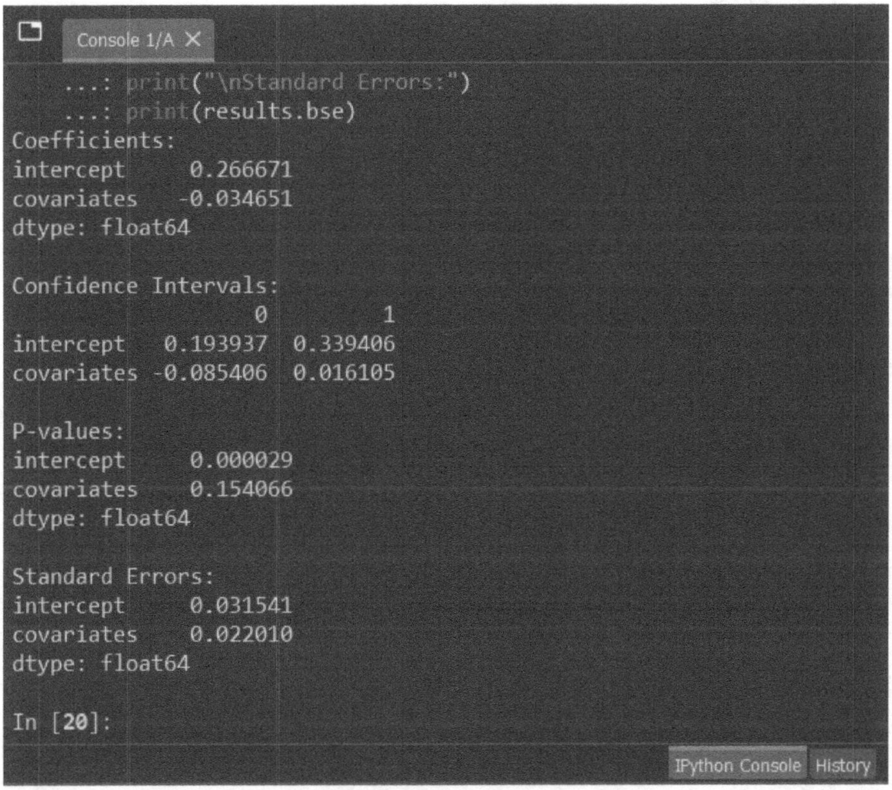

```
   ...: print("\nStandard Errors:")
   ...: print(results.bse)
Coefficients:
intercept     0.266671
covariates   -0.034651
dtype: float64

Confidence Intervals:
                   0         1
intercept   0.193937  0.339406
covariates -0.085406  0.016105

P-values:
intercept     0.000029
covariates    0.154066
dtype: float64

Standard Errors:
intercept     0.031541
covariates    0.022010
dtype: float64

In [20]:
```

Figure 13.16 – Meta-regression output

How do we interpret this output? We have the coefficient (under `covariates`) which is `-0.034`. It's negative, which means an inversely proportional association. We have the confidence intervals (the lower and upper bounds being named `0` and `1`, respectively), and we have the p-value. They all indicate a non-significant result (a p-value greater than 0.05). Even though the result is not statistically significant, sometimes the trends can be seen visually, and no interpretation should ever be based only on statistical significance.

Now, let us create a bubble plot that explains the meta-regression results:

```
# Create a bubble plot without confidence intervals
plt.figure(figsize=(10, 6))
plt.scatter(df['covariates'],
    df['effect_sizes'], s=100 * df['weights'],
    alpha=0.5, edgecolors='black')
plt.plot(df['covariates'], results.fittedvalues, 'r-')
plt.xlabel('Covariate')
plt.ylabel('Effect Size')
plt.title('Bubble Plot with Trendline')
plt.grid(True)
plt.show()
```

Here is the resulting plot for the code:

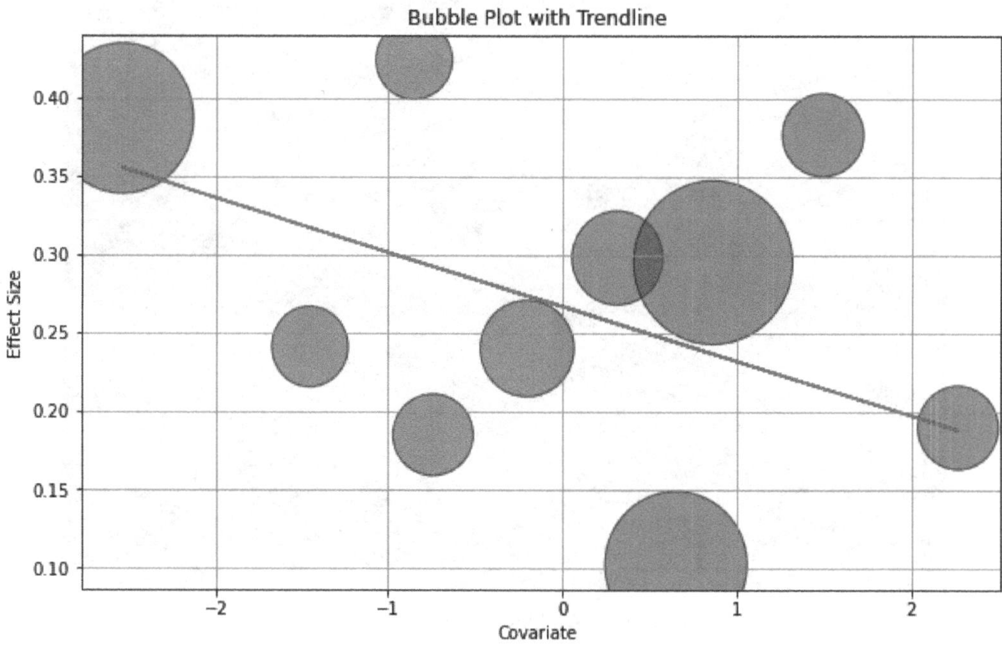

Figure 13.17 – Meta-regression data visualization using a bubble plot

You might be wondering how to interpret this plot. First, we have two axes, *x* and *y*, plotting **Covariate** and **Effect Size**, respectively. Essentially, this plot explains the association between a moderator (the covariate, in this case) and the effect size, which is the main effect outcome of the meta-analysis. The trendline shows the potential association, if present, and it can be used to explain how the main effect varies across different values of the covariate (e.g., age). This kind of plot is called a **bubble plot** because of the bubbles used to represent individual studies. These bubbles are used to describe the level of confidence we have in specific studies and are typically inversely proportional to confidence intervals or variances. In other words, if the uncertainty around individual studies is lower, the bubble will be larger, and vice versa.

In this specific plot, we can see that the covariate value decreases as the effect size increases and that some studies were more credible than others. There will be more on this in the next chapter, with a real-world example.

Summary

In this chapter, we learned how to perform meta-analysis using a randomly created practice dataset and the `PythonMeta` package.

We learned how to create meta-analysis forest plots and funnel plots and how to interpret them. We also learned how to perform and interpret subgroup analysis, which is very important in biomedical research. We learned how to evaluate individual study effects, pooled effects, and publication bias.

Furthermore, we used `statsmodels` to implement meta-regression as a weighted linear regression model.

Finally, we learned how to plot different types of plots for the interpretation of meta-analysis results such as forest plots, funnel plots, and meta-regression bubble plots.

In the next chapter, we will be implementing what we learned in this chapter using a real-world meta-analysis with real-world publications.

14
Part 3 Exemplar Project – Meta-Analysis of Survival Data in Clinical Research

In this chapter, we will be implementing a real-world exemplar project using data from real clinical trials to perform the Meta-analysis. We will be using **Randomized Controlled Trials** (**RCTs**) and survival data (**Progression-Free Survival** (**PFS**)), calculating risk ratios, constructing the forest plots and funnel plots, and interpreting the results. We will be discussing both biological and statistical aspects of the project. The main biological topic will be oncology and a comparison of specific treatments regarding the therapy effect on survival of non-small cell lung carcinoma (a type of lung cancer). Later, it will be shown how to implement the DerSimonian and Laird and Mentel-Haenzel estimator in a Meta-analysis and relate these to the biological research questions.

In this chapter, we're going to cover the following main topics:

- About the project and the dataset
- Implementing DerSimonian&Laird inverse variance method in Python
- Making the forest plots for oncology meta-analysis
- Making the funnel plots – publication bias analysis
- Implementing a Mentel-Haenzel estimator in a Meta-analysis

About the project and the dataset

This real-world exemplar project focuses on oncology research. Oncology research is one of the most important areas of life science, with a significant portion of clinical research being done in this field (around 50% of all clinical research). An even more important aspect is that mortality worldwide from oncological diseases is counted approximately at 10 million as reported by the **World Health Organization** (**WHO**); you can learn more at `https://www.who.int/news-room/fact-sheets/detail/cancer`.

Different treatment strategies have been developed for decades in the oncology area and the two most common types include chemotherapy and radiotherapy. In this project, we will focus on two types of chemotherapy: traditional and targeted therapies.

Traditional chemotherapy involves chemical agents that can interfere in the proliferation of cancer cells and target molecules involved in cancer cell cycles. (Note: cancer cells can divide and spread through the body because of the fact that their cell cycles are not controlled as normal cells' cycles are.)

Traditional chemotherapy agents work by binding to DNA or proteins in cancer cells and preventing them from dividing further by disrupting their cell cycles. However, the main problem with this approach is that chemotherapy typically affects the healthy cells too. For this reason, the adverse effects of traditional therapies are a big problem. Furthermore, it is known that the immune system plays a significant role in cancer immunity so another big question that arises is whether the chemotherapies decrease the immune potential of organisms toward cancer. All these aspects can be considered in relation to the overall effectiveness and safety of traditional chemotherapies.

Another approach that has been considered in the past decade is targeted chemotherapies. In this case, only a specific subset of cells and their molecules are targeted and the main goal for researchers is to decrease the adverse events while increasing the effectiveness through a targeted approach. The main weakness of this approach is that since it's a targeted approach, it may not be effective for all cancers but only for specific subtypes.

One such example is **Non-Small Cell Lung Cancer** (**NSCLC**), where a lot of research is being done around using targeted therapy called **Tyrosine Kinase Inhibitors** (**TKI**). These are known to inhibit the molecules expressed in large amounts on cells of this type of cancer.

In this project, we will be conducting Meta-analysis focusing on one main question: is the TKI approach more effective than the traditional chemotherapy approach in subjects with advanced NSCLC type of cancer in its advanced stage?

Research question:

Is the TKI approach more effective in relation to their PFS than traditional chemotherapy approach in subjects with advanced NSCLC type of cancer in its advanced stage evaluated at the 12 months?

The main question is how we perform such a Meta-analysis. The first step is to introduce yourself to the relevant literature and define the inclusion criteria for your meta-analysis.

In this case, we will define the following inclusion criteria:

Inclusion criteria	**Exclusion criteria**
Subjects with NSCLC	Subjects with multiple cancers
Subjects with III/IV stage cancer	Subjects with I/II stage cancer
Studies with RCT trial design	Studies with designs other than RCT
Studies that report PFS	Studies that don't report PFS
Studies that report the relevant statistical metrics	Studies that show the relevant statistical metrics.

Table 14.1 – Inclusion/exclusion criteria for studies

Next, we need to define the dataset and describe the data.

To define the dataset, we first need to define a string search strategy.

The following strings are defined:

Search strings	**Databases used as sources**
"NSCLC"	Scopus, Pubmed, Google search, and Cochrane search
"III/IV stage "	
"RCT"	
"Cancer"	
"Progression-free"	
"Survival"	
"PFS"	
"Hazard ratio"	
"Risk ratio"	
"TKI "	
"EGFR"	
"Chemo"	
"Targeted"	
"Therapy"	
"Clinical trial"	

Table 14.2 – Search strings for Meta-analysis

This dataset is a real-world dataset (*Table 14.4*) containing data extracted by me using Scopus, PubMed, Google search, and Cochrane search, which are the search platforms that can be used to find the relevant studies for meta-analysis.

To do this, I defined the following search terms and input them in the search platforms:

```
SEARCH TERMS:

NON-SMALL CELL LUNG CANCER
METASTATIC LUNG ADENOCARCINOMA
PHASE III CLINICAL TRIAL
BIOMARKER ANALYSIS
OVERALL SURVIVAL
CLINICAL TRIAL
EFFICACY
LUNG
CANCER
STUDY
RANDOMIZED
```

I also set specific inclusion/exclusion criteria such as that only Phase III RCTs will be included. The population with lung cancer, a specific type (NSCLC) will be included, and they may also have EGFR mutation. The studies that don't meet these criteria will be excluded.

Here are the studies that match this criteria :

- **Study 1**: Shi Y, Wang L. First-line Icotinib versus cisplatin/pemetrexed plus pemetrexed maintenance therapy in lung adenocarcinoma patients with sensitizing EGFR mutation (CONVINCE): a phase 3, open-label, randomized study. Ann Oncol 2017; 28: 2443-2450. `https://pubmed.ncbi.nlm.nih.gov/28945850/`
- **Study 2**: Mok TS, Wu YL, Thongprasert S, Yang CH, Chu DT, Saijo N, Sunpaweravong P, Han B, Margono B, Ichinose Y, Nishiwaki Y, Ohe Y, Yang JJ, Chewaskulyong B, Jiang H, Duffield EL, Watkins CL, Armour AA, Fukuoka M. Gefitinib or carboplatin-paclitaxel in pulmonary adenocarcinoma. N Engl J Med. 2009 Sep 3;361(10):947-57. doi: 10.1056/NEJMoa0810699. Epub 2009 Aug 19. PMID: 19692680. `https://pubmed.ncbi.nlm.nih.gov/19692680/`
- **Study 3**: Masahiro Fukuoka et al., Biomarker Analyses and Final Overall Survival Results From a Phase III, Randomized, Open-Label, First-Line Study of Gefitinib Versus Carboplatin/Paclitaxel in Clinically Selected Patients With Advanced Non–Small-Cell Lung Cancer in Asia (IPASS). JCO 29, 2866-2874(2011). DOI:10.1200/JCO.2010.33.4235

- **Study 4**: Sequist LV, Yang JC, Yamamoto N, O'Byrne K, Hirsh V, Mok T, Geater SL, Orlov S, Tsai CM, Boyer M, Su WC, Bennouna J, Kato T, Gorbunova V, Lee KH, Shah R, Massey D, Zazulina V, Shahidi M, Schuler M. Phase III study of afatinib or cisplatin plus pemetrexed in patients with metastatic lung adenocarcinoma with EGFR mutations. J Clin Oncol. 2013 Sep 20;31(27):3327-34. doi: 10.1200/JCO.2012.44.2806. Epub 2013 Jul 1. Corrected and republished in: J Clin Oncol. 2023 Jun 1;41(16):2869-2876. PMID: 23816960.

As you can see, the four RCTs are identified, which contain all the relevant data as defined in the inclusion/exclusion protocol and are relevant to the research question of interest.

From these aforementioned studies, here is the data extracted using the Kaplan-Meier curves:

Name of the study and year	Study type	Population	Variable of interest	Variable used as a therapy effect summary
Shi Y et al 2017	RCT	Subjects with III/ IV stage NSCLC	PFS (12 months)	Risk Ratio (PFS)
Fukuoka M et al 2011	RCT	Subjects with III/ IV stage NSCLC	PFS (12 months)	Risk Ratio (PFS)
Mok T et al 2009	RCT	Subjects with III/ IV stage NSCLC	PFS (12 months)	Risk Ratio (PFS)
Sequist L et al 2013	RCT	Subjects with III/ IV stage NSCLC	PFS (12 months)	Risk Ratio (PFS)

Table 14.3 – Studies descriptive table

In *Table 14.3*, the names of the authors, types of study designs, population studies, and (importantly) the metrics used to summarize the treatment effects are described. Every Meta-analysis should have such a descriptive table.

In the next table, the actual dataset to be summarized in another table with event data is presented. This table will contain event counts for experimental and control groups.

Name of the study and year	Events in the experimental group (TKI)	Number of subjects in the experimental group (TKI)	Events in the control group (traditional chemo)	Number of subjects in the control group (traditional chemo)
Shi Y et al 2017	91	138	106	122
Fukuoka M et al 2011	76	96	90	94
Mok T et al 2009	533	609	586	608
Sequist L et al 2013	153	230	104	115
Follow-up period – 12 months				

Table 14.4 – The dataset

This table shows events (progression) and the number of subjects within TKI and control (chemo) groups. We can use this data to calculate Risk Ratios and perform the meta-analysis of these four RCTs. Keep in mind that there is PFS as a metric here and not overall survival as this was the main research question we set at the beginning of the study. Some meta-analyses may set other variables of interest such as overall survival or recurrence-free survival, but in this specific project, this is out of the scope of the main research question.

In this section, we learned how to extract and structure the required data for meta-analysis.

Implementing DerSimonian and Laird inverse variance method in Python

In this section, we will utilize the knowledge from the previous chapter and use the `PythonMeta` package, as well as the same coding principles and functions learned so far, to create an actual meta-analysis using the four RCTs identified during the data search.

Now to the coding part. Open your Python editor (Jupyter or Spyder). In parallel, make sure you have installed the `PythonMeta` package using the following command:

```
pip install PythonMeta
```

Then, once this is complete, you can import the `PythonMeta` package and give it a shorter name:

```
import PythonMeta as PMA
```

Then load the `Data()`, `Meta()`, and `Fig()` classes we will need later to load the data, conduct meta-analysis, and visualize the results:

```
# Load classes
data = PMA.Data()
model = PMA.Meta()
figure = PMA.Fig()
```

Now we need to define the settings of the meta-analysis as we learned in the previous chapter.

Since we are performing the meta-analysis based on events data, which is binary (presence or absence of event), we will select the `CATE` (argument for categorical data) setting and the fixed effects method first. Set the `IV` algorithm and `RR` for the risk ratio effect metric. Refer to the following code snippet:

```
# Define settings
settings = {
    "datatype": "CATE",   # 'CATE' for binary data or 'CONT' for
continuous data
    "models": "Random",   # 'Fixed' or 'Random' effects
    "algorithm": "IV",   # Inverse variance based
    "effect": "RR" # RR/OR/RD for binary data; SMD/MD for continuous
data
}

# Set data type
data.datatype = settings["datatype"]
```

Here is the data written as an array from the data table presented in the dataset description:

```
# Provided data with subgroup
samp_categorical = [
    "Shi_Y_et_al, 91, 138, 106, 122",
    "Fukuoka_M_et_al, 76, 96,  90, 94",
    "Mok_T_et_al, 533, 609, 586, 608",
    "Sequist_L_et_al, 153, 230, 104, 115"
]
```

Now we can load the data and implement `settings`:

```
# Load data
studies = data.getdata(samp_categorical)

# Set the subgroup, data type, models, algorithm, and effect for meta-
analysis
model.subgroup = data.subgroup
model.datatype = data.datatype
model.models = settings["models"]
model.algorithm = settings["algorithm"]
model.effect = settings["effect"]
```

Here is how to implement the meta-analysis method:

```
# Perform the analysis
results = model.meta(studies)
```

In this section, we learned how to prepare the data and write code for implementing the DerSimonian and Laird inverse variance method in Python and store results for further interpretation and data visualization. With this information, we will now move to the next section, where we will learn how to make forest plots for oncology meta-analysis.

Making forest plots for oncology meta-analysis

In this section, we will use what we learned so far about the study design and the data and combine that with Python programming to make the forest plots. Forest plots are one of the main data visualization tools in Meta-analysis.

Here is how you can visualize the results using the forest plot:

```
# Show forest plot
figure.forest(results).show()
```

Here is the resulting forest plot:

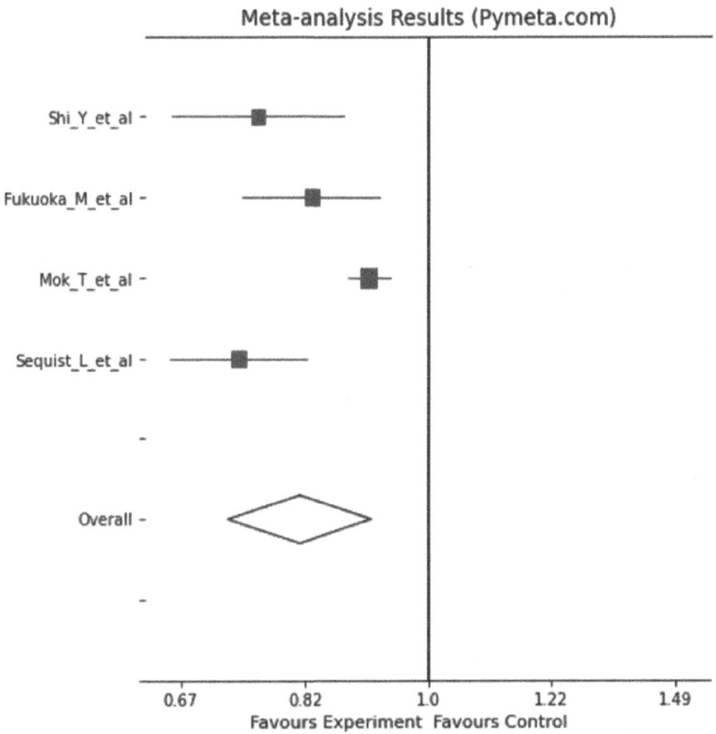

Figure 14.1 – Basic forest plot using Pymeta

Notice how I wrote what the values on the plot represent. So, all the values here are **Progression-Free Survival Risk Ratios** (**PFS RRs**) as calculated using the IV method specified before. This metric is used as a treatment efficacy measure.

The forest plot shows interesting results. First, it can be seen that all the studies favor the experimental treatment, which is the targeted TKI therapy over the traditional chemotherapy (control) in this case. Study 3 (Mok_T_et_al) has the smallest confidence interval, which means it adds significant weight to the meta-analysis. The most important aspect in this case is the diamond or the overall effect. We can see that the overall effect point estimate is **0.82** but the confidence interval (left and right edges of the diamond) are in full interval below **1.0**, which indicates a statistically significant result.

This first implementation included IV or inverse variance, also called DerSimonian and Laird method implementation, and the conclusion was that the overall effect indicates better treatment efficacy for TKI. How does the DerSimonian and Laird method work? It assigns weights to studies so that the studies with lower variance have more impact on the overall result. That is also why it is called the inverse variance approach; the studies with less variance will have more impact on the meta-analysis.

In the next section, we will learn how to make funnel plots, a special kind of plot that can be used to assess the publication bias in a meta-analysis.

Making funnel plots – publication bias analysis

Quantifying publication bias is one of the most important segments of Meta-analysis in general. Publication bias is an occurrence that may happen in certain life science fields, mostly due to publishers and authors favoring "positive" results over "negative" ones. If not assessed, the results could be biased due to publication bias. One of the main goals of any Meta-analysis project is to assess this aspect.

Let's explore how one can use Python and specific plots called *funnel plots* to evaluate publication bias.

We can also visualize the results using the funnel plot. Refer to the following code:

```
# Show funnel plot
figure.funnel(results).show()
```

This is how the plot appears:

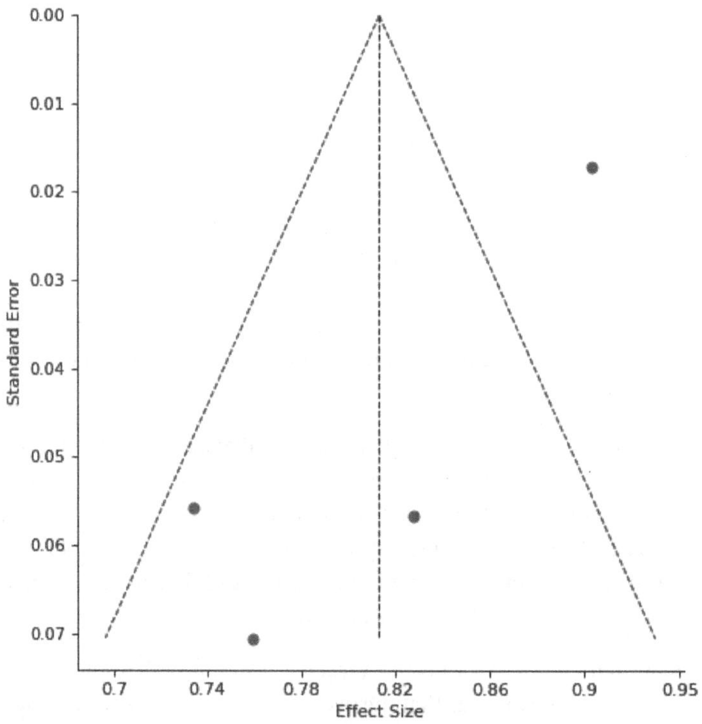

Figure 14.2 – The funnel plot

The funnel plot is used to investigate potential publication bias. If the plot is symmetric and within the triangle, there is little evidence for publication bias. For this specific case, we can see a level of symmetry and only one study is out of the triangle. Overall, there is little chance for publication bias in this situation. This is a good validation of previous results seen on the forest plot.

In this section, we learned how to make funnel plots using Python and use them to assess the publication bias as a validation step in a Meta-analysis. In the next section, we will learn how to use another estimator, the **Mantel-Haenszel** (**MH**) estimator, as the main Meta-analysis implementation.

Implementing the Mantel-Haenszel estimator in a Meta-analysis

The M-H estimator can be used in situations with less than 10 studies in a Meta-analysis. We will use a similar Python implementation as before, but this time, we will use a specific argument to change the estimator. If you change the setting of the algorithm to MH, we can implement the MH method which is more appropriate for this meta-analysis (having four studies):

```
# Define settings
settings = {
    "datatype": "CATE",  # 'CATE' for binary data or 'CONT' for
continuous data
    "models": "Random",  # 'Fixed' or 'Random'
    "algorithm": "MH",  #  Mantel-Haenszel method
    "effect": "RR"  # RR/OR/RD for binary data; SMD/MD for continuous
data
}
# Perform the analysis
results = model.meta(studies)
```

Now we can visualize the results using the forest plot as follows:

```
# Show forest plot
figure.forest(results).show()
```

Here is the plot for it:

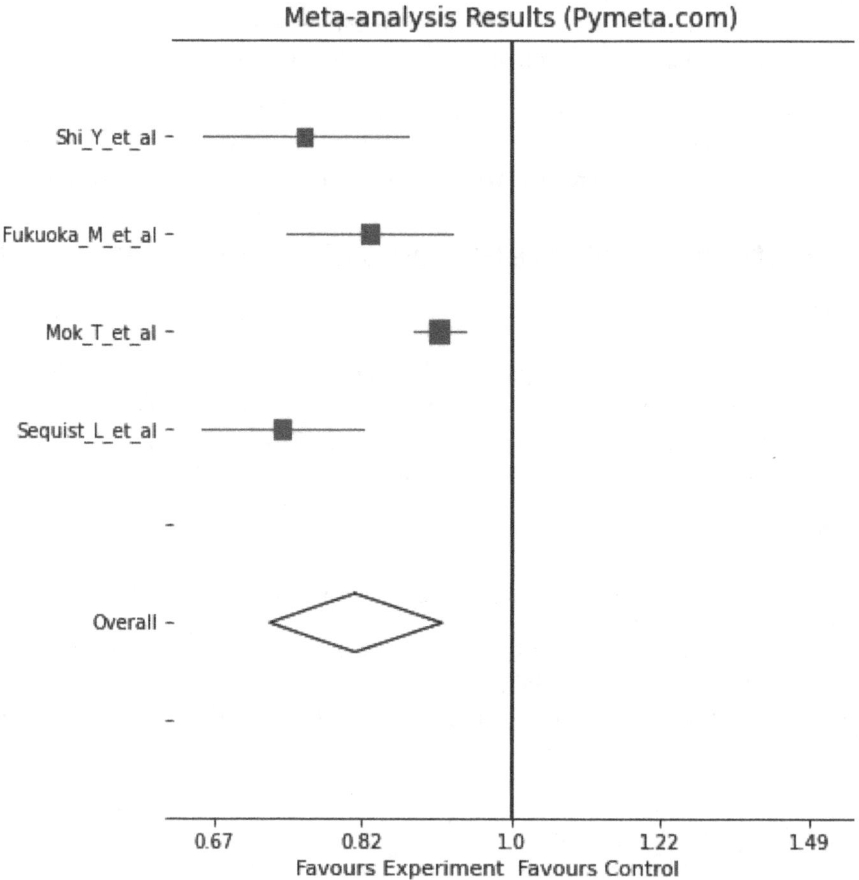

Figure 14.3 – Forest plot for the meta-analysis of the MH method

The forest plot using the MH method is very similar to the one implementing the DerSimonian and Laird method, and the conclusions are similar. The overall effect diamond is in its complete shape below the level of **1.0** for PFR RR; we can conclude that the overall effect indicates better PFS for subjects receiving targeted chemo in the period of 12 months. The next step is to evaluate the publication bias. This is done using the funnel plot code and this is how it can be done, similarly to the previous section:

```
# Show funnel plot
figure.funnel(results).show()
```

Here is the resulting Funnel plot, but this time for meta-analysis using the MH estimator.

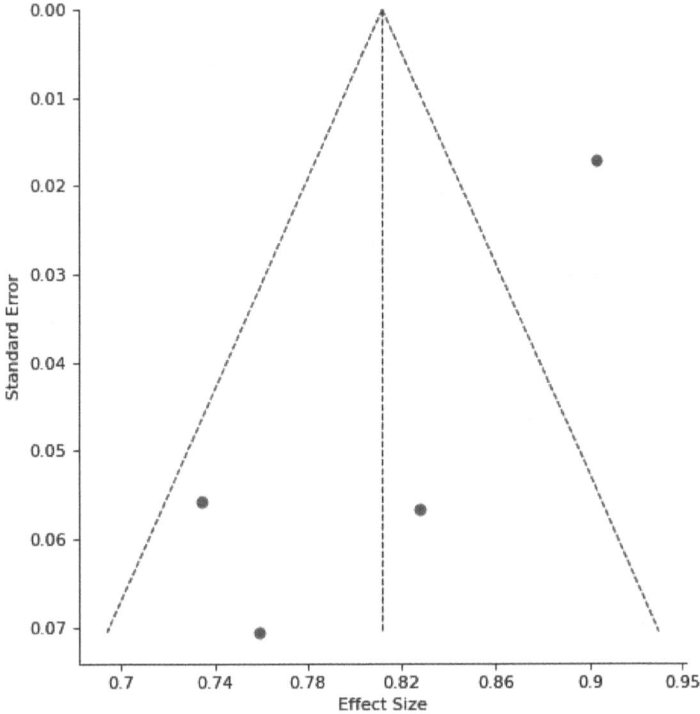

Figure 14.4 – Funnel plot for MH Meta-analysis

As you can see, three out of four studies are within the confidence interval triangle, two studies are to the left, and two studies are to the right on the effect size vertical line. The plot looks symmetric but we can never confirm there is no publication bias with four studies, even though there is a lower chance in this case. Further, since only RCTs are included, their reporting criteria are more standardized and more trustworthy and that can be considered as another factor that contributes to having more credibility for this specific project.

In this section, we learned how to approach the meta-analysis from the perspective of assessing the publication bias. This is a very important step to validate the meta-analysis results. We also learned how to use Python for the implementation and make funnel plots as publication bias visualizations.

Summary

In this chapter, we learned how to design and implement a Meta-analysis. We learned about the basic biology of cancer treatments and different chemotherapy strategies used, such as TKI and traditional chemotherapies. We learned how to form a research question based on cancer biology. Further, we learned how to design inclusion/exclusion criteria, describe the data, and create event tables for Meta-analysis. We learned how to implement DerSimonian and Laird inverse variance and MH methods. Finally, we learned how to visualize and interpret the results.

In the next chapter, we will gain a deeper understanding of biological variables. We will also learn how to relate biological variables to biological experiments and how to form research questions in life science.

Part 4: Biological and Statistical Variables and Frameworks, and a Final Practical Project from the Field of Biology

Part 4 is about biology and biotech research combined with Python. You will learn about the structure of biological variables and how to combine them with Python biostatistics implementations. Another important topic of *Part 4* is understanding of statistical frameworks in life sciences, frequentist and Bayesian methods, and their differences.

At the end of *Part 4*, there is a hands-on exercise that involves performing structural equation modeling for biology research. The main topic of the project is molecular biology data, and it involves analyzing protein relations in mice.

This part contains the following chapters:

15
Understanding Biological Variables

In this chapter, you will learn how to assess biological variables, as well as different biological aspects of biostatistics and one very important aspect, which is combining biological and statistical knowledge. You will learn about biological variables and experiments such as those in biotech research and clinical trials. You will learn about biological confounders and latent variables in biology research. Thus, you will be exploring the context and uncertainties around specific clinical and biological domains in clinical research. Finally, you will see how to validate biological data, which is one of the most important parts of biostatistics.

In this chapter, we're going to cover the following main topics:

- Understanding biological variables and experiments
- Confounders and latent variables in biology research
- Validating biological data

Understanding biological variables can be as important as understanding statistical segments of the analysis. This is especially the case in life science areas such as biology research statistics and clinical biostatistics. In this chapter, we will delve into biological system constructs in relation to statistical constructs and also learn how to evaluate the integrity of the biological data one receives as a biostatistician.

Understanding biological variables and experiments

Before delving into biostatistics professionally, it's important to understand biological variables, biological experiments, and their relation. One of the most important parts of biostatistics is understanding how biological measurements and observations transpose onto biological data.

We use the term *biological variable* to define a quantity, property, or measurement that can vary among species or populations, or even within a single species population. Examples were discussed in previous chapters, such as height, **BMI** (short for **body mass index**), and HbA1c (glycosylated hemoglobin), but there are many more that we will discuss in this chapter.

But before that, let's try to understand where biological data comes from. One of the largest sources of data for biological variables is experimental biology. Many different types of experiments are conducted with the main goal of answering biological research questions. These typically involve making measurements and observations that are specific to their experimental background. Experimental design provides scientists with the means to control what is being tested and isolate unwanted bias and noisy data from the scientific data they want to collect. For this reason, biological data arising from experiments has the highest value in terms of data quality.

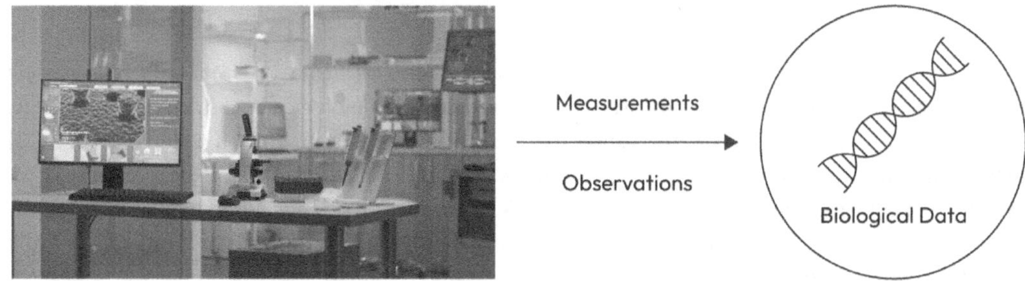

Figure 15.1 – Data collection in experimental biology (https://www.freepik.com/free-photo/modern-empty-biological-applied-science-laboratory-with-technological-microscopes-glass-test-tubes-micropipettes-desktop-computers-displays-pc-s-are-running-sophisticated-dna-calculations_17762122.htm#fromView=search&page=1&position=21&uuid=c5b6a66b-fd25-4f6f-8952-a9720939bdd1)

In the preceding figure, you can see how data is collected or observed in an experimental biology lab. All the measurements from biological experiments produce certain types of data. Different values on instrument colors, reactions, and observations can be used to create biological data for biological variables.

In biostatistics, we frequently relate biological experiment data with statistical methods to create a framework that helps us understand how biological variables vary, what they tell about the result of an experiment, and how this translates into scientific and research value.

Experiment design tells a researcher how to understand the data derived from the experiment, but this data needs to be quantified or categorized in a standardized and accurate way to derive value from it. This is where statistical methods are crucial.

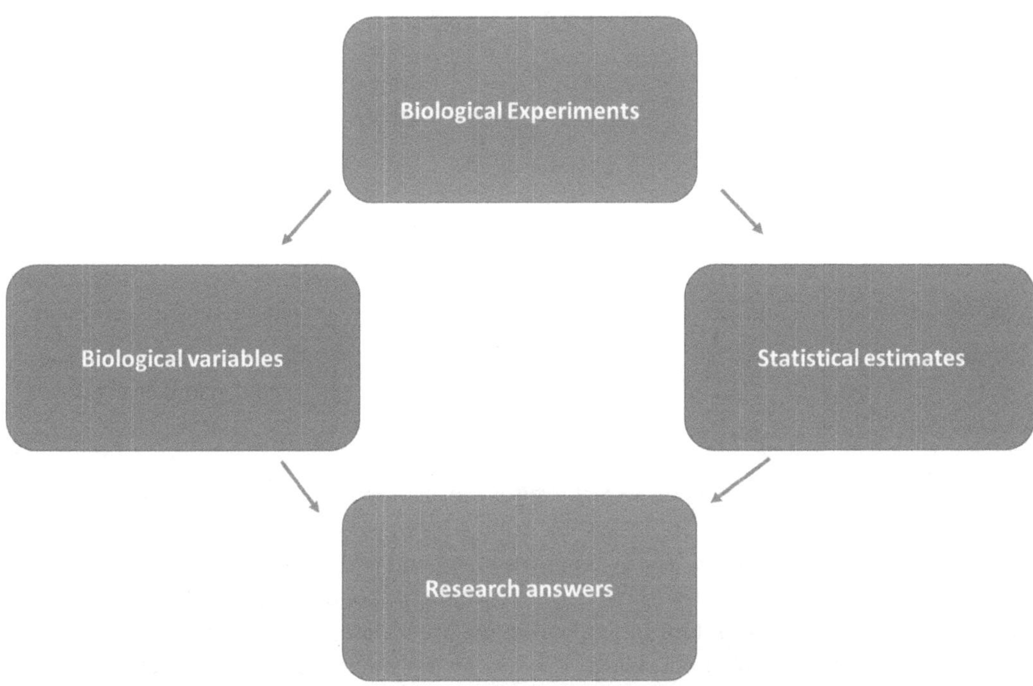

Figure 15.2 – Relation between biological variables and research answers

Here, you can see how biological experiments are combined with statistical estimates to provide a scientific answer to biological research questions. Note how biological experiments are at the core of all the previously mentioned biostatistical segments.

What does this mean? The quality of biostatistical data analysis heavily depends on the quality of biological experiments.

A good biostatistician is one who can relate biological context with the statistical methodology.

One very important feature of biological variables is their variability type. In this sense, there are two types of biological variables: static and dynamic.

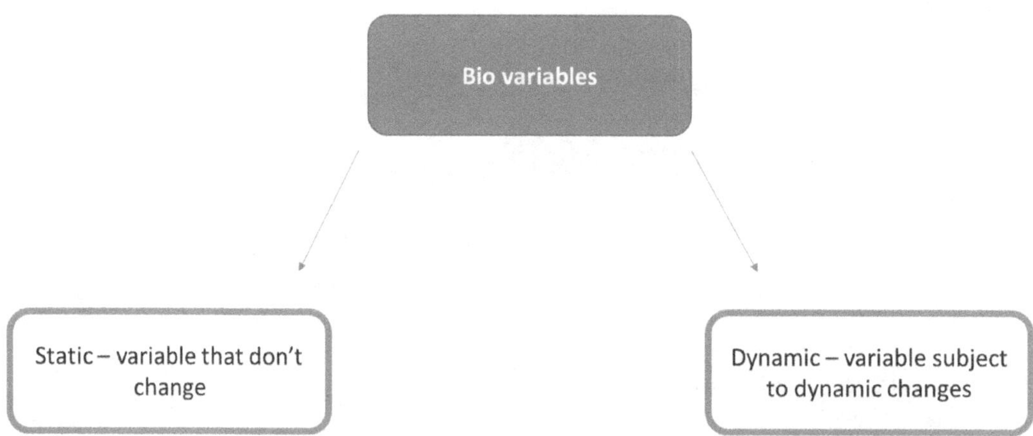

Figure 15.3 – Static versus dynamic variables

Some biological variables, such as blood groups (A, B, AB, and O), are defined by molecules on the cell surfaces of human erythrocytes. These may vary within the population, so different people may have different blood groups. But for individuals in the population, their blood group is the same throughout their lifetime; it does not change. If we test an individual's blood group at different times of the day, or months or years across their lifetime, it will always be the same; it is a static variable.

Now, let's consider another type of biological variable: blood pressure. This variable varies within the population, but it also varies for all individuals. It changes from minute to minute, from day to day, on a monthly basis, and so on. It is dependent on other variables, such as stress level or medication use, and is highly dynamic. For this reason, we call this type of variable a dynamic biological variable.

Let's explore some other biological variables and see which of these two classes they belong to (static versus dynamic). In addition to blood type, which is a genetic trait, let's explore a similar variable: eye color. It is also genetically defined and it does not change, so we can say it is generally static.

If an organism (humans and most other organisms) carries a specific hereditary mutation on a gene, it will also not change; it will be static. There is similar to the fact that if no hereditary mutation is present for a certain organism, the organism will not have this mutation in all cells. But mutations can sometimes be dynamic. Single cells or tissue may acquire mutations in organisms due to environmental variables. A good example is the sun's UVA and UVB rays causing mutations in the skin, even though they were not present before. So, sometimes the environment may cause specific parts of the body to acquire mutations. Also, in this case on a tissue or cellular level, these mutations may be dynamic biological variables and the values or status may change over time. But mutations that are hereditary are usually static variables.

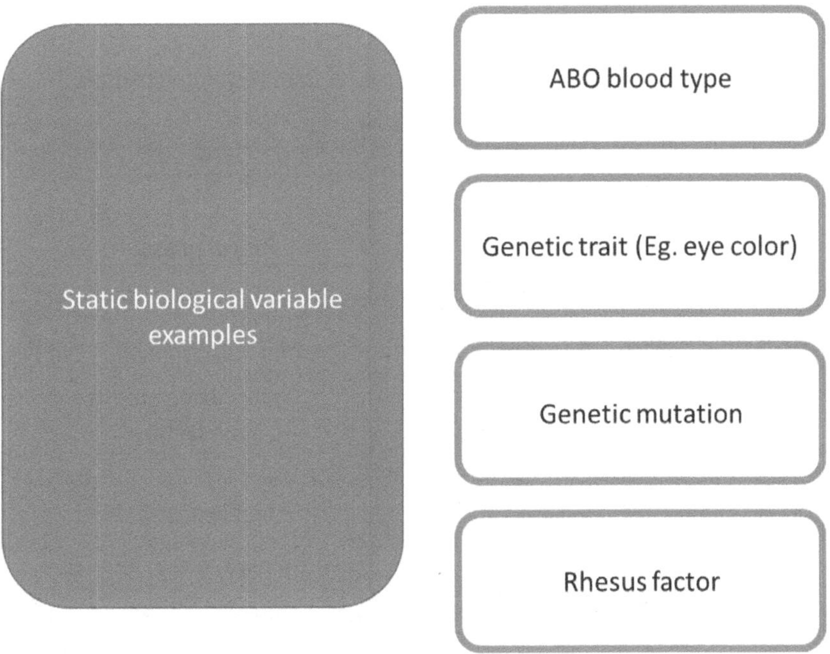

Figure 15.4 – Static biological variables

The preceding figure lists several static variables in the biological domain, such as blood type, genetic traits, and mutations. Now, let's explore some typical dynamic variables that may exist in a biological system. We mentioned blood pressure before, but there are many other dynamic biological variables, such as BMI, population size, and gene expressions. In fact, the majority of biological variables are dynamic as biological systems tend to change and vary.

For example, populations of organisms in almost all ecosystems vary all the time. Population size is dependent on many factors, such as food availability, temperature, potential relations between organisms, and even the diseases that organisms may or may not have. Even though the factors I mentioned vary at different times, they change in different geographical areas and cause the population size to change. This is a good example of a biological system that is composed of mostly dynamic biological variables. Dynamic biological variables are listed in the following figure:

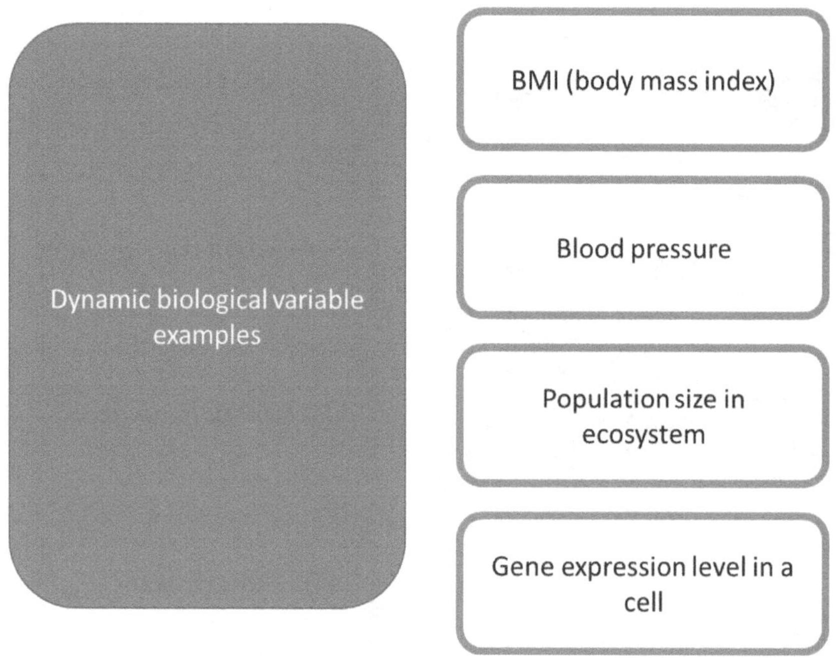

Figure 15.5 – Dynamic biological variables

Let's take another example of how a biological system may be composed mostly of dynamic biological variables. The BMI of human subjects may change relative to their nutritional habits, metabolism, and even aging. While BMI does not change as dynamically as blood pressure (hourly), it may change in time periods of weeks, months, or years. Another very important set of dynamic variables in most organisms is the gene expressions of cells. This set of variables, by studying the gene expression transcripts in bioinformatics, has also been found to be highly dynamic. Gene expressions may also be different, even on a time scale of milliseconds on a molecular scale. Gene expressions also tend to be controlled via molecular mechanisms, so even if they are dynamic, they have a level of stability and may be associated with different diseases, such as cancer, genetic diseases, or diabetes.

What does all this mean to a biostatistician? Well, we must have a clear methodology on how to deal with static or dynamic variables, and this methodology may differ based on the biological variables' characteristics.

Now, let's try to relate experiment areas, biological variables, and statistical estimates in a systematic way, which will enable analyzing and interpreting biological variable insights effectively.

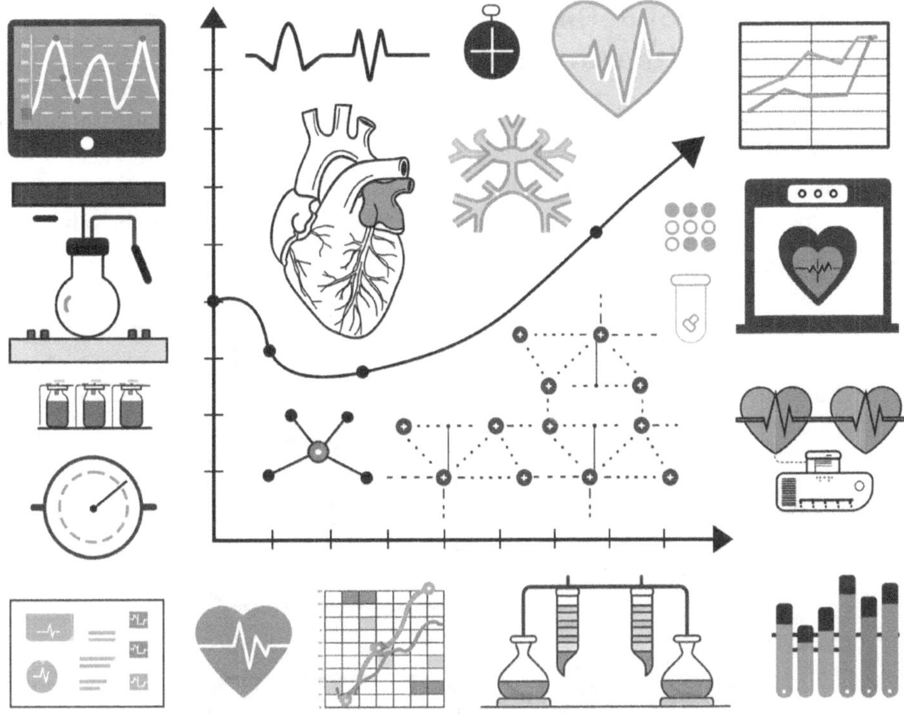

Figure 15.6 – Relating experiments and biological variables

Let's consider an experiment that is conducted in the cardiology research area, such as a clinical trial, which is effectively an experiment. In such an experiment, the main biological variable may be ejection fraction, so the researcher would be studying the amount of blood the heart pumps in each heartbeat. How can this be connected to the statistical estimate? Well, the mean difference between different biological states or treatments received could be measured to evaluate how each of those affects the ejection fraction of the heart.

> **Note**
>
> **Ejection fraction percent** (**EF%**) is calculated using the following formula:
>
> $EF\% = SV / DV * 100$
>
> Here, SV is systolic (heart muscle contraction) volume and DV is diastolic (heart muscle relaxation) volume.

In this section, we learned about biological variables in relation to biological experiments. We also learned how to differentiate and identify static and dynamic biological variables. In the next section, you will see practical examples of defining biological variables and associating them with statistics.

Practical examples of defining biological variables and associating them with statistics

Here, we will explore several practical examples of defining biological variables and associating them with statistics. The main goal of defining variables in biostatistics is to systematically relate the biology/life science area, the specific biological variable that is to be studied/analyzed, and the statistic we can relate to the biological variable. All these aspects need to be integrated in order to be valid.

To further explore this systematic definition, let's look at a few examples.

To begin, we can understand the field of life sciences with a simple example, focusing on ecology where species' richness is commonly examined. Here, species richness serves as a key ecological descriptor. We can then show how to define the biological variable and statistical estimate for it in a simplified way:

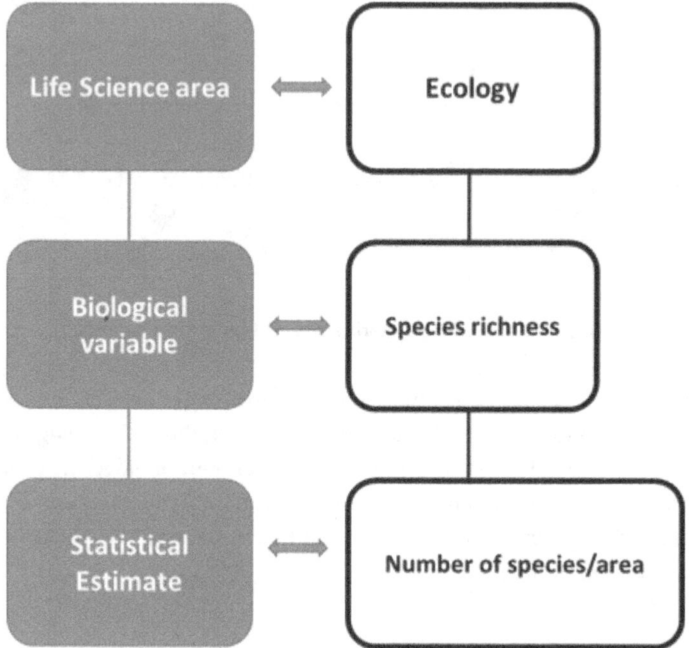

Figure 15.7 – Biological variable to evaluate species richness in a biological habitat

As you can see, integrating biological variables with the life science area and relevant statistics can be done easily in a three-step process.

Now, let's explore a slightly more complex example:

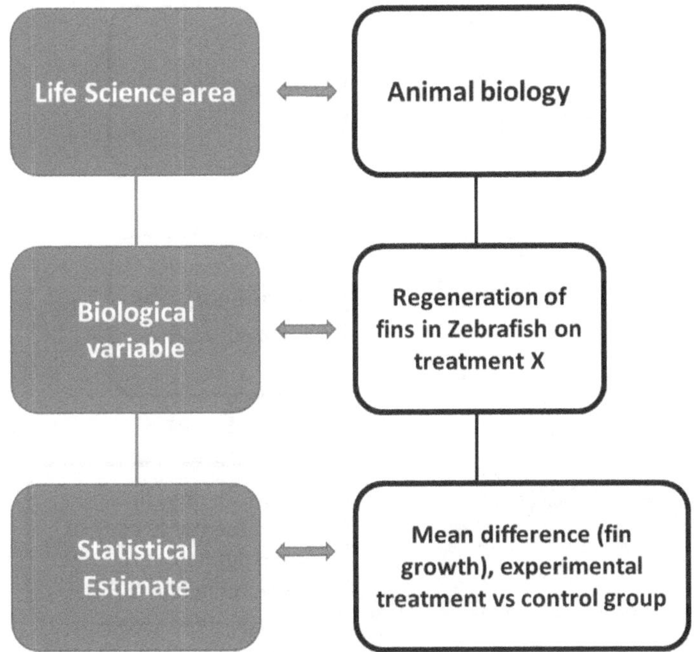

Figure 15.8 – Animal biology variables and statistical estimates example

Here is an example of **animal biology** as a life science area. This defines a specific organism, called the zebrafish, and the biological variable is associated with the fin regeneration of the zebrafish (zebrafish, or *danrio rerio*, is known to have fin regeneration potential).

This could be an example of studying zebrafish fin regeneration under different conditions, such as temperature. The final integration with a statistical estimate is the mean difference and experimental condition (low temperature), versus control group (normal body temperature).

The next variable to define is very specific to pharmaceutical research and is part of one of the most frequent workflows in biostatistics/research today. The goal of this research is often to evaluate the efficacy of different drugs or treatments. In this case, we will be exploring how to systematically define a biological variable to estimate vaccine efficacy.

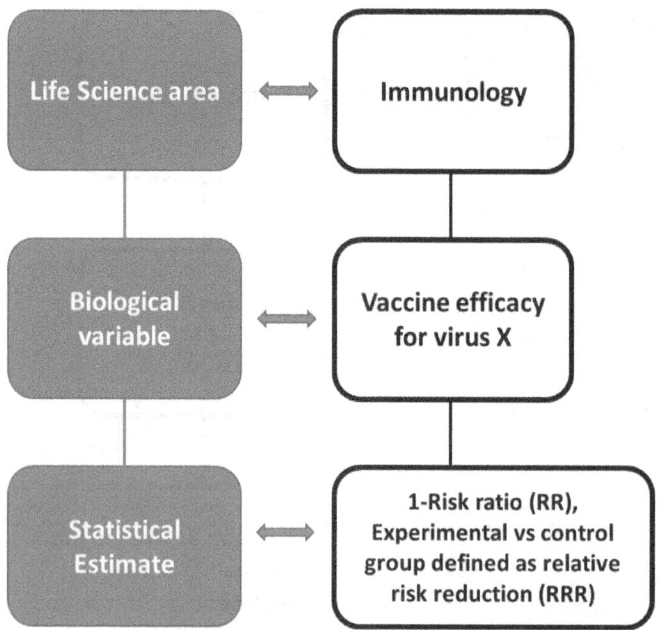

Figure 15.9 – Immunology biological and statistical variables example

As you can see, life sciences falls under the area of **immunology**. Why? Well, because vaccines work by introducing the immune system to viral or bacterial particles so they can develop antibodies to fight the disease.

We can define the biological variable as vaccine efficacy for some virus we will call *X*.

Now for the most complicated part: how to define the statistical part. One way is to consider risk reduction in developing a viral disease for subjects who have received the vaccine versus subjects who have not received it. We call this estimate **relative risk reduction** (**RRR**) and it is calculated by deducting the **risk ratio** (**RR**) from 1.

This example is the most complicated one we have seen so far, but it is also very important as it is frequently used in other areas of biostatistics, such as cardiology and oncology biostatistics.

Now, let's switch to anthropology as a life science area. We can define the biological variable as the BMI of subjects (a variable typically assessed in anthropology studies). To define a biological variable, we also need to input the definition of the subjects, such as the improvement of BMI in subjects in a specific exercise program (this could be daily or weekly).

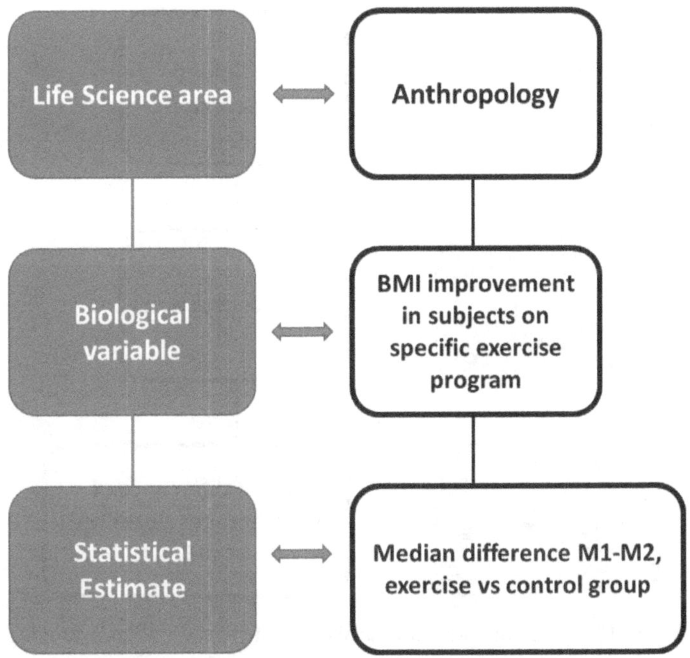

Figure 15.10 – Anthropology biological and statistical variables example

A statistical estimate can be the median difference, so M1 (median 1) – M2 (median 2), for exercise and control groups. Such a definition would cover all the aspects needed for biological variable estimation, but keep in mind that other statistical estimates, such as variability estimates, can be added later in the statistical analysis part.

Now, in this cardiology research-specific example, you can see how a biological variable can be connected to a simple statistical estimate such as the mean difference of percentages to evaluate how different conditions affect the ejection fraction of the heart:

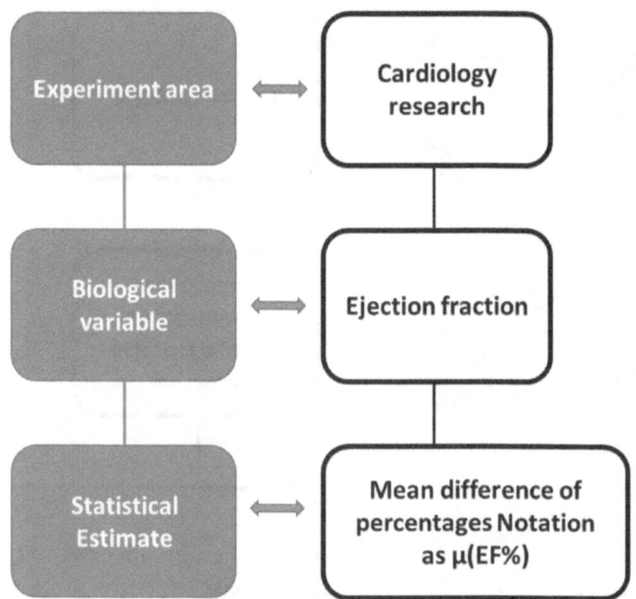

Figure 15.11 – Cardiology biological and statistical variables example

Now, let's define another systematic definition of a biological variable associated with its statistical counterpart, but this time in a hypothetical oncology research experiment.

Confounders and latent variables in biology research

What differentiates experiments and simple observations in the scientific world? One of the most important aspects to consider is **confounding**. Most observations we see around us are prone to being affected by many other variables, both those we see and those we don't see. Those we see and measure are called **covariates**, which we frequently use to create models (independent variables), but these variables can also alter the results, which may introduce bias into the models. Frequently, we can't observe or measure all the variables that may affect the data we collect; these variables are called **latent variables.** Covariates and latent variables that may affect the data we collect and influence the results meaningfully are called confounders. Have a look at the following figure:

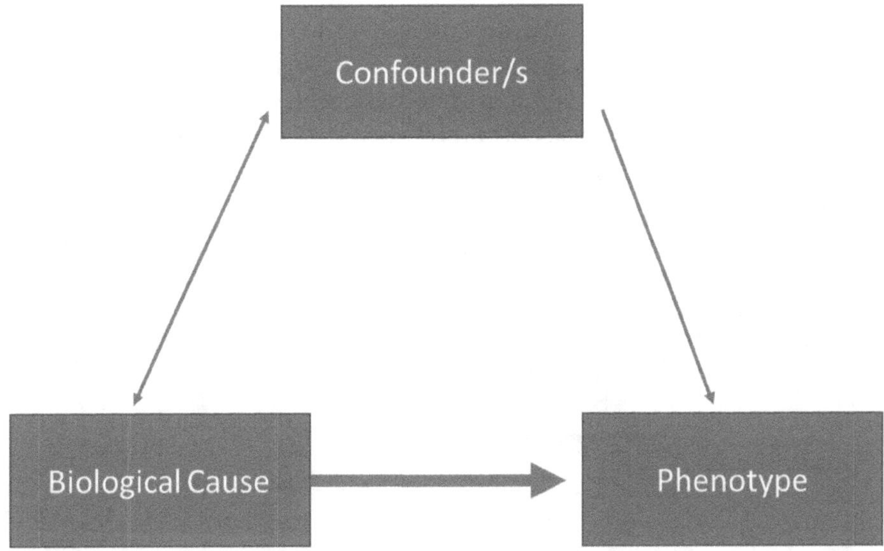

Figure 15.12 – Confounders

In this diagram, you can see an example of how biological causes may be affected by confounders to produce what we observe as biological characteristics called **phenotypes**. For example, genetics may affect the risk of obesity in some people. In this case, genetics is the biological cause and obesity is the phenotype. But the phenotype may affect the genetic background in a different way if confounded by other variables, such as lifestyle and nutrition.

Experiments are frequently designed to avoid confounding and isolate them from biological causes and phenotypes. By using experiments and isolating the confounders, one can more accurately assess the causes of many biological variables.

Biological experiments enable control of the confounders. One of the most basic principles of experiments is controlling for different variables.

Researchers can, for example, control which groups receive treatments and which don't and design the experiment in such a way as to show the actual treatment effects in statistics.

In the next section, we are going to discuss the validation of biological data, why it is important, and how it affects biostatistical analysis.

Validating biological data

Validating biological data involves multiple steps and principles. But first, let's discuss why data validation is important. Every result we get in a biostatistical analysis depends on the validity of the data used.

The first, and one of the most important, parts of the data validation workflow is checking the source of the data—a biological experiment, specific instrument, biological observation, or other sources of data. Checking the validity of the data means first checking the validity of the data source. As an example, the experiment documentation can be checked. By checking the documentation, the analyst is sure that the instruments are validated, which helps them to verify the data. Ensuring that they can trust the data collected is essential.

Figure 15.13 – Biological data validation

Replicates are another way of validating the data during its creation process. This is specific to dynamic biological variables. If variables such as gene expression that vary from measurement to measurement are replicated in multiple measurements, their validity is higher and more reproducible.

Biological controls are also used to make sure that data does not come from contaminated or biased sources. Many biochemical reagents can be used as controls in biological research to make sure that a comparison using them provides a valid reference for the measurements made in biostatistics.

Programmatic validation is another way of checking data after it's stored, such as checking Excel tables, CSV files, or other files used in data analysis. This can also be done for software such as R and Python by creating scripts that, once run, check for duplicates, missing data, data out of the possible reference range for biological variables, and other invalid data forms.

To assess all these aforementioned data validity aspects, data documentation should be read, and then each of the segments validated individually. Data-related documentation may be found in the experiment/study protocols, manuscripts, files, and folders associated with the data.

It is very important to know that without data validation, many sources of bias and invalid results may be introduced into biostatistical analysis.

Data validation should be performed before the actual data analysis starts and typically involves a multidisciplinary approach including biologists, clinicians, and statisticians to confirm that the data is valid before proceeding with the analysis.

In this section, we learned how to check and validate the data, data formats, and documentation in a biostatistical project.

Summary

In this chapter, we learned how to understand biological variables and experiments. These are some of the most important aspects of biostatistics, since most of the analysis relates to them. Also, we learned about confounders and latent variables in biology research, why it's important to control them, and how to improve the interpretation of the results. Finally, the last topic we learned was how to validate biological data, which is probably one of the most important parts of any biostatistical analysis, this time including both biological and statistical knowledge.

In the next chapter, you will learn how to implement actual biostatistical analysis in specific areas of biology, utilizing both the knowledge you gained in this chapter and Python coding.

16
Data Analysis Frameworks and Performance for Life Sciences Research

In this chapter, you will learn how to relate Biostatistics to study designs in Biology Research. Further, learning how to relate Biological experiments with different aspects of Data analysis is covered. Finally, learning how to interpret Biological data analysis is featured in this chapter. This chapter is one of the most important in the book as Biostatistical frameworks, Frequentist and Bayesian which are not easy to understand will be simplified and explained in detail.

You will learn about the differences between these frameworks and how to choose specific statistical framework for your analysis and study.

In this chapter we're going to cover the following main topics:

- Creating biology study designs
- Understanding the statistical frameworks
- Learning the Frequentist framework statistics
- Learning the Bayesian framework statistics
- Choosing a Statistical framework
- Connecting experiments to data analysis
- Latent variables and Causal inference

- Sensitivity and how to how to interpret biological data analysis

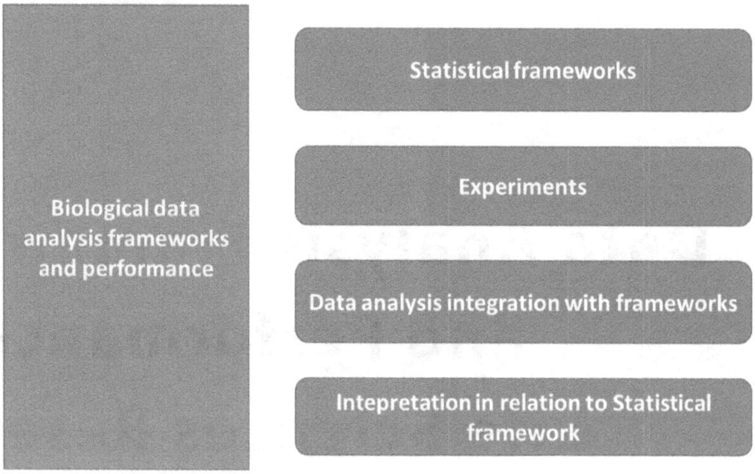

Figure 16.1 –Statistical frameworks and data analysis

Creating study designs in Life Science projects is one of the most important skills for Biostatisticians. In *Chapter 10*, we learned how to design studies in Clinical Research and in this chapter we will expand on this topic by adding a more detailed perspective, taking into account both biological domain specific knowledge and statistical domains such as Frequentist vs Bayesian frameworks.

One of the most important aspects of this chapter is learning how to take into account multiple angles, domain specific biological or clinical research angle, statistical framework angle, regulatory and study design angle, combine them and include in the overall interpretation of biostatistics.

We will learn how to connect the experimenter perspective to statistical frameworks. Finally we will learn how to combine all the above mentioned into an integrated skillset of interpreting the biological data analysis.

Biological / Life Science data analysis frameworks are specific and are dependent on both biological and statistical domains. In this chapter you will learn how to integrate these aspects, but you will learn how to adapt the interpretation principle to specific areas such as Sensitivity analysis, Latent variables and Causal inference. These aspects are of special importance, because most research questions in life science area are indeed related to statistical frameworks, causality and sometimes to latent variables too.

Creating biology study designs

To create a study design within any biological or other life science, we start with specific biological knowledge. That knowledge can be expanded, hypothesized, and tested and may lead to the acquisition of new knowledge within the studies. Sometimes this knowledge may be directly related to the data or we may have limited level of knowledge around the data and may want to expand it further.

Once we have enough Biological knowledge, we can identify and formulate research problems and later turn this information into the study design.

To create the study design, clear endpoints of the study must be defined. The endpoints are used to define exactly what are we estimating and how will these estimates be used to interpret the success and interpretation of the study. Biological knowledge may be used to create a variety of hypotheses and this type of the study is hypothesis based, but the studies may also be without a specific hypothesis and this type of the study is called **explorative type** of study.

Here is a list of steps which is essential in creating Biological study designs which can be used in various Life Science areas, such as Clinical Research, Ecology, Genetics studies and many other fields.

1. **Identify the Research problem**: In this step the author/researcher needs to study all the relevant details about the specific research domain and identify potential problems that could be addressed by conducting a study. Of course, consulting references and the already conducted research in the area of interest is the first step.
2. **Identify potential scenarios in which a study could be designed to answer the questions about the research problem**: In this step, the creativity of the author/researcher and the innovative approach is very important. Research has the highest weight if it brings new previously unexplored information and conclusions, but confirming the previously created theories and conducted studies has equal importance in biological research and especially in clinical research.
3. **Identify the statistical framework relevant to the study**: In this step, identifying the statistical framework that can help address the problems and answer the questions asked in the previous steps is the main focus. This is one of the most important aspects not just for the data analysis part, but also for the whole study design. A statistical framework is used also to frame the study in the specific design and is most often part of most study protocols.
4. **Identify the biological aspects of the study specific domain**: This means being able to understand the biological segments of the study in highest possible detail. This is one area that is specific for Biostatisticians and I would emphasize a special focus on this segment. Biostatisticians need to understand both biological and statistical aspects of the study. Both design and later interpretation of the study results would not be optimal if both biological and statistical segments were not understood in detail and integrated into the study design, data analysis and the expected result interpretation.
5. **Design the study documentation**: Why is this step important? Most studies in life science are conducted with specific regulations in place. Animal studies, observational studies, clinical trials and other types of studies are all implemented with specific regulations in mind.

   In some areas these regulations are less strict, such as observational studies and in others the regulations are very strict, such as those in clinical trials field.

For example Clinical trials are implemented with respect to regulatory bodies such as **Food and Drug Administration** (**FDA**) which have specific requirements in terms of transparency, documentation and validation required. For this reason it's important to understand and plan which types of documentation are needed and how to adhere to specific regulatory requirements in that respect.

Study documentation may include, protocols, data collection plans, data management plans, ethical consideration documents, legal documents and software documentation. In the case of Python implementations, it's important to include the Python documentation and the relevant Python package documentation. Finally, it's important to document the data and the code used to implement the analysis in Python.

Next important step is to define the Statistical framework, but before that one needs to understand them in detail.

Understanding the statistical frameworks

Every statistical analysis and especially its interpretation is dependent on the statistical framework. During the last several decades, two main statistical frameworks have evolved, the **Frequentist statistical framework** and the **Bayesian Statistical framework**. We can define these as two main Statistical philosophies and their application in Biological and Clinical research can determine the interpretations of results, and also define the decision making processes in life science.

Statistical frameworks make the assumptions and the foundation for the statistical analysis being implemented. They contain both the theoretical foundations and the relevant interpretation principles which will be used later in the analysis.

Frequentist framework has been here for decades and a majority of Biological data analysis, especially in Clinical research is conducted within frequentist framework. Bayesian framework has been popularized in the past years and is also applicable for many clinical research aspects. In this chapter we will focus on these two frameworks in detail and show how they can be applied to life science biostatistics and research overall.

Frequentist and Bayesian frameworks are dominantly used in today's Biostatistics. The choice of framework depends on the type of study, goals and the types and the amount of data being analyzed. In the rest of the chapter we will discuss how and when to use these two frameworks, but first lets try to understand them better.

Bayesian framework also became popular and is used in many areas of clinical research in the past few decades. Today, it is used in life science manufacturing area and clinical trials area. It is also used in many adaptive clinical trial and dose-finding studies.

Let's learn the foundations and principles of Frequentist and Bayesian statistical frameworks, their foundations, differences and similarities, and very importantly their philosophies. Also we will learn when to use them in specific research areas.

Learning the Frequentist framework statistics

So what is the Frequentist framework? We discussed the p values before, they are a segment of the Frequentist framework. But the Frequentist framework is much more. As its name says, the Frequentist framework is based on frequencies or repetitions of sampling in order to estimate the parameters we are looking for, in this case biological parameters. The main aspect of the frequentist framework is that what is inferred statistically is founded on the principle of repeated sampling. This ensures that the main focus of results is on their reproducibility and scientific defensibility.

Frequentist framework is the most commonly used framework in Biostatistics. Most analyses in biology, clinical research, bioinformatics and other life science areas are indeed implemented in the frequentist framework. It has been a standard for decades and it still is in most research areas. This is mainly due to the nature of the Frequentist framework. It is tailored for reproducibility, objectiveness and scientific defensibility overall. For this reason it remains the gold standard in clinical research and any life science areas where trustworthiness of statistical analysis is essential.

In Frequentist framework, the focus is on the data from the sample and priors are not included. The context around the data is also included in the result interpretation. The reproducibility is in the focus of statistical evaluation, so the statistical significance parameters such as p values and confidence intervals are based on the potential long term sampling instead of being focused on a single sample or a fewer number of samples. One of the main goals of a Frequentist is to try to estimate the uncertainty not around the single sample, but around the potential data generation process from a long-term frequency of samplings.

Have a look at the following figure explaining the sampling distribution from a larger population and how it converges to a bell shaped normal distribution

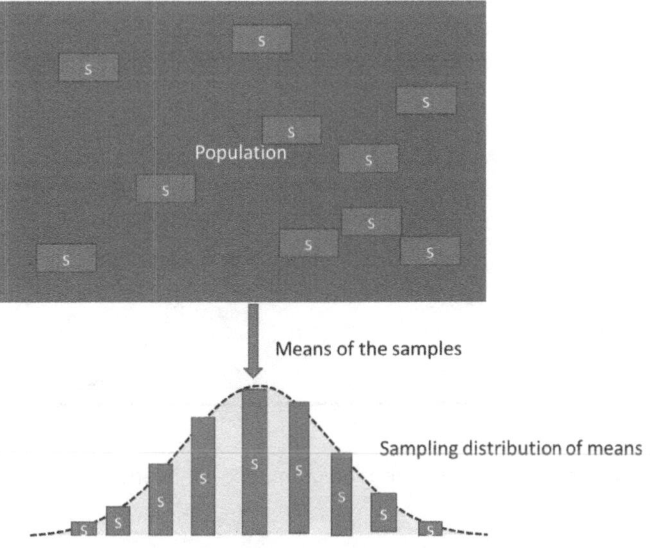

Figure 16.2 – Samples and sampling distribution

If we think of most scientific phenomena, we can indeed see that long term frequency of repetitions is indeed the one which adds trustworthiness into the results. From simple coin toss to complex experiments, reproducibility of statistical estimates is indeed highly dependent on the long-run frequencies of repetitions.

Experiment design in biology, clinical research and other life science areas is also meant to be as reproducible as possible and many regulatory agencies such as **Food And Drug Administration (FDA)** and **European Medical Agency (EMA)** insist on the reproducibility of results as one of the most important segments of clinical research.

In the Frequentist framework, probabilities are also estimated relative to the frequencies and repetitions. This means that estimated probabilities would occur if the observer or experimenter were to repeat the study many times. What does this mean?

Imagine conducting a biological study by making 100 measurements, or a sample of $n=100$. In the sample there are two groups, A and B. By comparing them a researcher gets a specific result, a mean difference of 1, confidence interval from 0.7 to 1.3 for a statistical significance level set at 0.05 (a hypothetical scenario).

For a Frequentist implementation, this would mean that if the researcher were to repeat the study many times, the mean difference would be a fixed number that is within the confidence interval 95% of the time, or better said in 95% of the studies/samples. When the inference is made about a specific interval of probability, we actually use the term *confidence* to better describe the Frequentist framework.

95% Confidence interval

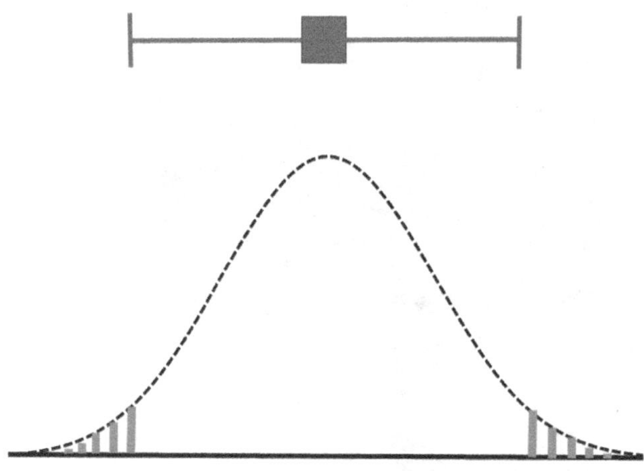

Figure 16.3 – The visualization of the 95% confidence interval

Another specific angle of the Frequentist framework is that the parameters are considered fixed. Parameter is a measure of specific characteristic we are studying. In Frequentist framework, every characteristic we study has a fixed, singular quantity that describes it. If we take the earlier example, if one were to measure the whole population, the parameter, *mean difference* would become known and have an exact point value and it would be a fixed value. The confidence interval in that case would not be needed to describe the parameter.

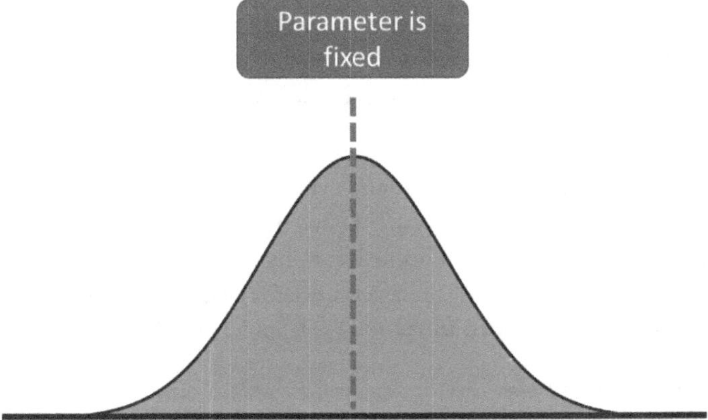

Figure 16.4 – Fixed parameter in a distribution

When the confidence interval for a biological parameter is calculated, it means that 95% of the sample confidence intervals would contain the fixed parameter and each sample can either contain or not contain the parameter, there is no probability perspective here, but rather a deterministic perspective.

This is another reason why Frequentists usually use the term confidence instead of probability, which would correspond to the random variable nature of the parameters instead of the fixed parameters.

The p values we mentioned in the previous chapters are an important segment of the Frequentist framework, but the confidence intervals are an even more important segment. Confidence intervals allow to interpret both the statistical significance and the magnitude of the effects in biological and biomedical research.

Most Frequentist approaches try to establish as objective as possible approach and for this reason, rarely combine the effects from different statistical methodologies. Instead the focus is on the data and the context around it but independently and objectively.

For this reason Frequentist statistics is a standard in most life science areas. Inability to incorporate prior knowledge in Frequentist frameworks is sometimes regarded as a weakness in the framework. Meta-analysis allows this when we have multiple studies about the same research endpoint. When interpreting the statistical significance in the Frequentist framework, we typically use the p values to interpret the results based on the statistical hypotheses. A null statistical hypothesis is set – typically no difference between distributions or no significant associations between variables.

Then a statistical method from the frequentist framework (frequentist statistical test) can be used to evaluate potential hypothesis set in the research project.

Another very important alternative to explore is the Bayesian framework.

Learning the Bayesian framework statistics

Bayesian statisticians, sometimes called 'Bayesians' base their approach to the Bayes theorem but also a variety of other approaches such as including prior data and calculating posterior probabilities for statistical estimates. Before learning about all these, lets learn about the Bayes theorem. Bayes theorem is about the conditional and joint probabilities.

Let's consider that there are two events, event A and event B. We can consider two scenarios, scenario one, what is the probability of an event A occurring given that event B occurred. An example would be what is the probability that thunder will occur if it rains. We can also consider a reverse scenario, what is the probability it rains given that we hear thunder. These are the conditional probabilities of events and we can write them as P (A|B) in scenario one and P (B|A),

Bayes rule (based on Bayes theorem)

$$P\left(A \middle| B\right) = \frac{P(B|A)*P(A)}{P(B)}$$

The Bayes theorem enables the calculation of P (A|B), the probability of an event A given B if we have the probability of an event B given A (B|A) and individual P(A) and P (B). While this is the Bayes theorem it enables the development of a whole branch of Statistical philosophy called Bayesian statistics or more accurately the Statistics with the Bayesian framework. It allows for calculating the posterior P (A|B) based on the prior and new evidence based on the preceding formula. Even in the situation where no prior data is present the Bayesian framework can be implemented.

Figure 16.5 – Updating the prior evidence with new evidence

As it can be seen, in all three scenarios. **Prior belief** (as evaluated by domain experts), **No prior data**, and actual **Prior data present** Bayesian analysis can be implemented. In scenario 1, the prior belief can be used to set prior based on expert opinion. But keep in mind that this approach may be subjective and this can be its weakness. In scenario 2, no prior data, a prior can be set as a non-informative prior, the prior which does not affect the model much. This approach is more objective and will often provide similar results as the ones from the Frequents framework. In scenario 3 when actual credible data is present as prior data, one can construct the Bayesian analysis in a highly trustworthy way and this is an especially useful approach in Clinical trials where trustworthiness is of high importance. Such an example is the use of data from one Clinical trial as a prior for data for another type of Clinical trial, for example, Bayesian Adaptive Clinical trial.

Based on these approaches the priors used in Bayesian analysis can be separated into these categories:

- Strong Informative subjective priors
- Strong Informative objective priors
- Weak subjective informative priors
- Weak objective informative priors
- Non-informative priors

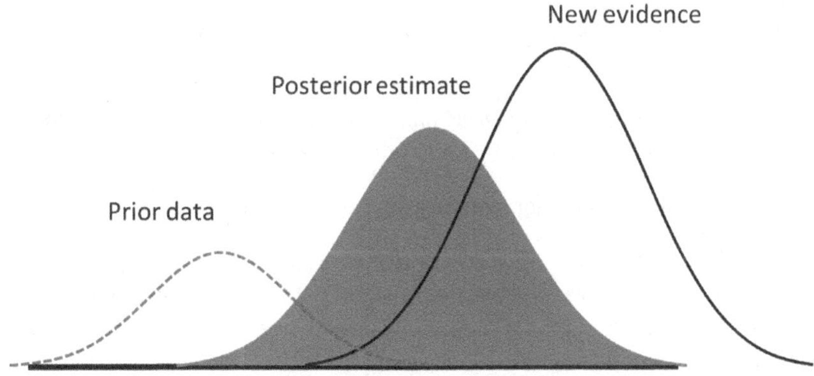

Figure 16.6 – Visualizing the prior and posterior distribution

This formula can be used to address the problem of using the prior knowledge about the scientific phenomenon and combine it with the new data into the posterior estimate of the evidence. As you can see from the preceding figure, the posterior estimate is in between the prior data and the new evidence. Bayesians would say for this principle that the prior estimate is updated with the new evidence and the posterior probability estimate is formed.

Bayesians try to take a different approach in terms of the hypothesis formation. Bayesians try to ask the following question, *given the data we have from the sample (new evidence) what is the probability estimate of obtaining a specific result as a probability interval?* But all this is complemented by a prior background information. As for a comparison, Frequentists would ask a different question for the same situation: *Given that the null hypothesis is true, what is the probability of obtaining the false positive result, and express this as the p value*?

To achieve the desired results, Bayesian implementations often include the simulations as part of the method implementation. Most often, a type of simulation called **MCMC** or **Markov Chain Monte Carlo** simulation is used as a part of estimating the Bayesian probability estimates.

Choosing between Frequentist and Bayesian framework is very important and is usually done in the initial phase of studies. Let's explore how to go through the selection process.

Choosing a Statistical framework

Choosing a Statistical framework depends on many factors. For the past few decades, there has been and still is a debate if the Frequentist framework is better than the Bayesian and vice versa. A much better approach is to try to differentiate in which specific study designs and life science areas it is better to use a Frequentist and in which a Bayesian approach and what is their level of scientific defensibility.

Even though these frameworks have similar roots in probability, their applications in Life Science research are different. The Frequentist framework is best used for studies in which we have enough data to be able to base strong claims based on that data and the context around it. Such examples are usually large retrospective studies, longitudinal studies, phase II, III, IV studies, and post marketing clinical studies.

The *Adaptive Designs for Clinical Trials of Drugs and Biologics, Guidance for Industry* document by the FDA that explains these aspects well : `https://www.fda.gov/media/78495/download`

Most meta-analyses are also well implemented in the Bayesian framework.

On the other hand Bayesian framework is specific for its ability to include priors in the data. While this approach may be considered as less objective by some Statisticians, in situations where the data is not abundant, such as smaller samples or in situations where prior data is objective and trustworthy, Bayesian designs can be a good solution. Such examples include smaller retrospective or smaller prospective studies, preclinical studies and phase I studies in which the data is scarce, dose escalation studies in which starting dose is not defined, and later phase clinical trials with well-founded data from the previous phase trials.

Both Frequentist and Bayesian implementations are included in today's life science, especially in clinical research and should be used according to a specific situation as explained earlier.

One specific example is the use of Bayesian approaches for the medical device statistics in clinical trials.

> **Note**
>
> You may find more information about the use of Bayesian framework in Bayesian implementations in clinical trials here, `https://www.fda.gov/regulatory-information/search-fda-guidance-documents/guidance-use-bayesian-statistics-medical-device-clinical-trials`.

This document is called Guidance for the use of Bayesian Statistics in Medical Device Clinical Trials. I strongly suggest reading this document for anyone interested in the use of Bayesian statistics in clinical research.

After learning how to reason the statistical frameworks, such as Bayesian and Frequentist, it's very important to learn how to connect those frameworks to the actual biological experiments in an integrated way. In the next section this will be discussed in more detail.

Connecting experiments to data analysis

Before proceeding with this topic, it's important to know why we need experiments and how they can improve the study. As its name states, experiments have an empirical nature. For thousands of years different types of experiments have been used to study natural phenomena. But experiments exist in life science too, a good example is the **randomized controlled trial** (**RCT**) which is performed in vivo, sampled from a study population. The experiments can also be implemented using in vitro biological experiments used in drug discovery and molecular biology areas which are of huge importance to understanding the biological data used in biostatistical analyses.

Experimentation is used to:

- A: Empirically evaluate phenomena and scientific theory.
- B. Isolate the variables of interest
- C. Make a controlled study in which the experimenter controls all the important aspects of the study

Have a look at the following figure:

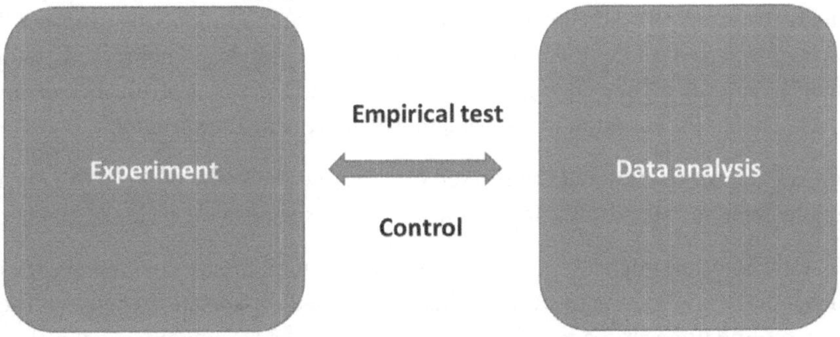

Figure 16.7 – Experiment vs Data analysis

So how do we connect the data analysis part with the experimental part?

First we need to understand the domain studied in detail and as a result how to reason various aspects of a biological or clinical experiment. These include the theoretical context, scientific reasoning, logic, and the current state of knowledge about the specific scientific domain relevant to the experiment.

Once the context, background and scientific reasoning about the research topic are integrated, the next step is to design the experiment. The experiment is typically designed to show clear empirical nature of the research phenomena studied. The main question in this field is often how to relate all these aspects with a clear and falsifiable research question. This means that the research question relevant to the experiment design should be answerable in a clear and distinct way. This also means that the research questions should be empirically testable. Research questions with multiple potential meanings should be avoided in the experiment design and data analysis and may lead to biased research interpretations.

Second goal of relating experiments to data analysis is trying to isolate the variables of interest from confounding. Whether we are designing a technical experiment or a clinical research experiment such as RCT ,one of the goals is to try to control the effects of unwanted variables, confounders that may cause the bias in the analysis. This is achieved through the design of the experiment and by strict control of the experiment workflows.

Another very important statistical approach and framework is the causal inference framework. This framework usually includes using a multi-angle approach to isolate the causal effects from non-causal ones and this will be discussed in the next section.

Latent variables and Causal inference

When Biological datasets are analyzed it's very important to understand that there are two types of variable constructs to keep in mind. Variables that are in the dataset, the variables that can be measured, and turned into numerical data are called the known or observed variables. We may include these in the models, perform statistical inference and other statistical methods. But there are always hidden, latent variable constructs which most often affect the results. These variables belong to the latent variable construct and are called the **latent variables**. They are important because most of the time in biological experiments and studies there are hundreds if not thousands of unobserved, latent variables which influence the results. They often introduce the bias and increase statistical errors in analyses. They also contribute to the variation in the data, mostly increasing the variation.

The question is what can we do about them? Well first, we must be aware of them. We must be aware that the variables we have in the dataset are not the only ones existing in the reality of biological studies.

We must be aware how they might affect the interpretation. Latent variables often prevent a researcher from making sound causal claims in studies. This is because the associations between the variables could be due to some unobserved variable and not to the ones we actually have in the data. This means that unless we deal with the latent variable constructs we should not make causal claims in result interpretation.

Most statistical frameworks on their own do not answer the main research questions in most life science areas. Causality is at the core of most research questions we ask. Statistical frameworks isolated from the context around the data and the study itself cannot answer these questions. For this reason statistical frameworks are combined with specific experimental designs and specific biological contexts to try to answer the Causal research questions.

In biostatistics some of the most important Research questions are actually related to the causality. One such area is diseases research and modeling.

The following Research questions is a standard in the disease research field:

- What is the cause of a specific disease?
- Can we identify a gene that causes the disease?
- Can we identify a protein modification which causes the disease?
- Can a researcher identify the lifestyle which may cause a specific disease?
- Does a specific virus or bacteria cause a specific infection?

We could also ask a question like this :

- Does a specific biological molecule cause the disease to disappear?
- Does a specific treatment cause the disease status to improve
- Does a specific treatment cause any side effects?
- Does a specific treatment cause any serious side effects?

As you can see both diseases and treatment research questions of highest importance to science are relevant to the causality principles. Does A cause B? Does A improve B?

The main principle of causality is to evaluate is a specific event, let's call it even B is caused by another event A. Simplified formulation would be: Does A cause B, mathematically expressed does A -> B? as you can see there is a direction in causality, and one of the main goals in the causal analysis is to avoid a problem called **reverse causation** which may cause the analysis to be biased.

Another aspect of causality is the time aspect. This means that if we call the event A the cause of the event B, the event A must also precede the event B. This makes them the two main foundation principles of causality, the direction and the times aspect .To summarize, if the event A causes the event B, there is a direction of A towards B and A precedes B.

Causation can be mathematically presented using a simple graph

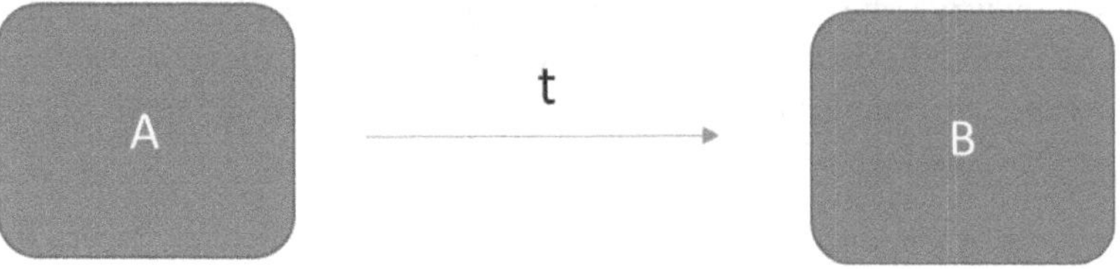

Figure 16.8 – Cause and effect in time

Notice how I used the directed edge to show the direction of causation and place **t** (time) on it to specify that the time aspect is present. These two conditions must be met in the causal analysis.

To study more about them, we will go to the following sections.

Counterfactual design

Counterfactual design of the studies is one of the most scientifically defensible ways of using study design to actually assume causality within the statistical interpretation later. This means the groups studies are often allocated in such a way that a researcher can study what would happen if the treatment is present and what if the treatment is absent in population without significant interference of other variables on the end result. Such design is often called **counterfactual design**. So what is the counterfactual design. We can imagine what would happened in scenario A and what in scenario B. Also this principle can be expanded into what would effect would occur for a standard situation and would there be any difference for some different scenario, or intervention. In Clinical research it is achieved by creating controlled trials, with treatment and control groups randomly allocated.

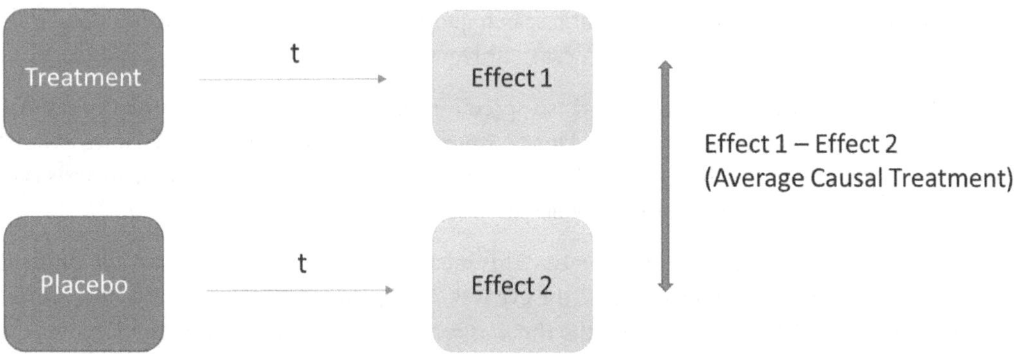

Figure 16.9 – The counterfactual scenario

Randomized controlled trials

Randomization is needed to shield the analysis both from known and unknown confounders. In other words, randomization decreases the bias that would originate from known and unknown latent variables.

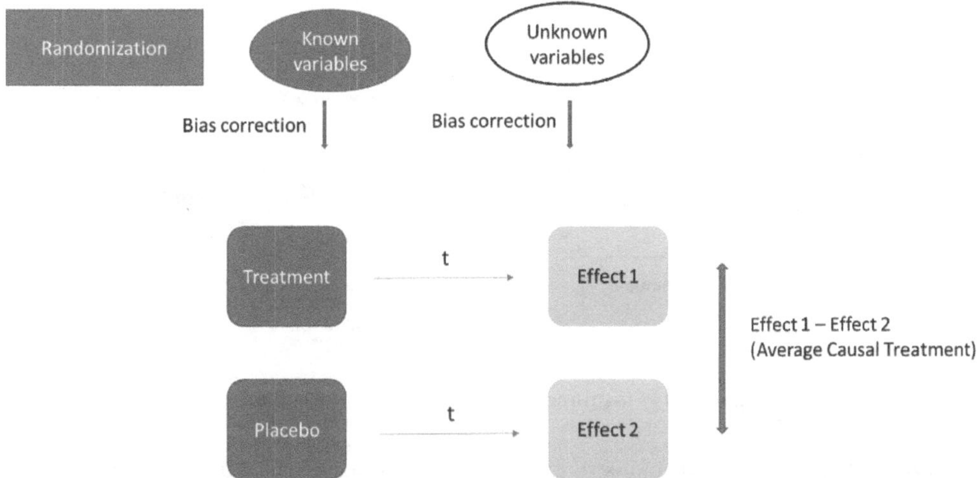

Figure 16.10 – Randomization and counterfactual design

This bias correction for the unknown variables is very important. Randomization is one of the rare statistical principles which can actually correct for the unknown bias. Most other methods and adjustments correct only for the known bias and now also for the latent variable which causes unknown bias. For this reason randomization can be used to isolate the variables of interest to a high degree and the effects estimate are actually causal treatment effects. Because they are averaged across the population, we also call them the **Average Causal Treatment Effects**.

Instrumental variables

When experimental analysis is not possible, an alternative approach to causal inference is called **instrumental variable analysis**. Instrumental variables are variables that can be used to correct the portion of the bias originating from the confounders. Variables are considered as instruments if they are not correlated with the target variable but influence the independent, explanatory variables. This characteristic makes them a potential '*shield*' from the external influence on the relation between target and explanatory variable association.

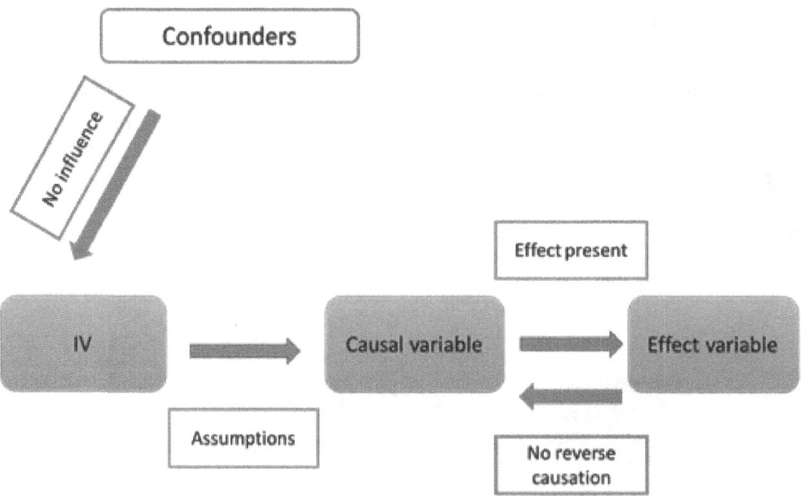

Figure 16.11 –Instrumental variable as a Causal variable

Another important characteristic of instrumental variables is that we need to have domain-specific factual knowledge about the nature of the association between instrumental variables and target/explanatory variables. Just statistical estimates are not enough, we need to logically conclude these types of associations from context and domain knowledge.

As it can be seen the biological assumptions need to be met to assume that the **instrumental variable (IV)** influences the causal variable but is not influenced itself by the confounders. This way the instrumental variable shields the analysis from the confounder bias and partially isolates it.

This approach enables combining biological and statistical knowledge in observational causal studies and for this reason it is one of the most advanced statistical methodologies in biostatistics. The most common form of instrumental variable analysis is called **instrumental variable regression** (IV regression) and is very frequently used in areas such as Oncology and Cardiology.

After learning the statistical frameworks, its very important to learn how to validate the use of those frameworks in any specific statistical analyses. One very important aspect in this field is called the Sensitivity analysis and will be discussed in more detail in the next section.

Sensitivity and how to how to interpret biological data analysis

One of the most important aspects of any biostatistical analysis is performing the sensitivity analysis. In fact FDA, the top regulatory body for Drugs in United States, defined a framework on how to conduct this very important segment of research. You may find the document here, `https://www.fda.gov/regulatory-information/search-fda-guidance-documents/e9r1-statistical-principles-clinical-trials-addendum-estimands-and-sensitivity-analysis-clinical`.

But, what is the sensitivity analysis? In fact there are many types of sensitivity analysis. Sensitivity is all about evaluating how sensitive the analysis is in terms of some assumption, method variable or even segment of the study. Generally the topics defined by the FDA are the most important ones and are relevant to the assumption and methodological sensitivity.

As you have probably seen so far in the book, most methods in statistics come with specific assumptions. For example, Student's t-tests, Chi square tests, ANOVA, Linear regression and other statistical methods and hypothesis tests have different assumptions. They work only when these assumptions are true. So it's always a good idea to try similar methods and methods with different assumptions to check if the results hold under different assumptions. Why is this done? Well to make sure that any assumptions in the methods do not influence the results in such a way that would make a researcher make wrong conclusions.

Also the methods are different in the way they are implemented and in the frameworks they are implemented. So Frequentist implementations and Bayesian implementations are not the same. The way they are interpreted is also different. So trying different frameworks and implementations is also a good idea to try to conclude that the results are not a result of methodological implementation of framework characteristics and that they would hold in different settings.

Here is a visualization of how confidence intervals in a method-wise sensitivity analysis may look like:

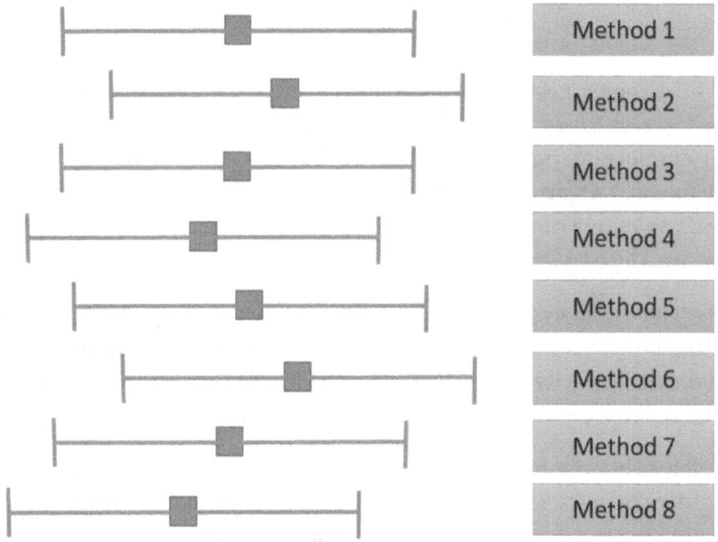

Method 1
Method 2
Method 3
Method 4
Method 5
Method 6
Method 7
Method 8

Figure 16.12 – Method sensitivity analysis.

When interpreting the biological data analysis, multiple segments must be taken into account. Of course, the data segment is very important, but one must also take into account the context around the data, sometimes previous knowledge, reference studies, the statistical framework and last but not least, specific life science domain knowledge. A Researcher should know what the requirement is which can be considered as meeting the primary endpoint of the study.

Meeting the primary endpoint doesn't necessarily mean a positive result for let's say a treatment effect in biostatistics. Meeting the primary endpoint means that the topic one was set out to study was actually answered. This means that the data and results were trustworthy enough to answer the research questions of the study. Secondary endpoints may not be a strict requirement but are generally needed to augment meeting the primary endpoints. This is the first aspect to interpret for a Biostatistician. Have the primary and secondary endpoints been met with the available data and results?

If not, the study implementation should be reevaluated. If the answer is yes, a Biostatistician can proceed with the next step. The next step is evaluating the statistical and research outcomes in terms of the research questions and hypotheses. Do we accept the hypotheses or do we reject them. Also it's important to interpret both Research and Statistical hypotheses. It's also important to differentiate these terms. In theory, one may accept the statistical hypothesis but reject the research hypothesis and vice versa.

Its important to evaluate statistical hypotheses, the magnitude of the effects, their relation to biological domain context, references, and specific clinical or other life science knowledge and integrate these aspects. Integrating these aspects should be done in a way that can clearly without any double meanings answer the research question in a way that is scientifically defensible and logical for the audience to reason scientifically.

So how do we interpret the results section?

There are always 3 segments to interpret in any Biostatistical analysis as explained in the following sections.

Trustworthiness

To interpret trustworthiness, attention should be paid to reproducibility, sampling, confidence, credibility, bias analysis and potential errors in the data. Obviously, the statistical significance is one of the main indicators of trustworthiness, keeping in mind that in the Frequentist framework, the p values are about reproducibility and sampling based statistical significance. The lower the p value, the more we can trust the result in terms of sampling and reproducibility. This is mainly attributed to lower p values having lower probability of Type I errors, false positive result. Another angle of trustworthiness interpretation is the Statistical power, it is a bit different from the statistical significance because it focuses on the type II errors or false negatives.

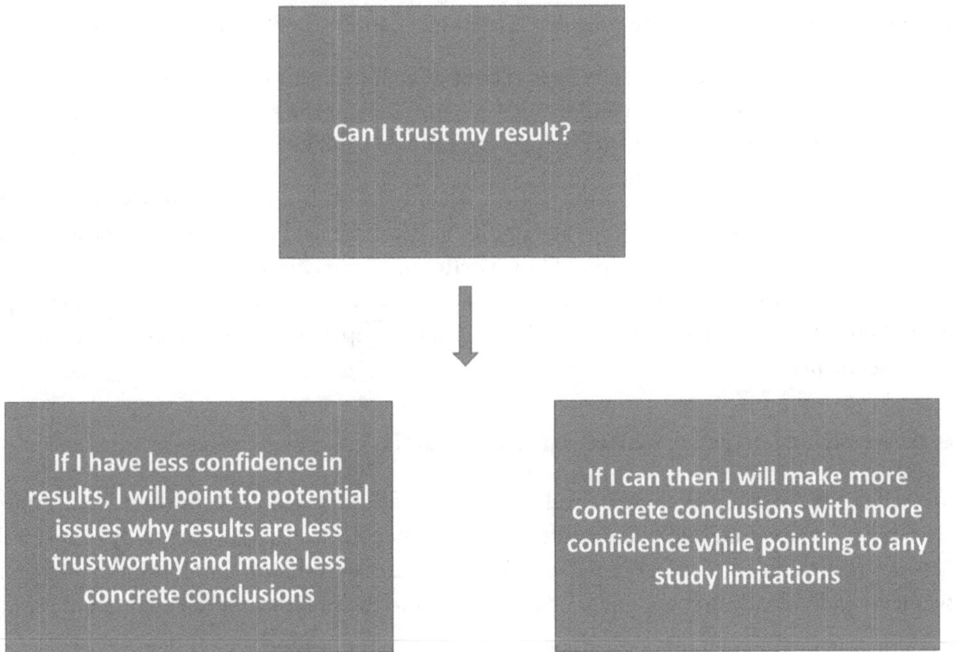

Figure 16.13 – Level of trust in results and how to position the interpretation

Confidence intervals in Frequentist statistics and credible intervals in Bayesian statistics are also used to interpret the uncertainty around the result. Statistical significance, statistical power, and uncertainty are the foundation of interpreting the trustworthiness segment of a statistical result. To summarize, lower p values (typically <0.05), higher statistical power and smaller uncertainty intervals will indicate a better level of trustworthiness and vice versa.

Magnitude perspective of the results and Biological context

Magnitude of the effect is one of the most important parameters of any biostatistical result. Magnitude is interpreted in terms of the efficacy, effectiveness and scale of the effect. Therapy efficacy is often expressed as a risk reduction. An example would be a risk reduction in **coronary artery disease** (CAD) development when taking a specific therapy. Another example would be vaccine effectiveness. A vaccine having an immunization effectiveness of 50% would be quite a different story to a vaccine having a 97% level of effectiveness. The magnitude of the effects is one of the most important aspects relating to statistical and biological/clinical interpretation of results in studies and research. In the example above the magnitude is the difference in percentages.

Two vaccines may have a similar effectiveness, e.g. 50% for control and 55% for experimental vaccine and the difference may reflect a $p<0.05$ but this difference is not large in magnitude and the clinical significance of such difference would be lower as regulatory agencies such as FDA would usually require larger vaccine efficacy to be considered for further studies and use in population.

Every result interpretation must have a biological context included. This means including the biological knowledge about a specific domain and using it to complement the statistical interpretation. Only when combined, biological and statistical angles can provide a full interpretation of statistical results.

Even once the interpretation of a specific statistical/study result is finalized internally, it's always a good idea to compare your result with other studies in the same field of biology or clinical research. This way, we can see if the results are providing a better or worse alternative to what is a standard in research in a given moment of analysis. Also making a result comparative with other studies, opens the door to potential meta-analysis which could be used to compare multiple studies, treatment effects and other segments.

The overall scientific value of the result

The overall scientific value of a study result depends on all the previous 4 points. Combining the trustworthiness, magnitude, biological context and reference studies in a specific study or research is the best way to create a composite biostatistical interpretation of results. It's important to interpret all the segments individually first but integrate them later into the overall conclusion about the results.

The overall conclusion should answer if the primary and secondary endpoints have been met, if potential beneficial, neutral or worsening effects (clinical trials) have been identified, what is their magnitude and potential use in the biomedical and pharmaceutical industry.

Finally it's important to pay attention to standards and regulations in specific areas of research and interpret the results in relation to them. The novelty of results is especially important in research, but even the results which are not novel can serve as confirmatory results for a specific field and should not be disregarded. Biologically and Clinically negative results also have value and should be published as well as positive results, as long as the results are trustworthy and accurate. Not publishing negative results can lead to publication bias and is a practice that should be avoided.

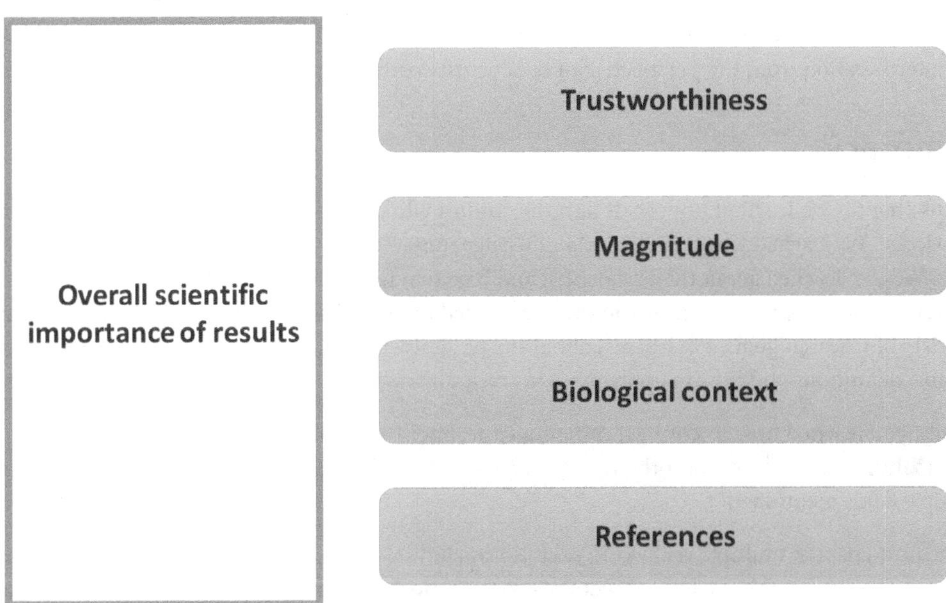

Figure 16.15 – Scientific importance of a result

As seen in the previous figure, the overall scientific result depends on many different fields and these fields are integrated into the overall interpretation. For this reason the author should have a specific domain knowledge and combine it with the results of statistical analysis to achieve a full interpretation. One of the biggest mistakes to avoid in this area is to base the overall significance of a result solely on statistical significance. As discussed before, statistical significance is just another segment of the overall statistical result.

Novelty value of a result

Another very important segment to consider when making overall interpretation of the result is its novelty. This is also considered by Life Science journals when publishing the research as most journals would rank the novel approaches higher that the approaches which have been established before. But it's very important to know that even if the approach is not entirely novel, its use in different areas may be a source of novelty. Also many researchers around the world may try similar approaches and if their results are in agreement, even without novelty, the agreement itself may be scientifically useful in terms of confirming the results.

To interpret the overall result, novelty is combined with the trustworthiness of the result, magnitude of the difference found in results, biological context and very importantly the references which represent the current state of a certain scientific topic. To relate this the previous vaccine example. If a vaccine exists on a market for a specific pathogen with 83% accuracy, references would be used to present that as a current state, and the novelty and magnitude of a new vaccine would be evaluated against that reference to see if a higher effectiveness can be achieved with the novel vaccine.

This concludes the section with main perspective on validation of the results trough sensitivity analysis and interpretation from the perspective of result trustworthiness.

Summary

In this chapter we learned how to design the studies while keeping both biological and statistical knowledge. We learned why is biological knowledge important when designing studies in life science areas. We also learned about the Frequentist and Bayesian frameworks in Biostatistics, what are their differences, philosophies and how can they be applied in different Biostatistical analyses and achieve a variety of research goals. We learned about the parameters in biostatistics, their fixed and random variable definitions and how to apply these in research.

Finally, you learned how to connect different experiment and study angles, connect them with the data analysis part and interpret the results of a biological domain studies by integrating the earlier multiple fields mentioned.

In the next chapter, multiple real world exercises including performing PCA (Principal Component Analysis), Latent variable analysis, Structural equation modeling and Bayesian analysis will be covered.

17

Part 4 Exercise – Performing Statistics for Biology Studies in Python

In this chapter, you will learn how to perform biostatistical analysis using advanced methods such as **Principal Component Analysis** (**PCA**), random forests, latent variable modeling, and others. Data dimensionality (having a large number of biological variables) is a common aspect of real-world biological datasets. This is often an advantage because we have more data and more insights as a result. But sometimes we want to reduce dimensionality to better summarize and understand the data from the perspective of having fewer dimensions than in the original data. This set of methods is called **data dimensionality reduction**. This is especially important in studies involving genetics and protein analysis. In this chapter, you will learn how to practically reduce dimensionality and perform PCA in Python using a real-world mice protein dataset with Down syndrome data.

Further, you will learn how to identify the unknown, latent factors that may influence the analysis. This is one of the most difficult, but most important segments of biostatistics, and in this chapter, you will see a practical exercise in this field.

We will be covering the following topics in this chapter:

- Understanding data dimensionality and resolving data complexity
- Learning how to identify latent factors in data and associate them with biological aspects

Understanding data dimensionality and resolving data complexity

In this section, we will be introduced to a specific exercise that is conducted to show the dimensionality of complex datasets in the biology field, a mice protein analysis, and how to address that dimensionality.

The mice protein dataset is a part of the UCI Machine Learning Repository and can be found at `https://archive.ics.uci.edu/dataset/342/mice+protein+expression`. This is licensed under Attribution 4.0 International for Clara Higuera, Katheleen Gardiner, and Krzysztof Cios (2015). Mice Protein Expression. UCI Machine Learning Repository. `https://doi.org/10.24432/C50S3Z`.

Additionally, the dataset may also be found here `https://www.kaggle.com/datasets/ruslankl/mice-protein-expression`.

Before we start with defining and exploring the variables, let's first set up the framework for biological variables.

What needs to be defined at this stage is the experiment area, a biological variable, and the statistical feature that we will use to explore the biological variables (multitude in this case):

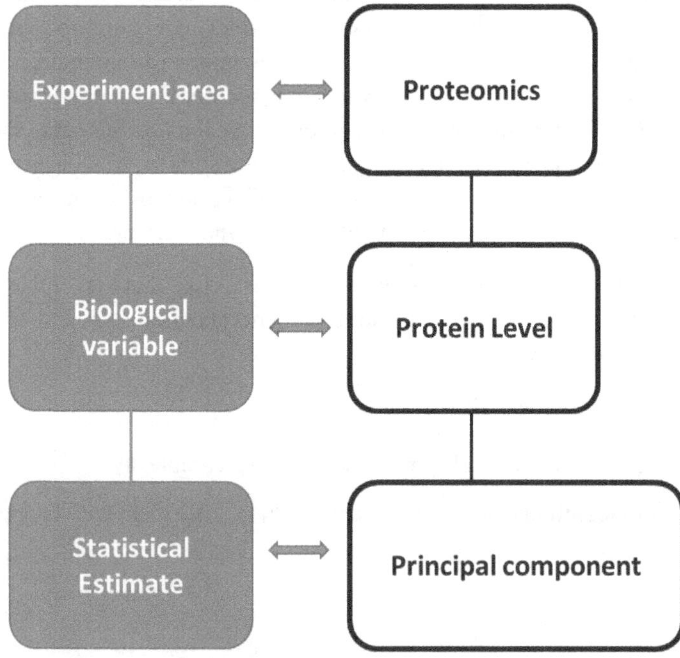

Figure 17.1 – Variable definitions for PCA

As you can see, the experiment area is proteomics, which is a field of biology in which the main topic is studying the protein levels in tissues and cells and finding associations between proteins and biological conditions of interest.

So, the biological variable will be a protein level, or the protein variable.

Finally, we will be using different estimates related to principal components, which serve to reduce dimensions in complex biological datasets such as this one. Principal components are new variables generated using PCA. These new variables simplify the dataset by segmenting the data into a number of data components that are not correlated.

Open Spyder (or Jupyter Notebook).

Now let's load the libraries needed for this project:

```
from ucimlrepo import fetch_ucirepo
from sklearn.decomposition import PCA
from sklearn.preprocessing import LabelEncoder
from sklearn.impute import SimpleImputer
import pandas as pd
import matplotlib.pyplot as plt

# fetch dataset
mice_protein_expression = fetch_ucirepo(id=342)

# data (as pandas dataframes)
X = mice_protein_expression.data.features
X = X.iloc[:, :-3] #remove - reduntant with 'class' col
y = mice_protein_expression.data.targets

# metadata
print(mice_protein_expression.metadata)

# variable information
print(mice_protein_expression.variables)
```

The target column to predict in this case will be in the `y` object and it's called the `class` column. The `Treatment`, `Genotype`, and `Behavior` columns are basically contained in that column, so they are redundant and are removed.

Let's see what the dataset looks like in the Variable Explorer (Spyder IDE):

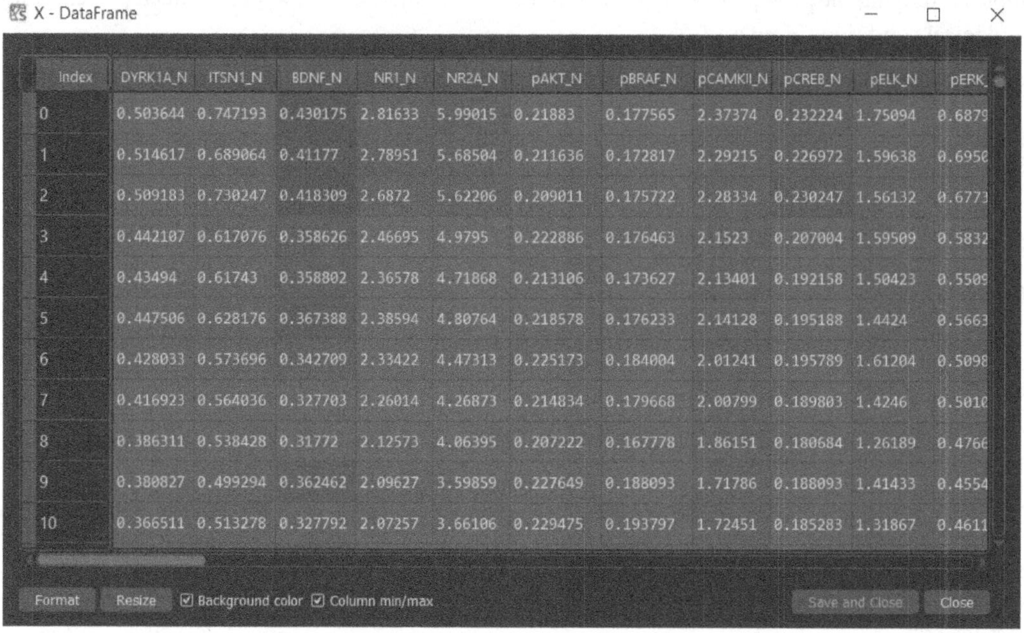

X - DataFrame

Index	DYRK1A_N	ITSN1_N	BDNF_N	NR1_N	NR2A_N	pAKT_N	pBRAF_N	pCAMKII_N	pCREB_N	pELK_N	pERK_
0	0.503644	0.747193	0.430175	2.81633	5.99015	0.21883	0.177565	2.37374	0.232224	1.75094	0.6879
1	0.514617	0.689064	0.41177	2.78951	5.68504	0.211636	0.172817	2.29215	0.226972	1.59638	0.6950
2	0.509183	0.730247	0.418309	2.6872	5.62206	0.209011	0.175722	2.28334	0.230247	1.56132	0.6773
3	0.442107	0.617076	0.358626	2.46695	4.9795	0.222886	0.176463	2.1523	0.207004	1.59509	0.5832
4	0.43494	0.61743	0.358802	2.36578	4.71868	0.213106	0.173627	2.13401	0.192158	1.50423	0.5509
5	0.447506	0.628176	0.367388	2.38594	4.80764	0.218578	0.176233	2.14128	0.195188	1.4424	0.5663
6	0.428033	0.573696	0.342709	2.33422	4.47313	0.225173	0.184004	2.01241	0.195789	1.61204	0.5098
7	0.416923	0.564036	0.327703	2.26014	4.26873	0.214834	0.179668	2.00799	0.189803	1.4246	0.5010
8	0.386311	0.538428	0.31772	2.12573	4.06395	0.207222	0.167778	1.86151	0.180684	1.26189	0.4766
9	0.380827	0.499294	0.362462	2.09627	3.59859	0.227649	0.188093	1.71786	0.188093	1.41433	0.4554
10	0.366511	0.513278	0.327792	2.07257	3.66106	0.229475	0.193797	1.72451	0.185283	1.31867	0.4611

Format Resize ☑ Background color ☑ Column min/max Save and Close Close

Figure 17.2 – Mice protein dataset

This is what the mice protein dataset looks like. At the top, you may see the protein names, and in the table their corresponding protein levels. The protein levels can be used as independent variables to try to create statistical models to assess different biological classes of mice.

But before we go there, let's explore the biological classes of mice in this dataset:

c-CS-s: Control mice who are stimulated to learn (saline treatment)

- **c-CS-m**: Control mice who are stimulated to learn (memantine treatment)
- **c-SC-s**: Control mice who are not stimulated to learn (saline treatment)
- **c-SC-m**: Control mice who are not stimulated to learn (memantine treatment)

t-CS-s: Trisomy mice who are stimulated to learn (saline treatment)

- **t-CS-m**: Trisomy mice who are stimulated to learn (memantine treatment)
- **t-SC-s**: Trisomy mice who are not stimulated to learn (saline treatment)
- **t-SC-m**: Trisomy mice who are not stimulated to learn (memantine treatment)

You may observe the class object by opening the `y` object in the Variable Explorer (`y` stands for the target class biological variable).

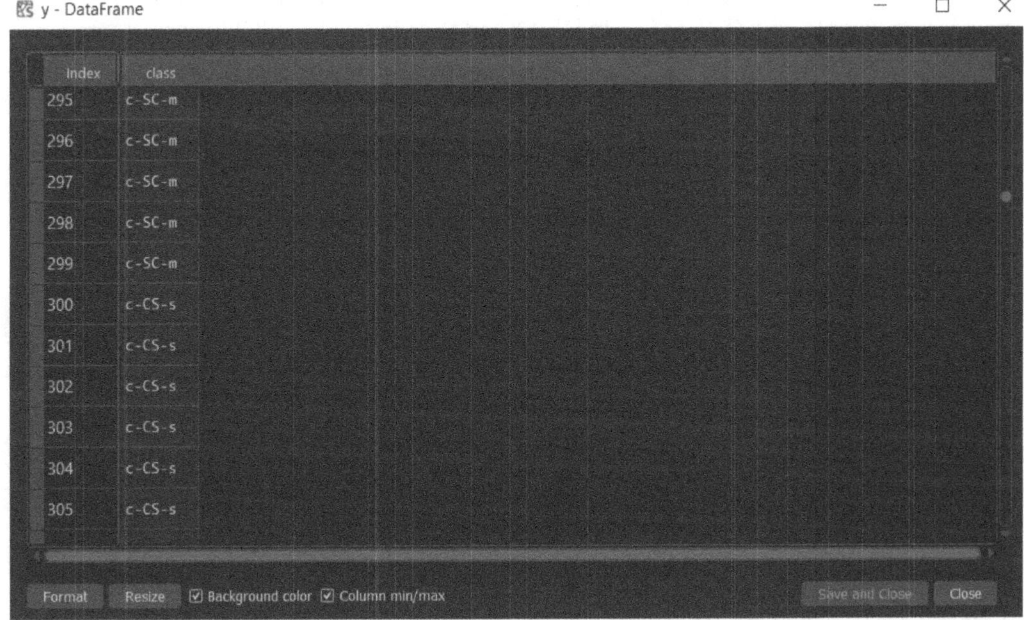

Figure 17.3 – Class object

The next step is to define the non-numeric columns in the dataset and convert them to numeric ones. This implementation (**SEM** – short for **Structural Equation Modeling**) would work much better if all columns were converted to numeric columns and this is one of the important segments to define:

```
non_numeric_columns = X.select_dtypes(include=['object']).columns

# Convert non-numeric columns to numeric using LabelEncoder
label_encoders = {}
#Add the for i loop, which will loop over #non_numeric_columns and use
label encoder
for column in non_numeric_columns:
    le = LabelEncoder()
    X[column] = le.fit_transform(X[column].astype(str))
    label_encoders[column] = le

# Impute missing values
imputer = SimpleImputer(strategy='mean')  # or you can use other
strategies like 'median', 'most_frequent', etc.
X_imputed = imputer.fit_transform(X)
```

The next step is to perform PCA, which will reduce the complexity (dimensionality) of the data to a lower dimensional space. The PCA will take data with high complexity (many variables) and turn it into a lower number of dimensions, enabling us to explore all the variables at the same time in, let's say, two dimensions (`X` and `Y`). Keep in mind that `X` and `Y`, or, in this case, PC1 (principal component 1) and PC2 (principal component 2) will contain the most variance explained in the data. Also, PC1 and PC2 are orthogonally projected to each other, which means that they are uncorrelated and explain different sources of variability.

Let's use the `sklearn` package implementation by simply using the `PCA()` function and specifying the number of components as `2`, as shown in the following code (`2` is chosen for visualization simplicity):

```
# Impute missing values
```

As you can see, we used `SimpleIpmputer()` to fill in the missing values. In this case, we used the mean as the most representative value to impute. Keep in mind that the mean is a very simple point to use for data imputation and can be biased, but for this exercise, we will keep the mean imputation:

```
# Perform PCA
pca = PCA(n_components=2)
X_pca = pca.fit_transform(X_imputed)

# Create a DataFrame for the PCA results
pca_df = pd.DataFrame(data=X_pca, columns=['PC1', 'PC2'])
```

This is depicted here:

Index	PC1	PC2
0	1.33784	2.74036
1	0.923475	2.50339
2	0.894646	2.5104
3	0.0196342	1.87768
4	-0.218342	1.71293
5	-0.17131	1.75436
6	-0.612545	1.46101
7	-0.803238	1.31066
8	-1.168	1.20594
9	-1.6545	0.907836
10	-1.64487	0.919657

Figure 17.4 – Dataframe containing PC1 and PC2 values

We now encode the target labels:

```
# Encode target labels to numeric values for plotting
target_le = LabelEncoder()
y_encoded = target_le.fit_transform(y)
pca_df['Target'] = y_encoded

# Print the explained variance ratio
print(f"Explained variance ratio: {pca.explained_variance_ratio_}")
```

You should get this as output:

Explained variance ratio: [0.3171965 0.2672305]

Now let's plot the PCA results using Matplotlib and create scatterplots to show each observation on the PCA plot along the PC1 and PC2 axes:

```
# Plotting the PCA results
plt.figure(figsize=(10, 8))

# Map each category to a different color
categories = target_le.classes_
colors = plt.cm.get_cmap('viridis', len(categories))

for category, color in zip(categories, range(len(categories))):
    subset = pca_df[pca_df['Target'] == color]
    plt.scatter(subset['PC1'], subset['PC2'],
        label=category, color=colors(color))

plt.xlabel('Principal Component 1')
plt.ylabel('Principal Component 2')
plt.title('PCA of Mice Protein Expression Dataset')
plt.legend(title='Category')
plt.show()
```

Here is the plot for it:

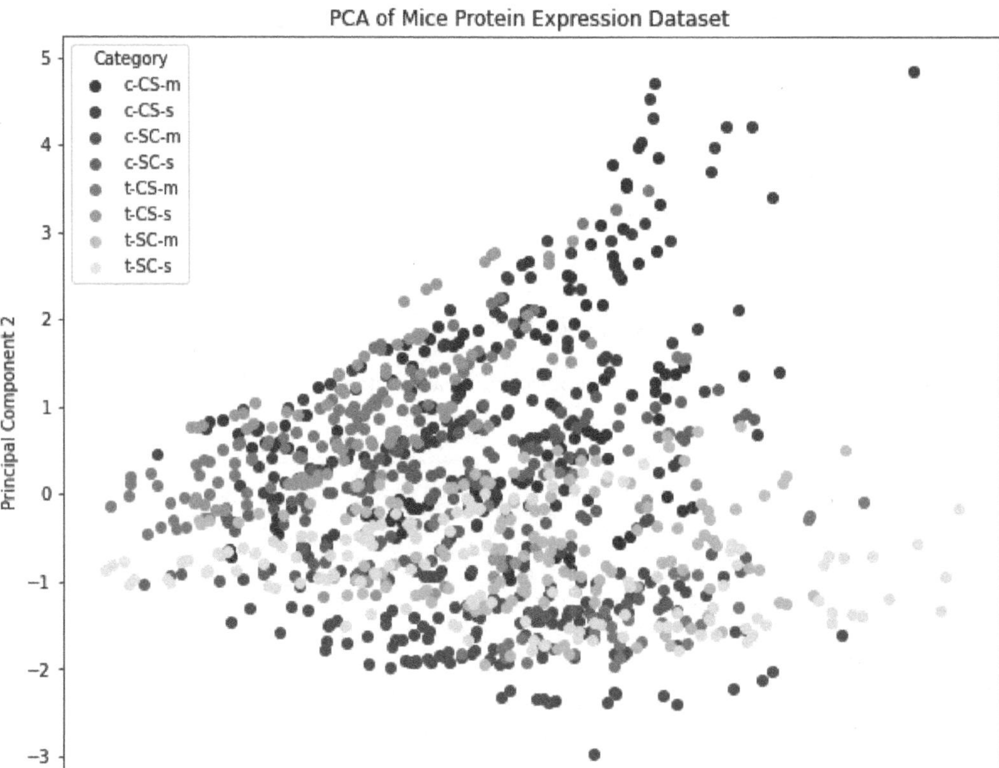

Figure 17.5 – PCA visualization

The preceding visualization represents two principal components, or the newly created variables that represent a bigger number of variables in the dataset. **Principal Component 1 (PC1)** is the component that contains the most variability in the data. **Principal Component 2 (PC2)** is the next component with the second largest variability in the data but it's a collection of compressed segments of variables that are not correlated to PC1. To summarize, PC1 and PC2 contain segments of variability from many variables in the dataset, but it's important to understand that PC1 and PC2 are chosen as they are the least correlated to each other.

To resolve the complexity around the data, we will use random forest, an ensemble algorithm designed to perform machine learning, but also the feature selection we need to select the variables that increase the separation between classes.

Random forest is a machine learning method used for classification tasks by constructing multiple decision trees (decision trees are based on classification thresholds for variables). Each tree is built from a random subset of the data (this selection is called bagging) and a random selection of variables. This generates diversity and enables random forests to learn complex patterns in the data.

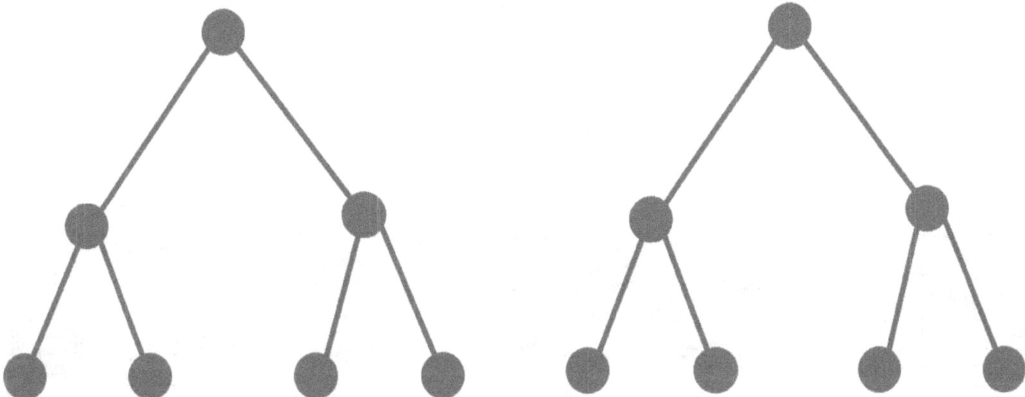

Figure 17.6 – Random forest

The random forest then aggregates the results from all individual trees – using majority voting. This method-specific use case, which is within the scope of this book, is called feature selection. We will be using the random forest to select the top 10 features that contribute the most to the classification of the data:

```
# Train a Random Forest model to select the top 10 features
from sklearn.ensemble import RandomForestClassifier
rf = RandomForestClassifier(n_estimators=100, random_state=42)
rf.fit(X_imputed, y_encoded)
```

The *n* estimators in the code mean that we are using 100 trees to construct the random forest model:

```
# Get feature importances and select the top 10 features
feature_importances = pd.Series(
    rf.feature_importances_, index=X.columns)
top_10_features = feature_importances.nlargest(10).index

# Filter the original data to include only the top 10 features
X_top_10 = pd.DataFrame(
    X_imputed, columns=X.columns
)[top_10_features]

# Write the top 10 features to a CSV file
X_top_10.to_csv('top_10_features.csv', index=False)
```

```
# Perform PCA on the top 10 features
pca = PCA(n_components=2)
X_pca_top_10 = pca.fit_transform(X_top_10)

# Create a DataFrame for the PCA results
pca_df_top_10 = pd.DataFrame(
    data=X_pca_top_10, columns=['PC1', 'PC2'])
pca_df_top_10['Target'] = y_encoded
```

Here is the output for it:

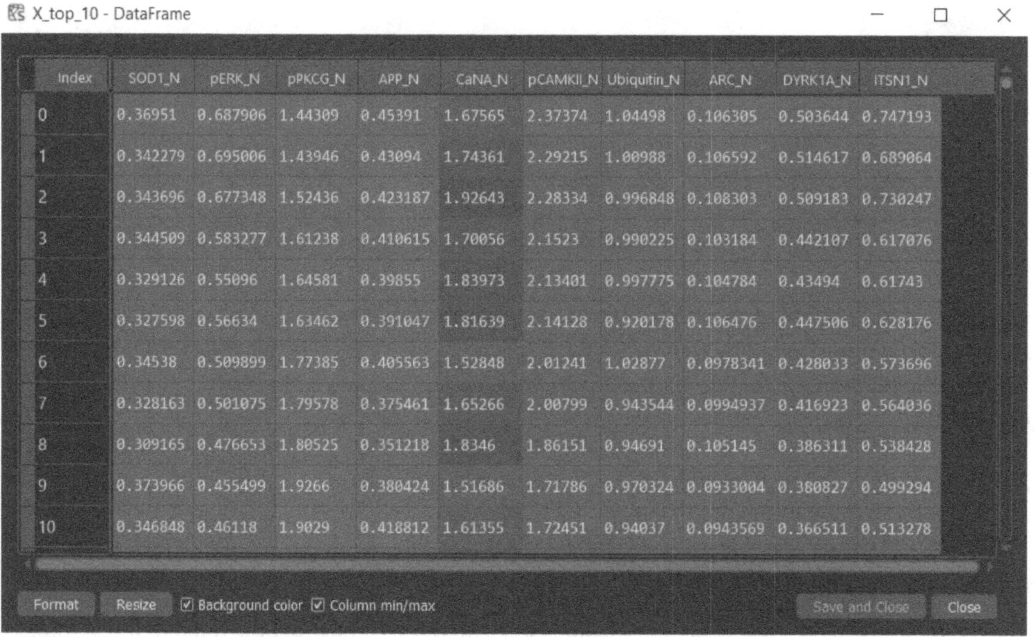

X_top_10 - DataFrame

Index	SOD1_N	pERK_N	pPKCG_N	APP_N	CaNA_N	pCAMKII_N	Ubiquitin_N	ARC_N	DYRK1A_N	ITSN1_N
0	0.36951	0.687906	1.44309	0.45391	1.67565	2.37374	1.04498	0.106305	0.503644	0.747193
1	0.342279	0.695006	1.43946	0.43094	1.74361	2.29215	1.00988	0.106592	0.514617	0.689064
2	0.343696	0.677348	1.52436	0.423187	1.92643	2.28334	0.996848	0.108303	0.509183	0.730247
3	0.344509	0.583277	1.61238	0.410615	1.70056	2.1523	0.990225	0.103184	0.442107	0.617076
4	0.329126	0.55096	1.64581	0.39855	1.83973	2.13401	0.997775	0.104784	0.43494	0.61743
5	0.327598	0.56634	1.63462	0.391047	1.81639	2.14128	0.920178	0.106476	0.447506	0.628176
6	0.34538	0.509899	1.77385	0.405563	1.52848	2.01241	1.02877	0.0978341	0.428033	0.573696
7	0.328163	0.501075	1.79578	0.375461	1.65266	2.00799	0.943544	0.0994937	0.416923	0.564036
8	0.309165	0.476653	1.80525	0.351218	1.8346	1.86151	0.94691	0.105145	0.386311	0.538428
9	0.373966	0.455499	1.9266	0.380424	1.51686	1.71786	0.970324	0.0933004	0.380827	0.499294
10	0.346848	0.46118	1.9029	0.418812	1.61355	1.72451	0.94037	0.0943569	0.366511	0.513278

Figure 17.7 – Variables with the most separating power for PCA

As can be seen, the random forest algorithm identified the top 10 variables that increase the separation and resolve the complexity in the dataset. Now we can use these 10 variables to plot a better PCA visualization:

```
# Plotting the PCA results for the top 10 features
plt.figure(figsize=(10, 8))

# Map each category to a different color
categories = target_le.classes_
colors = plt.cm.get_cmap('viridis', len(categories))
```

```
for category, color in zip(categories, range(len(categories))):
    subset = pca_df_top_10[pca_df_top_10['Target'] == color]
    plt.scatter(subset['PC1'], subset['PC2'],
        label=category, color=colors(color))

plt.xlabel('Principal Component 1')
plt.ylabel('Principal Component 2')
plt.title('PCA of Mice Protein Expression Dataset (Top 10 Features)')
plt.legend(title='Category')
plt.show()
```

This is shown in the following figure:

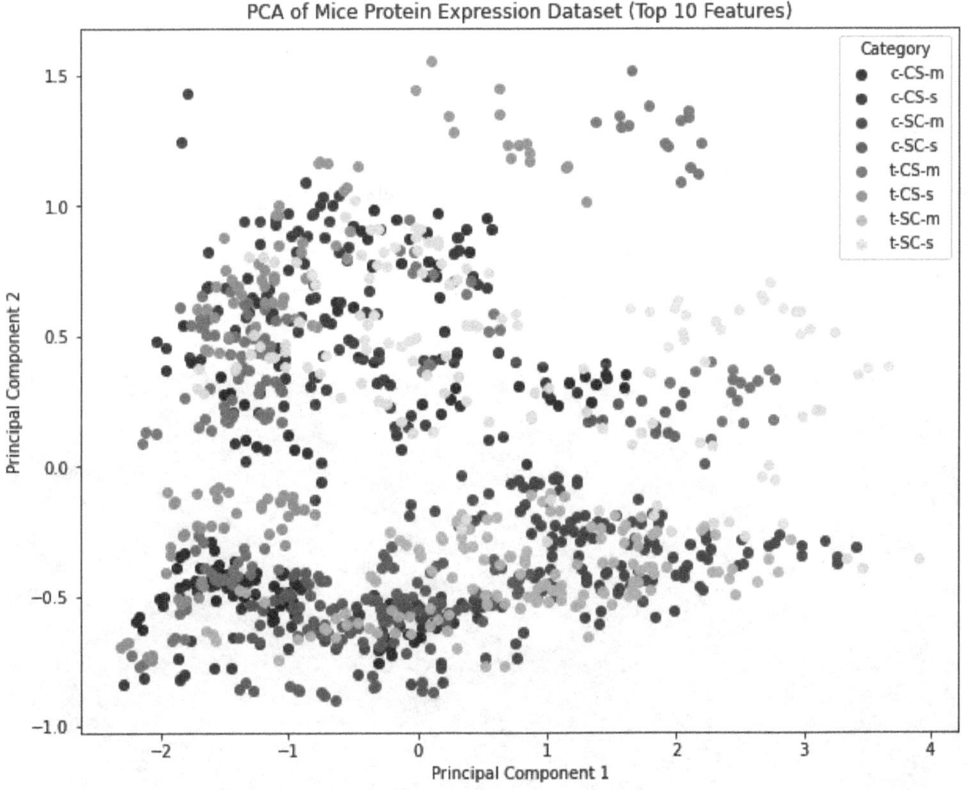

Figure 17.8 – Clear PCA separation

Now we can see that the classes are separated more clearly and there are very interesting results for interpretation. It can be seen that the classes of mice with trisomy that were treated with memantine tend to be positioned lower in the plot. This can now serve as very important input for which variables to focus on (the 10 identified by the random forest) and which classes to focus on, such as those simulated to learn. Such exploratory analysis would be ideal.

In this section, the main learning topic was how to reduce dimensions in the dataset using PCA, but there is much more to learn about complex data. In the next section, you will learn about practical implementations to identify the latent factors in the data.

Now, we can explore the important features for the classification of the data based on the `class` variable:

```
feat_imp = rf.feature_importances_
# Create a pandas Series with feature names as the index
feat_imp_series = pd.Series(feat_imp, index=X.columns)

# Sort the feature importances in descending order
sorted_feat_imp =feat_imp_series.sort_values(ascending=False).head(10)
```

Now explore the `sorted_feat_imp` object using the Variable Explorer or use the `print` statement:

```
print(sorted_feat_imp)
```

This is depicted here:

Figure 17.9 – Identified features for further studies

As can be seen, random forest was more powerful compared to PCA in identifying the most important features for the classification of the data according to the `class` variable. The 10 most important protein levels can now serve as a basis for further experimental research. We have narrowed the features down from more than 70 to 10. Further experimental research will now be more targeted on a real-world scenario.

Learning how to identify latent factors

In biological research, latent variables are almost everywhere. These are the variables that may affect the results and we might not even have them in our data. Variables could be latent because they are unmeasurable, which is a frequent phenomenon in fields such as neuroscience. But most of the time, latent variables are there because of the high complexity of biological systems, often including thousands of variables that are associated with each other. In most experiments, it's not feasible to measure every single variable associated with a biological system. This means that, in most experiments and studies, there will be a lot of latent variables and sometimes we need to understand them to make sure we understand the results of our research.

Studying latent variables can also help uncover new mechanisms and new variables of interest for biological research.

Let's define the experiment domain and the biological variable for latent factor analysis:

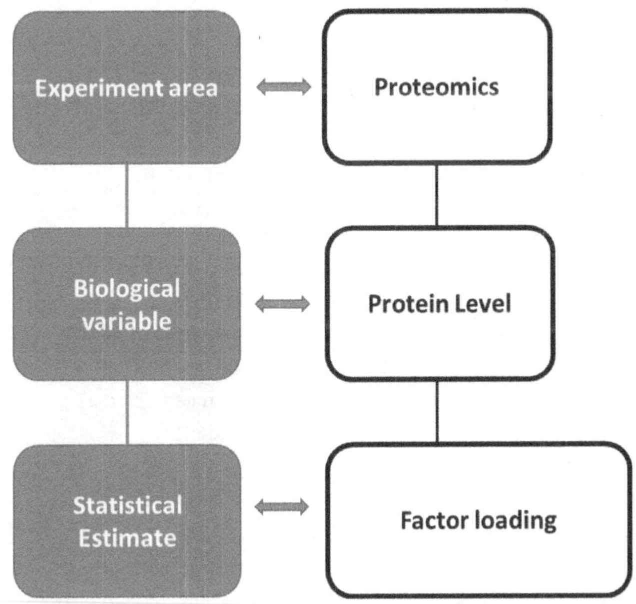

Figure 17.10 – Defining the biological variable for the mice dataset exercise

The first steps for the latent variable analysis exercise are the same as in the PCA exercises.

Let's load the libraries and the data. But this time we will be using some additional libraries, such as `semopy` and `graphviz`.

Make sure you have installed these using `pip install semopy` and `pip install graphviz` in your command prompt and in the environment for Python you are using. For `graphviz`, you may need to install it through the `conda` prompt:

```
import pandas as pd
from ucimlrepo import fetch_ucirepo
from sklearn.preprocessing import LabelEncoder, StandardScaler
from sklearn.experimental import enable_iterative_imputer  # noqa
from sklearn.impute import IterativeImputer
from semopy import Model, calc_stats, semplot
import graphviz
```

As can be seen, the `semopy` library will be used to perform the SEM and `graphviz` will be used for the visualization of SEM diagrams once loaded.

Now we can fetch the dataset using the `fetch` function as follows:

```
 # Fetch dataset
mice_protein_expression = fetch_ucirepo(id=342)

# Data (as pandas dataframes)
X = mice_protein_expression.data.features
y = mice_protein_expression.data.targets

# Data preprocessing
# Separate numeric and categorical columns
numeric_cols = X.select_dtypes(include=['float64', 'int64']).columns
categorical_cols = X.select_dtypes(include=['object']).columns

# Impute missing values for numeric columns using MICE
imputer = IterativeImputer(random_state=222)
X[numeric_cols] = imputer.fit_transform(X[numeric_cols])

# Encode categorical columns
label_encoders = {}
for col in categorical_cols:
    label_encoders[col] = LabelEncoder()
    X[col] = label_encoders[col].fit_transform(X[col])

# Standardize numeric data
```

```
scaler = StandardScaler()
X[numeric_cols] = scaler.fit_transform(X[numeric_cols])
# Prepare data for SEM
data = pd.concat([X, pd.DataFrame(y, columns=['Target'])], axis=1)
```

Before running the latent variable SEM analysis, we need to describe the SEM model.

SEM is a statistical approach that allows examining complex relationships among observed and non-observed (latent) variables within a theoretical construct (adding a theoretical knowledge perspective).

In this example, a theoretical foundation for the SEM model will be created using the molecular biology knowledge of the author.

The main goal is to evaluate the theoretical construct of two potential latent factors that are not in the data. These two factors are cell division signaling and apoptosis signaling. Cell division signaling may be involved in many processes in neural tissues, such as cells reproducing to generate new neural tissues. Another process that occurs in neural tissues, as in other tissues in mouse organisms, is apoptosis. Apoptosis is needed to eliminate cells that have faults in their biological systems, such as those that divide abnormally or are in some other way dysfunctional.

Using `https://reactome.org/`, I have identified that pAKT_N, pBRAF_N, and pCREB_N are associated with cell division signaling pathways. The MTOR_N and P38_N proteins are associated with apoptosis signaling pathways. Cell division and apoptosis signaling pathways are often associated with each other in many molecular biology processes.

We will use this theoretical foundation to form an SEM model and test it statistically. Let's create the model as `model_desc`, which will be the description for a basic structural equation model for this project:

```
# Define SEM model
# """ for lengthy text, code, or data embedded within  code
# Define the theoretical SEM model
model_desc_theoretical = """
# Measurement Model

Cell_division_signaling =~ pAKT_N + pBRAF_N + pCREB_N
Apoptosis_signaling =~ MTOR_N + P38_N
#+ EGR1_N

# Structural Model
Cell_division_signaling ~  Apoptosis_signaling
Apoptosis_signaling ~ Cell_division_signaling
"""
```

The =~ sign is a novel combination of signs in this book and it's used to define the latent variables' relation with the observed variables – in this case, protein level variables.

Notice how I also set the potential association between three latent variables using the ~ and + signs.

Next, let's run the analysis and fit the latent variable SEM model:

```
# Initialize and fit the model
model = Model(model_desc_theoretical)
model.fit(data)

# Retrieve and print fit statistics
fit_stats = calc_stats(model)
print("Theoretical Model Fit Statistics:")
print(f"CFI: {fit_stats['CFI']}")

parameter_estimates = model.inspect()
parameter_estimates.to_csv('parameter_estimates.csv', index=False)
```

Here is the output for the analysis:

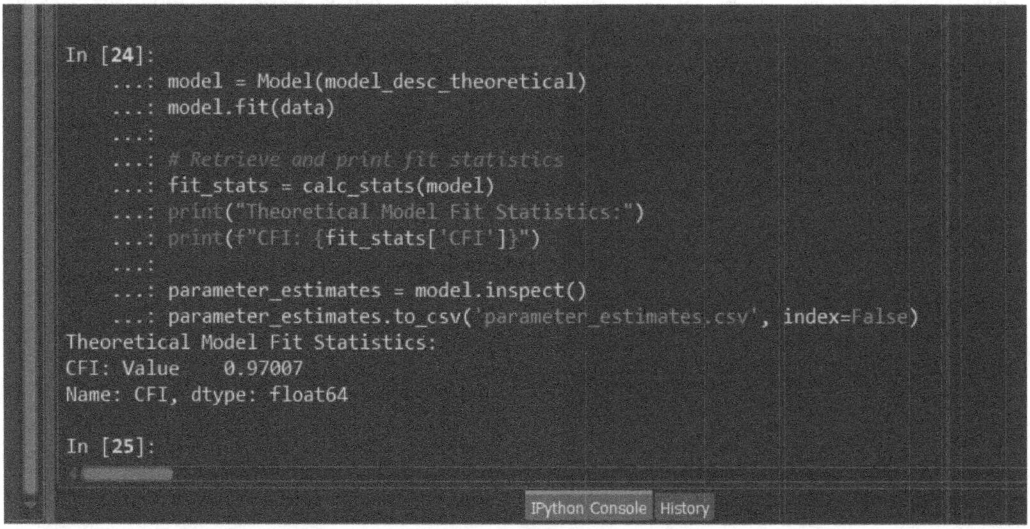

```
In [24]:
    ...: model = Model(model_desc_theoretical)
    ...: model.fit(data)
    ...:
    ...: # Retrieve and print fit statistics
    ...: fit_stats = calc_stats(model)
    ...: print("Theoretical Model Fit Statistics:")
    ...: print(f"CFI: {fit_stats['CFI']}")
    ...:
    ...: parameter_estimates = model.inspect()
    ...: parameter_estimates.to_csv('parameter_estimates.csv', index=False)
Theoretical Model Fit Statistics:
CFI: Value    0.97007
Name: CFI, dtype: float64

In [25]:
```

IPython Console History

Figure 17.11 – Inspecting the SEM model CFI

Detailed evaluation of the SEM model can now be performed using the sempplot() function. The next step is to explore the theoretical model we defined in more detail:

```
# Plot the path diagram
semplot(model, "theoretical_path_diagram.png", show='estimates')
```

Here is the output plot:

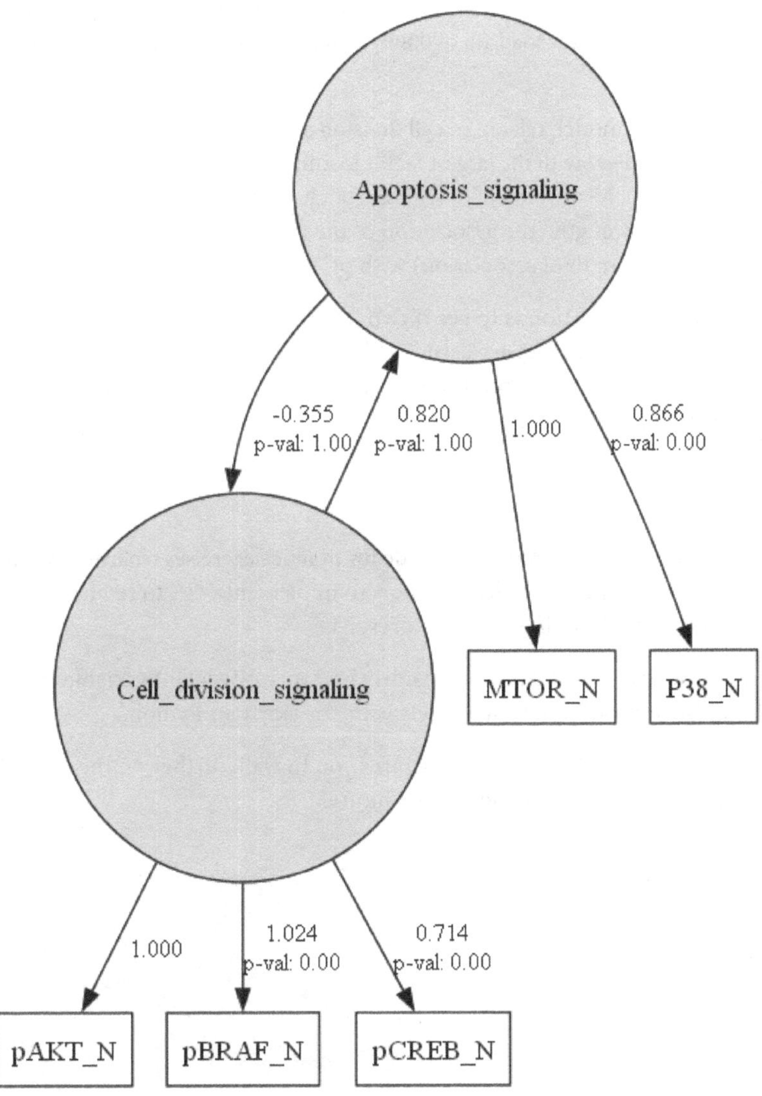

Figure 17.12 – Structural equation model visualization

In the structural equation model, we can now observe individual associations between variables and the latent factors. Notice how each latent factor (`Apoptosis_signaling` and `Cell_division_signaling`) has an association edge of 1.00. These factor loadings are automatically set by the SEM model for reference. So, `Apoptosis_signaling` association with MTOR_N and `Cell_division_signaling` association with pAKT_N are reference paths. We can compare them to other items in the SEM model.

It can be seen that the factor loading of 0.866 for p38_N in relation to apoptosis signaling is lower compared to MTOR_N. The conclusion is that MTOR_N is strongly affected by apoptosis signaling in this SEM model. But the factor loading of 0.866 is not small either and it has a significant p value (0.00, which is $p<0.05$).

The other part of the SEM model, related to cell division signaling is more complex. The cell division signaling influences pBRAF_N with the largest factor loading of 1.024. This is very similar to pAKT_N-related factor loading, which is a reference. We can conclude that cell division signaling affects these two proteins at a similar strength. The association of the cell signaling pathway is present but with a smaller factor loading (strength of association) with pCREB_N.

The final part of the interpretation is to see if two latent factors (`poptosis_signaling` and `Cell_division_signaling`) are associated with each other. Two factor loadings for edges between them are -0.355 and 0.820, but both p values are 1.00, which means that the associations are not statistically significant.

Summary

In this chapter, we learned how to implement code for practice exercises regarding studies in different biological and biostatistical fields. We learned how to implement PCA to reduce the dimensionality of biological datasets and how to interpret PCA results.

In one of the exercises, we learned how to identify and quantify latent variables and interpret the results of **SEMs** or **Structural Equation Models** using `semopy` in Python.

I hope you enjoyed this journey and that it inspires you to walk further on the path of Biostatistics and Python programming, a combination of the future.

Index

Symbols

A

B

E

F

G

H

I

T

W

X

Other Books You May Enjoy

If you enjoyed this book, you may be interested in these other books by Packt:

15 Math Concepts Every Data Scientist Should Know

David Hoyle

ISBN: 978-1-83763-418-7

- Master foundational concepts that underpin all data science applications
- Use advanced techniques to elevate your data science proficiency
- Apply data science concepts to solve real-world data science challenges
- Implement the NumPy, SciPy, and scikit-learn concepts in Python
- Build predictive machine learning models with mathematical concepts
- Gain expertise in Bayesian non-parametric methods for advanced probabilistic modeling
- Acquire mathematical skills tailored for time-series and network data types

Building Statistical Models in Python

Huy Hoang Nguyen, Paul N Adams, Stuart J Miller

ISBN: 978-1-80461-428-0

- Explore the use of statistics to make decisions under uncertainty
- Answer questions about data using hypothesis tests
- Understand the difference between regression and classification models
- Build models with stats models in Python
- Analyze time series data and provide forecasts
- Discover Survival Analysis and the problems it can solve

Packt is searching for authors like you

If you're interested in becoming an author for Packt, please visit `authors.packtpub.com` and apply today. We have worked with thousands of developers and tech professionals, just like you, to help them share their insight with the global tech community. You can make a general application, apply for a specific hot topic that we are recruiting an author for, or submit your own idea.

Share Your Thoughts

Now you've finished *Biostatistics with Python*, we'd love to hear your thoughts! Scan the QR code below to go straight to the Amazon review page for this book and share your feedback or leave a review on the site that you purchased it from.

`https://packt.link/r/1837630968`

Your review is important to us and the tech community and will help us make sure we're delivering excellent quality content.

Download a free PDF copy of this book

Thanks for purchasing this book!

Do you like to read on the go but are unable to carry your print books everywhere?

Is your eBook purchase not compatible with the device of your choice?

Don't worry, now with every Packt book you get a DRM-free PDF version of that book at no cost.

Read anywhere, any place, on any device. Search, copy, and paste code from your favorite technical books directly into your application.

The perks don't stop there, you can get exclusive access to discounts, newsletters, and great free content in your inbox daily

Follow these simple steps to get the benefits:

1. Scan the QR code or visit the link below

`https://packt.link/free-ebook/978-1-83763-096-7`

2. Submit your proof of purchase
3. That's it! We'll send your free PDF and other benefits to your email directly

Made in United States
Cleveland, OH
06 July 2025